AF676011

Moving toward Biomimetic Tissue Engineered Scaffolds

Moving toward Biomimetic Tissue Engineered Scaffolds

Guest Editors

Costantino Del Gaudio
Silvia Baiguera
Lucy Di Silvio

MDPI

Basel • Beijing • Wuhan • Barcelona • Belgrade • Novi Sad • Cluj • Manchester

Guest Editors

Costantino Del Gaudio
Italian Space Agency
Rome
Italy

Silvia Baiguera
Department of Clinical Science and Translational Medicine
University of Rome "Tor Vergata"
Rome
Italy

Lucy Di Silvio
Centre of Oral, Clinical & Translational Sciences
King's College London
London
UK

Editorial Office

MDPI AG
Grosspeteranlage 5
4052 Basel, Switzerland

This is a reprint of the Special Issue, published open access by the journal *Nanomaterials* (ISSN 2079-4991), freely accessible at: https://www.mdpi.com/journal/nanomaterials/special_issues/Tissue_Eng_Biomimetics_Nanomed.

For citation purposes, cite each article independently as indicated on the article page online and as indicated below:

Lastname, A.A.; Lastname, B.B. Article Title. *Journal Name* **Year**, *Volume Number*, Page Range.

ISBN 978-3-7258-4103-5 (Hbk)
ISBN 978-3-7258-4104-2 (PDF)
https://doi.org/10.3390/books978-3-7258-4104-2

© 2025 by the authors. Articles in this book are Open Access and distributed under the Creative Commons Attribution (CC BY) license. The book as a whole is distributed by MDPI under the terms and conditions of the Creative Commons Attribution-NonCommercial-NoDerivs (CC BY-NC-ND) license (https://creativecommons.org/licenses/by-nc-nd/4.0/).

Contents

nanomaterials

MDPI

Editorial

Moving Toward Biomimetic Tissue-Engineered Scaffolds

Silvia Baiguera [1], Lucy Di Silvio [2] and Costantino Del Gaudio [3,*]

1 Independent Researcher; silvia.baiguera@gmail.com
2 Centre for Oral, Clinical & Translational Sciences, Faculty of Dentistry, Oral & Craniofacial Sciences, King's College London, London SE1 9RT, UK; lucy.di_silvio@kcl.ac.uk
3 Italian Space Agency, Via del Politecnico snc, 00133 Rome, Italy
* Correspondence: costantino.delgaudio@asi.it

Advancing experimental methodologies to accurately replicate the physiological and pathological characteristics of biological tissues is pivotal in tissue engineering. Such advancements enable the development of precise in vitro and in vivo models for therapeutic research. This Special Issue, titled "Moving Toward Biomimetic Tissue-Engineered Scaffolds", emphasizes the importance of biomimicry in designing customized scaffolds for tissue engineering applications.

A key focus in tissue engineering is replicating the microenvironment of target tissues to guide and enhance cellular responses. Ideally, scaffolds should mimic the tissue-specific extracellular matrix (ECM), be able to provide structural and biochemical cues, and support effective tissue regeneration [1]. Designing such scaffolds requires a comprehensive analysis of materials, fabrication techniques, and biological signals. Despite the challenges in replicating the hierarchical organization and intricate biochemical interactions of native ECM, a multidisciplinary approach can identify the critical features that are essential for regeneration. Scaffolds that incorporate biologically relevant properties and exhibit appropriate mechanical characteristics can potentially direct cellular behaviour to provide a conducive regenerative environment [2]. Decellularized ECM-based scaffolds are particularly promising, as combining natural biomaterials with advanced manufacturing and surface modification technologies can significantly enhance regenerative potential. This strategy broadens their therapeutic applications, offering substrates that support cellular viability and function while promoting tissue-specific regeneration [3].

Technologies play a pivotal role in the fabrication stage, leading to the development of functional microarchitectures. Key features such as porosity, pore size, interconnectivity, biosorption, and mechanical properties represent fundamental requirements for initiating a biomimetic strategy. This initial step must be further refined, for instance, by incorporating bioactive compounds through specific chemical treatments to achieve a stable binding to the scaffold or a controlled release with predefined kinetics, ensuring a precise dosage delivery. Moreover, the external environment must be considered within a comprehensive experimental design. The signals arising from culture conditions can significantly contribute to replicating physiological niches and provide additional, necessary inputs to achieve more realistic outcomes.

This rationale underpins the present Special Issue, which aims to deliver a focused, albeit partial, overview of the research conducted thus far in tissue engineering and regenerative medicine. Additionally, it seeks to provide perspectives on potential future projects in the field.

A biomimetic strategy remains one of the principal drivers for designing innovative investigative protocols. To this end, an effective approach must appropriately integrate the properties of biomaterials, cellular responses, and signalling pathways. This Special Issue offers a balanced mix of original research and review articles, alongside a perspective study addressing critical aspects of this topic.

Citation: Baiguera, S.; Di Silvio, L.; Del Gaudio, C. Moving Toward Biomimetic Tissue-Engineered Scaffolds. *Nanomaterials* **2024**, *14*, 2028. https://doi.org/10.3390/nano14242028

Received: 22 November 2024
Revised: 29 November 2024
Accepted: 5 December 2024
Published: 17 December 2024

Copyright: © 2024 by the authors. Licensee MDPI, Basel, Switzerland. This article is an open access article distributed under the terms and conditions of the Creative Commons Attribution (CC BY) license (https://creativecommons.org/licenses/by/4.0/).

Notably, three research articles focus on electrospun scaffolds as effective tools to replicate tissue-specific characteristics or promote tailored cellular responses. Rivera-Torres et al. [4] developed composite scaffolds composed of poly(vinyl alcohol) and bioglass type 58S to investigate the spontaneous activity of chick embryonic cardiomyocytes, examining how the inorganic component content affects synchronous beating patterns. Sousa de Almeida et al. [5] explored the role of polyurethane electrospun nanofibers in nanoparticle uptake by human fibroblasts and epithelial cells, revealing that substrate properties significantly influence cellular behaviour. These findings provide critical insights for designing disease models and nanoparticle-based therapies. Niu and Galluzzi [6] studied layered tubular hyaluronic acid-collagen scaffolds that mimic the intima and media layers of blood vessels. Their results demonstrate that the complete endothelialisation of mouse vascular endothelial cells along the inner wall facilitates vascular smooth muscle cell infiltration and alignment, conforming to the hierarchical architecture of the nanofibrous scaffold This structure supports the maintenance of contractile phenotypes and natural growth behaviour. Sassi et al. [7] expanded the scope of tissue engineering by investigating decellularized rat livers. Their study employed a custom-designed bioreactor to sustain dynamic culture conditions, enabling non-invasive analyses of cell viability, distribution, and function.

The review articles critically evaluate the existing data, emphasizing the importance of biomimetic procedures in achieving functional tissue regeneration. Górnicki et al. [8] present a survey on various scaffolds that facilitate three-dimensional cell cultures, a well-established prerequisite for accurately mimicking biological tissues. Roato et al. [9] analyze the role of scaffolds in periodontal tissue engineering, with an emphasis on mechanical stimulation to enhance the reliability of the results. Cun and Hosta-Rigau [10] examine cellular responses to scaffold topography, focusing on the behaviour of mesenchymal stem cells. They emphasize how surface features, including roughness, patterns, and porosity, regulate cell fate. Although a definitive correlation between topographical features and cellular responses remains unestablished, this approach offers a complementary or alternative strategy to the biochemical modulation of stem cell differentiation. An essential aspect of implantable devices is their immunological impact. Corsi et al. [11] address this by investigating macrophage polarization through bioactive metallic and ceramic nanoparticles, which hold promise as tools for promoting skeletal muscle regeneration. Politi et al. [12] extend this discussion by reviewing biofunctionalized electrospun mats and their potential for developing advanced smart scaffolds.

Furthermore, Baiguera et al. [13] present a forward-looking perspective on 3D-printing biomimetic scaffolds for skeletal muscle regeneration. They propose an experimental approach, utilizing additive manufacturing techniques, specifically stereolithography, to process hydrogels enriched with a decellularized extracellular matrix.

In summary, the contributions presented in this Special Issue highlight the immense potential of biomimetics for designing and fabricating scaffolds that advance tissue engineering. Enhancing the performance of biomedical devices to drive functional tissue regeneration is a pressing need to achieve the goal of improving the outcomes of novel therapeutic protocols.

Conflicts of Interest: The authors declare no conflict of interest.

References

1. Baiguera, S.; Urbani, L.; Del Gaudio, C. Tissue engineered scaffolds for an effective healing and regeneration: Reviewing orthotopic studies. *Biomed Res. Int.* **2014**, *2014*, 398069. [CrossRef] [PubMed]
2. Tonti, O.R.; Larson, H.; Lipp, S.N.; Luetkemeyer, C.M.; Makam, M.; Vargas, D.; Wilcox, S.M.; Calve, S. Tissue-specific parameters for the design of ECM-mimetic biomaterials. *Acta Biomater.* **2021**, *132*, 83–102. [CrossRef] [PubMed]
3. Zhang, S.; Guo, Y.; Lu, Y.; Liu, F.; Heng, B.C.; Deng, X. The considerations on selecting the appropriate decellularized ECM for specific regeneration demands. *Mater. Today Bio.* **2024**, *29*, 101301. [CrossRef] [PubMed]

4. Rivera-Torres, F.; Maciel-Cerda, A.; González-Gómez, G.H.; Falcón-Neri, A.; Gómez-Lizárraga, K.; Esquivel-Posadas, H.T.; Vera-Graziano, R. In Vitro Modulation of Spontaneous Activity in Embryonic Cardiomyocytes Cultured on Poly(vinyl alcohol)/Bioglass Type 58S Electrospun Scaffolds. *Nanomaterials* **2024**, *14*, 372. [CrossRef] [PubMed]
5. Sousa de Almeida, M.; Lee, A.; Itel, F.; Maniura-Weber, K.; Petri-Fink, A.; Rothen-Rutishauser, B. The Effect of Substrate Properties on Cellular Behavior and Nanoparticle Uptake in Human Fibroblasts and Epithelial Cells. *Nanomaterials* **2024**, *14*, 342. [CrossRef] [PubMed]
6. Niu, Y.; Galluzzi, M. Hyaluronic Acid/Collagen Nanofiber Tubular Scaffolds Support Endothelial Cell Proliferation, Phenotypic Shape and Endothelialization. *Nanomaterials* **2021**, *11*, 2334. [CrossRef] [PubMed]
7. Sassi, L.; Ajayi, O.; Campinoti, S.; Natarajan, D.; McQuitty, C.; Siena, R.R.; Mantero, S.; De Coppi, P.; Pellegata, A.F.; Chokshi, S.; et al. A Perfusion Bioreactor for Longitudinal Monitoring of Bioengineered Liver Constructs. *Nanomaterials* **2021**, *11*, 275. [CrossRef] [PubMed]
8. Górnicki, T.; Lambrinow, J.; Golkar-Narenji, A.; Data, K.; Domagała, D.; Niebora, J.; Farzaneh, M.; Mozdziak, P.; Zabel, M.; Antosik, P.; et al. Biomimetic Scaffolds-A Novel Approach to Three Dimensional Cell Culture Techniques for Potential Implementation in Tissue Engineering. *Nanomaterials* **2024**, *14*, 531. [CrossRef] [PubMed]
9. Roato, I.; Masante, B.; Putame, G.; Massai, D.; Mussano, F. Challenges of Periodontal Tissue Engineering: Increasing Biomimicry through 3D Printing and Controlled Dynamic Environment. *Nanomaterials* **2022**, *12*, 3878. [CrossRef] [PubMed]
10. Cun, X.; Hosta-Rigau, L. Topography: A Biophysical Approach to Direct the Fate of Mesenchymal Stem Cells in Tissue Engineering Applications. *Nanomaterials* **2020**, *10*, 2070. [CrossRef] [PubMed]
11. Corsi, F.; Carotenuto, F.; Di Nardo, P.; Teodori, L. Harnessing Inorganic Nanoparticles to Direct Macrophage Polarization for Skeletal Muscle Regeneration. *Nanomaterials* **2020**, *10*, 1963. [CrossRef] [PubMed]
12. Politi, S.; Carotenuto, F.; Rinaldi, A.; Di Nardo, P.; Manzari, V.; Albertini, M.C.; Araneo, R.; Ramakrishna, S.; Teodori, L. Smart ECM-Based Electrospun Biomaterials for Skeletal Muscle Regeneration. *Nanomaterials* **2020**, *10*, 1781. [CrossRef] [PubMed]
13. Baiguera, S.; Del Gaudio, C.; Carotenuto, F.; Di Nardo, P.; Teodori, L. Information-Driven Design as a Potential Approach for 3D Printing of Skeletal Muscle Biomimetic Scaffolds. *Nanomaterials* **2020**, *10*, 1986. [CrossRef] [PubMed]

Disclaimer/Publisher's Note: The statements, opinions and data contained in all publications are solely those of the individual author(s) and contributor(s) and not of MDPI and/or the editor(s). MDPI and/or the editor(s) disclaim responsibility for any injury to people or property resulting from any ideas, methods, instructions or products referred to in the content.

nanomaterials

Article

In Vitro Modulation of Spontaneous Activity in Embryonic Cardiomyocytes Cultured on Poly(vinyl alcohol)/Bioglass Type 58S Electrospun Scaffolds

Filiberto Rivera-Torres [1], Alfredo Maciel-Cerda [2], Gertrudis Hortensia González-Gómez [3], Alicia Falcón-Neri [3], Karla Gómez-Lizárraga [4], Héctor Tomás Esquivel-Posadas [1] and Ricardo Vera-Graziano [2,*]

[1] Facultad de Química, Universidad Nacional Autónoma de México, Circuito Escolar de Ciudad Universitaria, Coyoacán, Ciudad de México 04510, Mexico; riveratf@unam.mx (F.R.-T.); esquivel.hector99@gmail.com (H.T.E.-P.)
[2] Instituto de Investigaciones en Materiales, Universidad Nacional Autónoma de México, Circuito Escolar de Ciudad Universitaria, Coyoacán, Ciudad de México 04510, Mexico; macielal@unam.mx
[3] Facultad de Ciencias, Universidad Nacional Autónoma de México, Circuito Escolar de Ciudad Universitaria, Coyoacán, Ciudad de México 04510, Mexico; hortecgg@ciencias.unam.mx (G.H.G.-G.); maryaliciafalcon@ciencias.unam.mx (A.F.-N.)
[4] Cátedra CONAHCyT/Instituto de Investigaciones en Materiales, Universidad Nacional Autónoma de México, Circuito Escolar de Ciudad Universitaria, Coyoacán, Ciudad de México 04510, Mexico; karla.gomez@materiales.unam.mx
* Correspondence: graziano@unam.mx

Abstract: Because of the physiological and cardiac changes associated with cardiovascular disease, tissue engineering can potentially restore the biological functions of cardiac tissue through the fabrication of scaffolds. In the present study, hybrid nanofiber scaffolds of poly (vinyl alcohol) (PVA) and bioglass type 58S ($58SiO_2$-$33CaO$-$9P_2O_5$, Bg) were fabricated, and their effect on the spontaneous activity of chick embryonic cardiomyocytes in vitro was determined. PVA/Bg nanofibers were produced by electrospinning and stabilized by chemical crosslinking with glutaraldehyde. The electrospun scaffolds were analyzed to determine their chemical structure, morphology, and thermal transitions. The crosslinked scaffolds were more stable to degradation in water. A Bg concentration of 25% in the hybrid scaffolds improved thermal stability and decreased degradation in water after PVA crosslinking. Cardiomyocytes showed increased adhesion and contractility in cells seeded on hybrid scaffolds with higher Bg concentrations. In addition, the effect of Ca^{2+} ions released from the bioglass on the contraction patterns of cultured cardiomyocytes was investigated. The results suggest that the scaffolds with 25% Bg led to a uniform beating frequency that resulted in synchronous contraction patterns.

Keywords: PVA/bioglass hybrid; electrospun scaffolds; cardiomyocytes; physicochemical characterization; Ca^{2+} bioactivity; cell contractility patterns

Citation: Rivera-Torres, F.; Maciel-Cerda, A.; González-Gómez, G.H.; Falcón-Neri, A.; Gómez-Lizárraga, K.; Esquivel-Posadas, H.T.; Vera-Graziano, R. In Vitro Modulation of Spontaneous Activity in Embryonic Cardiomyocytes Cultured on Poly(vinyl alcohol)/Bioglass Type 58S Electrospun Scaffolds. *Nanomaterials* **2024**, *14*, 372. https://doi.org/10.3390/nano14040372

Academic Editor: Jean-Marie Nedelec

Received: 11 January 2024
Revised: 12 February 2024
Accepted: 13 February 2024
Published: 17 February 2024

Copyright: © 2024 by the authors. Licensee MDPI, Basel, Switzerland. This article is an open access article distributed under the terms and conditions of the Creative Commons Attribution (CC BY) license (https://creativecommons.org/licenses/by/4.0/).

1. Introduction

In medicine, myocardial infarctions and other diseases of the cardiopulmonary system are considered high-risk diseases due to the high mortality rate in the world population [1]. In 2021, 20.5 million people died from cardiovascular diseases (CVDs): "High blood pressure, air pollution, tobacco use, and elevated LDL cholesterol were among the leading contributors to CVD deaths" [2]. In 2018, more than 202 million people worldwide suffered cardiac ischemia [3], so there is an urgent need to develop various strategies to prevent and treat them. Heart valves from metal alloys, as well as stents for the repair of defects in the coronary vessels, were developed [4]. Damage to cardiac tissue due to arterial lesions, infections, or myocardial infarctions is also being investigated and is expected to be repaired by tissue engineering [5]. Tissue engineering for the treatment of myocardial infarction involves the fabrication of three-dimensional scaffolds that promote cell proliferation and adhesion based on biocompatible natural or synthetic polymers. These scaffolds can be porous or dense patches. The human heart has a very low regeneration rate because the cell loss that occurs during infarction is replaced by fibrous avascular tissue that alters the heart rhythm. For these reasons, there is interest in developing biopolymers and bioglasses that regulate cardiac rhythm and regenerate the infarcted area of the myocardium [6]. It is also necessary to measure the effectiveness of the developed biomaterials on the activity patterns of the heart to restore it.

To develop biopolymers for cardiac muscle tissue repair, several polymers from natural sources were tested [7]. To mimic the mechanical and surface properties of the cardiac tissue to be regenerated, hybrid organic/inorganic scaffolds were developed using natural biopolymers; the most commonly used are gelatin [8], collagen [9], alginates [10,11], chitosan [12,13], and fibrin [13], whereas synthetic biopolymers such as poly(l-lactic acid) (PLLA) [14], polyurethane (PU) [15], poly(lactic-co-glycolic acid) (PLGA), poly (ε-caprolactone), (PCL), their respective copolymers, and, more recently, carbon nanofibers are less commonly used [16]. Poly (ethylene glycol) (PEG) alginate hydrogels have been used for the controlled release of drugs into cardiac tissue [17]. Hydrogels and PVA membranes have also been developed for the treatment of vascular emboli due to their biocompatibility and low protein adsorption [18]. The degree of hydrolysis of PVA and its molecular weight determine its mechanical properties, the degree of water solubility, viscosity, and chemical activity of the surface. The latter is due to the ability of the hydroxyl groups on the surface to interact with the surrounding medium via hydrogen bonds, which is mediated by the degree of hydrolysis of PVA. Based on these parameters, the properties of a biomaterial for cardiac tissue regeneration can be developed using a hybrid of PVA and a bioglass [19].

Bioglasses have been used since the second half of the last century as bioactive components in tissue engineering to stimulate angiogenesis. The most successful was Bioglass 45S5®, invented by Larry Hench in 1969, thanks to its good bioactivity and biocompatibility [20]. Bulk bioglasses are rigid, but in the form of microparticles mixed with ad hoc organic polymers, they enhance the mechanical properties of the polymer. Kargozar, S. et. al. [1] used the well-known bioglass 45S5® [$45SiO_2$-$24.5CaO$-$24.5Na_2O$-$6P_6O_5$] mol%, obtained by casting, for cardiac applications. Still, since grinding produces particles with very different diameters, this leads to poor results for soft tissue engineering. Some bioglasses of this type can be obtained as nanoparticles by using the sol-gel method, such as 58S, whose composition is [$58SiO_2$-$36CaO$-$6P_2O_5$] mol% [21]. In the present work, the synthesis, and properties of a type 58S bioglass (Bg) obtained by using the sol-gel method from tetraethyl orthosilicate (TEOS), triethyl phosphate (TEP), calcium chloride dihydrate ($CaCl_2 \cdot 2H_2O$), and deionized water were studied to investigate the effect of Ca^{2+} ions on cardiac functions [22].

Ca^{2+} ions in cardiac tissue affect functions that enable blood circulation in the body. It is known that the contractile work performed by these tissues is due to the tension and contractile efforts of myocytes and fibroblasts, which decrease with age, and to the contractile proteins surrounding cardiomyocytes. Muscle contractions are a function of intracellular Ca^{2+} levels in cardiomyocytes, and a decrease in these ions reduces myocardial conductivity and the development of cardiovascular disease [23,24]. One type of cardiovascular disease is ischemia, in which the loss of cardiomyocytes is later replaced by fibroblasts and scar tissue. These anatomical changes alter the electrophysiological properties of the myocardium and can lead to arrhythmias and heart failure [25]. The infarcted zone becomes asynchronous to the healthy tissue, and the heart expands and remodels to compensate, eventually leading to heart failure [26]. In this sense, the design of a tissue scaffold for the growth of cardiomyocytes in vitro must consider its conductive property, a property necessary for the expression of proteins and specific regulators that allow the reproduction of the vital functions of the tissue to be repaired [27,28]. Therefore, it is necessary to consider the release of Ca^{2+} ions from the scaffold to promote myofilament activation. Previous literature has not examined the effect of the amount of calcium provided by the bioglass nanoparticles on adhesion or the electrical activity of cardiomyocytes by using fluorescence patterns in electrospun hybrid scaffolds.

On the other hand, chick embryonic cardiomyocytes seeded on electrically conducting scaffolds show contractile activity and increased expression of several proteins involved in contractility and electrical intercellular coupling [29,30]. Therefore, tuning the electrical conductivity of scaffolds to achieve the desired scaffold properties and cell responses is of great interest. To date, electrospun scaffolds made of conventional polymers lack suitable mechanical and electrical properties. Therefore, the synchronous beating rate of cardiac cells cultured on these conventional materials has not been achieved [29,31].

Thus, this work aimed to estimate the contractile activity response of cardiomyocytes from chicken embryos on organic/inorganic electrospun hybrid scaffolds of poly(vinyl alcohol) with bioglass ($58SiO_2$-$33CaO$-$9P_2O_5$) by stepwise addition of bioglass nanoparticles containing Ca^{2+} ions obtained by sol-gel.

2. Materials and Methods

2.1. Materials

Tetraethyl orthosilicate TEOS [$Si(OC_2H_5)_4$]; Triethyl phosphate TEP [$(C_2H_5)_3PO_4$]; calcium chloride dihydrate ($CaCl_2 \cdot 2H_2O$); Hydrochloric acid (HCl); and PVA with an average molecular weight between 31,000 and 50,000 g/mol and degree of hydrolysis of 98–99%. Glutaraldehyde grade I, 25% in H_2O, was used for the chemical crosslinking of PVA. All of the above chemical reagents were purchased from Sigma Aldrich (Burlington, MA, USA).

The following reagents were used for the culture of embryonic cardiomyocytes: Hanks' saline solution composition (mM) 125 NaCl, 0.9 KCl, 3.6 $NaHCO_3$, 0.3 Na_2HPO_4, 0.4 KH_2PO_4, 0.5 $MgCl_2$, 0.4 $MgSO_4$, 10 glucose, 2.9 sucrose, 9.9 HEPES, and 2.2 $CaCl_2$, pH 7.2. Cell culture media M199 Gibco BRL 10468-018 with inhibitors of cell proteolytic activity were used to prevent cell damage. Fetal bovine serum (FBS), horse serum (HS), gentamicin (Gibco Life Tech., New York, NY, USA), and 17 mM HEPES (Sigma Aldrich) were added to enhance cell growth. A local supplier provided fertilized eggs according to USDA quality guidelines (Memo 800.65). Calcium Green^{TM}-1 AM Cell Permeant, Invitrogen Thermo Fisher Scientific (Waltham, MA, USA), was used as a fluorochrome, with emission at 506–531 nm in a Leica MZ75 fluorescence stereomicroscope, Leica Microsystems, Wetzlar, Germany, equipped with a lamp with a 506–531 nm wavelength filter.

2.1.1. 58S Bioglass Sol-Gel Synthesis

The synthesis of bioglass 58S was realized according to the report by Bui, X.V. et al. [22] with slight modifications. Bioglass (Bg) was synthesized in a closed vessel by a sol-gel reaction at room temperature using the following procedure: 100 mL of hydrochloric acid solution (HCl, 0.1 M) and 100 mL of TEOS were mixed in a reactor and stirred for 45 min. A volume of 23 mL TEP was poured into it with stirring. Finally, after 45 min, 36.68 g $CaCl_2 \cdot 2H_2O$ was added with stirring for another 45 min to complete the reaction. The phases of polycondensation, solution gelation, and aging were carried out for ten days at room temperature in a closed container. Then, it was stabilized at 70 °C for three days and then at 120 °C for two more days. The product was carefully ground in an agate mortar to obtain nanometric particles. The obtained bioglass was characterized physicochemically.

2.1.2. Electrospinning of PVA and Hybrid Scaffolds

As described elsewhere, a customized experimental electrospinning setup was used to fabricate the scaffold [32]. The conditions for electrospinning PVA and hybrid scaffolds were spinneret needle size and thickness of 0.7 × 30 mm 22 G (1/4), a spinneret-collector distance of 10 cm, a volumetric flow rate of 0.2 mL/h, an applied voltage of 15 kV, an electrospinning time of 5 h, and a flat collector area of 8 × 8 cm. Typical environmental conditions were 45% relative humidity and a temperature of 25 °C. PVA scaffolds co lected on aluminum sheets were dried for 24 h at 70 °C in a vacuum oven to remove water residues.

In this study, the Bg composition was kept unchanged. However, to analyze the effect of calcium concentration on scaffold properties and cell response, scaffolds with different PVA/Bg ratios were prepared according to Table 1. To obtain scaffolds with a certain concentration of PVA/Bg, it was considered to keep the mass of PVA constant in all hybrid scaffolds and determine the amount of Bg that allows obtaining the concentrations indicated in Table 1. An aqueous solution of PVA and an amount of Bg were mixed at 70 °C for 1 h with continuous vigorous stirring and then poured into the electrospinning syringe. Table 1 also shows the concentration of CaO and Ca^{2+} based on the fact that each gram of Bg contains 0.099 g of calcium.

Table 1. Composition of PVA/Bg solutions and calculated calcium concentration.

Hybrid Scaffold Name	PVA/Bg Ratio (*w/w*)	CaO Concentration in Bg	Ca^{2+} Concentration in Bg
PVA	100/0	0	0
5%	95/5	0.1590	0.0114
10%	90/10	0.3180	0.0227
15%	85/15	0.0477	0.0341
20%	80/20	0.0635	0.0454
25%	75/25	0.0794	0.0567
30%	70/30	0.0953	0.0681

2.1.3. Chemical Crosslinking of PVA/Bg Scaffolds

The PVA/Bg scaffolds were crosslinked as follows to prevent dissolution in water. In a separating funnel, 5 mL of HCl at pH = 2 and 500 μL of glutaraldehyde (GA) 25% in water were mixed; then, 25 mL of toluene was added. The mixture was stirred vigorously, and after the chemical reaction, the phases were separated, leaving the residue of glutaraldehyde dissolved in toluene in the upper part of the funnel. At the same time, the HCl migrated to the aqueous phase [33]. The scaffolds were cut with a round punch and immersed for 24 h in beakers containing 4.5 mL of the toluene-GA phase. Finally, the crosslinked samples were dried in a vacuum oven for 24 h at 70 °C.

2.2. *Methods*

2.2.1. Characterization of Non-Crosslinked and Crosslinked PVA/Bg Hybrid Scaffolds

Scanning electron microscopy (SEM) images of crosslinked and non-crosslinked hybrid PVA/Bg scaffolds were acquired using a JEOL JSM-7600F, JEOL Ltd., Tokyo, Japan, field emission scanning electron microscope. A voltage of 10 kV and a current of 4.5×10^{-11} A were employed. The PVA/Bg scaffolds were sputter-coated with gold before being studied by SEM.

The diameters of the fibers were measured using the ImageJ software, Java 8. The obtained data were statistically analyzed using the program "R", and finally, a histogram (Figures S1 and S2) was generated for each micrograph of the electrospun scaffolds.

In addition, to study the distribution of bioglass elements (silicon, calcium, and phosphorus) on the PVA matrix, a mapping of the elements that compose the bioglass was performed using JEOL-ARM-200F, JEOL Ltd., Tokyo, Japan, transmission electron microscopy (TEM) in the mode of STEM. A voltage of 200 kV and a current of 1.5×10^{-11} A were employed.

The nanofibers loaded with 20% bioglass were dispersed in ethanol with the help of ultrasound equipment (60 W/min), and a representative fiber of the tissue scaffold was coated with carbon and stacked on a 100-mesh copper grid (Cat # GD1010-Cu). Subsequently, the surface was analyzed by energy dispersive X-ray spectrometry (EDX) to determine the presence of the elements Si, Ca, and P.

The chemical structure of PVA and PVA/Bg hybrid scaffolds were analyzed in a wavelength range of 4000–400 cm^{-1} at a resolution of 4 cm^{-1} using a Fourier transform infrared spectroscopy with attenuated total reflectance (FTIR-ATR) spectrometer, Thermo Fisher Scientific Nicolet 6700, Waltham, MA, USA. Samples of scaffolds of 1 cm^2 size were placed in the holder and adjusted with a screw.

Thermal transitions (glass transition temperature, Tg, and melting temperature, Tm) of the PVA/Bg hybrid scaffolds were determined by differential scanning calorimetry (DSC) using a TA Instruments DSC 2910 modulated DSC instrument, New Castle, DE, USA. Also weight loss was determined by thermal gravimetric analysis (TGA) using a TA Instrument TGA 2950 high-resolution thermogravimetric analyzer, New Castle, DE, USA. The DSC analysis was performed under a nitrogen atmosphere in a heating/cooling/heating cycle. A first run from 20 °C up to 119 °C at 10 °C/min to remove residual solvent was followed by cooling to 20 °C at 10 °C/min and a second run from 20 °C up to 320 °C at 10 °C/min.

2.2.2. Cardiomyocyte Culture on PVA/Bg Scaffolds

After physicochemical characterization, isolated primary embryonic chick cardiomyocytes were cultured on the PVA/Bg scaffolds to analyze their effects on cardiomyocyte survival, adhesion, and activity patterns. The pure PVA and hybrid scaffolds were cut into 0.25 cm^2 pieces and sterilized under UV-C light for 25 min in 7.5 cm diameter Petri dishes into which dissociated cardiomyocytes were seeded. Primary cell culture was performed according to previously established procedures [34–38]. Ventricular cells collected from embryonic chick hearts in ovo at 7–8 days were enzymatically dissociated to obtain a density of 4×10^4 cells/mL. An aliquot of approximately 20 μL was added to the scaffolds for 20 min to allow the cells to precipitate and adhere to the scaffolds. The remainder of the M199 medium supplemented with 5% FBS and 10% HS was added. Dishes containing cardiomyocyte-seeded scaffolds with media were incubated in a 5% CO_2 atmosphere for up to 48 h at 37 °C. The cell media was replaced and replenished every 24 h under sterile conditions. Previous work has shown that the survivability of cardiomyocytes on different substrates is determined by their contractile status after a toxicity assay with a vital dye plus detergent [32,39]. Spontaneous cell contractility was observed at 36 °C $\pm$ 0.5 as the first sign of cell survival by ocular inspection on a microscope stage; this activity may persist for several days. Calcium Green-1 fluorophore was used to detect intracellular calcium fluctuations (ΔCa^{2+}). The effect of calcium ion availability on the contractility of seeded cardiomyocytes was also investigated.

2.2.3. Fluorescence Analysis

An aliquot of Calcium-Green-1 fluorophore (with a light absorption/emission frequency of 485/517 nm) was diluted in 10 μL of dimethyl sulfoxide (DMSO) and Pluronic acid, sonicated for 15 min, and made up to 500 μL with Hanks' solution. The cell culture medium was carefully replaced with the fluorophore dilution and the dish was incubated at 37 °C for 25 min. Finally, the fluorophore solution was replaced again with Hanks' solution and analyzed using a Leica MZ75 fluorescence stereomicroscope. Images were captured using a Leica DFC360FX, (Leica Microsystems, Wetzlar, Germany). All camera and exposure settings were made using the freely available Micro-Manager software 2.0, (National Institutes of Health). For offline analysis, one to four-minute recordings were made with an exposure time of 50–130 ms. Video images were analyzed by extracting fluorescence amplitude fluctuations in different regions of interest (ROI). All image and video sequences were analyzed using ImageJ 1.6.0_20 software (National Institutes of Health, Bethesda, MD, USA).

Cell images showing contractile activity were outlined with circles. The different scaffold regions were analyzed separately, and fluorescence intensity was plotted as a function of time. ROIs were selected using ImageJ Time Series Analyzer V3.2 tool based on the propagation patterns of the contraction. Finally, the generated data matrix was analyzed using OriginPro 9.0 software to obtain plots over time of contractile activity.

3. Results and Discussion

3.1. Characterization of Electrospun PVA/Bg Hybrid Scaffolds

3.1.1. Nanofiber Diameter and Morphology by SEM

Fiber morphology and the effect of Bg concentration on fiber diameter were analyzed using SEM. Figure 1 shows the micrographs of the electrospun PVA/Bg scaffolds before and after chemical crosslinking. The morphology and diameter show a correlation with the Bg concentration as well as with the immersion of the hybrid scaffolds in the GA/toluene crosslinking solution. The average fiber diameter before crosslinking ranged from 0.127 to 0.338 μm, while after crosslinking ranged from 0.129 to 0.938 μm. Before PVA crosslinking, the fibers have similar fiber diameters and homogeneous surfaces at Bg concentrations below 15%, Figure 1a–c. However, at higher Bg concentrations, Figure 1d–f, the fibers do not have a cylindrical shape, with fibers containing 30% Bg having the largest diameter. The fibers have irregularities and a curved profile with Bg nanoparticles on their surface.

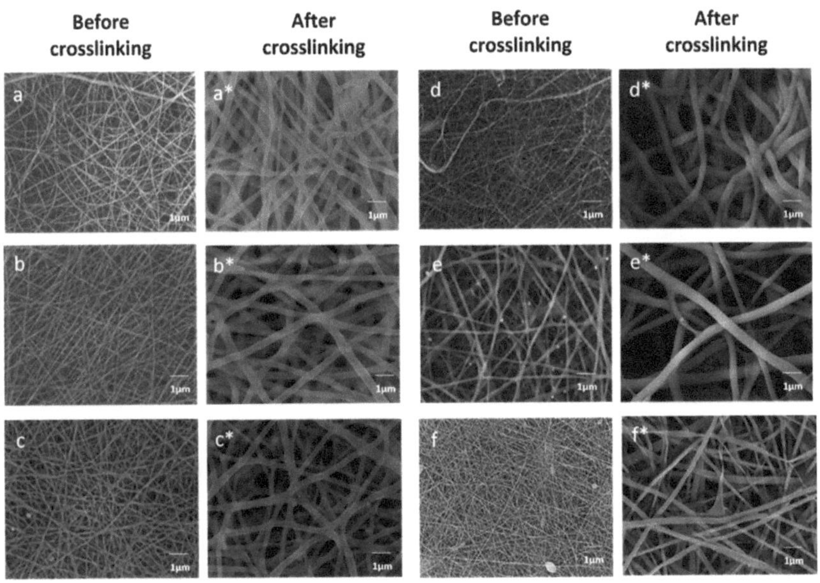

Figure 1. SEM micrographs of PVA scaffolds with different Bg concentrations, before and after (*) chemical crosslinking: (**a**,**a***) 5% hybrid, (**b**,**b***) 10% hybrid, (**c**,**c***) 15% hybrid; (**d**,**d***) 20% hybrid; (**e**,**e***) 25% hybrid and (**f**,**f***) 30% hybrid.

On the other hand, the increase in fiber diameter is also explained by the fact that glutaraldehyde and toluene molecules are introduced into the fibers, while some GA molecules participate in the chemical crosslinking reaction with PVA molecules. The absorbed toluene molecules cause swelling of the fibers and are interstitially located at the interface between the polymer and the bioglass particle.

The calculated uncertainties, given as direct measurements in Figure 2, include both statistical and nonstatistical uncertainties due to the resolution of the measurement system.

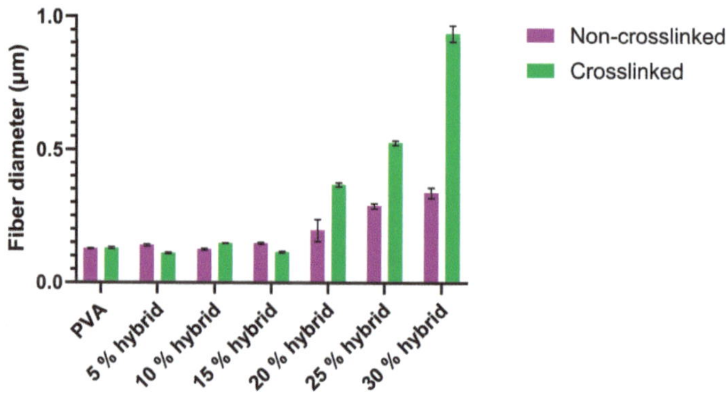

Figure 2. Shows the mean fiber diameters as a function of Bg concentration before and after chemical crosslinking with GA, measured with ImageJ software.

3.1.2. Elemental Mapping of Bg Particles by TEM

The elements of interest in the Bg composition (Ca, Si, and P) were mapped by TEM to determine their spatial distribution within the fibers (Figure 2). It should be noted that the microscopic image of pristine PVA fiber was taken during a short exposure time to avoid the incident electron beam damaging the sample. In contrast, hybrids with a Bg concentration higher than 15% resisted the effect of the electron beam, which allowed a longer exposure time and better image resolution. The fibers with 20% Bg are a representative example. Figure 3a shows the nanofiber without highlighting the Bg elements. A regular surface can be seen without the formation of Bg agglomerates.

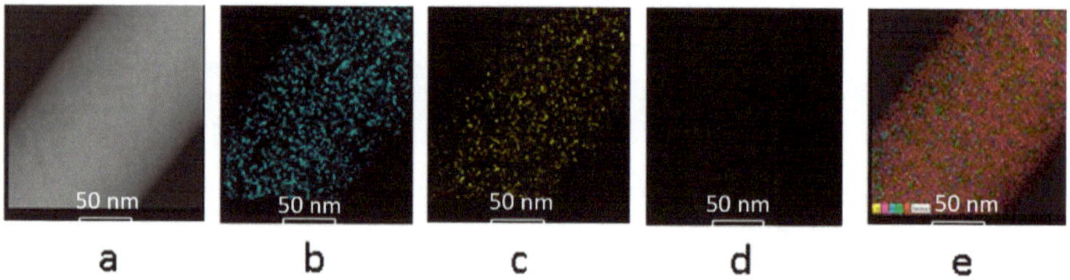

Figure 3. Elemental mapping in hybrids scaffold at 20% Bg concentration: (**a**) electron micrograph of PVA nanofiber, (**b**) Si (cyan), (**c**) Ca (yellow), and (**d**) P (green), and (**e**) merge image overlaying all elemental distributions on the nanofiber micrograph. Specific colors were added afterward.

Figure 3b–d show a uniform distribution of the bioglass elements (silicon, calcium, and phosphorus). Figure 3e shows a uniform distribution of the superimposed elements in the nanofiber of the scaffold. As expected, the distribution of each element was homogeneous along the fiber segment. It is also observed that the degree of element coverage in the fiber is proportional to the molar percentage concentration, namely 58 mol% SiO_2, 33 mol% CaO, and 9 mol% P. The analysis of TEM was performed on the scaffold whose average fiber diameter ranges from 196 to 369 nm, while the diameter of the particles is roughly estimated to be 20 nm or less.

3.1.3. Functional Group Analysis by FTIR-ATR

The FTIR-ATR spectra of the PVA/Bg hybrid scaffolds with different Bg concentrations show the characteristic functional groups of PVA and Bg (Figure 4). Table S1 shows the wavenumber presented in the different scaffolds. Figure 5 illustrates the chemical crosslinking mechanism between the molecules of PVA and GA in an acidic medium, which was confirmed by FTIR-ATR analysis. According to this reaction mechanism, the OH- groups of PVA react mainly with the aldehyde groups of GA, forming O-C-O bonds associated with the vibration of the acetal bridges (900–1150 cm^{-1}) and reducing the intensity of the OH- signal and therefore the hydrophilicity of the fibers [39–41]. PVA and Bg FTIR-ATR spectra are shown in Figure S3.

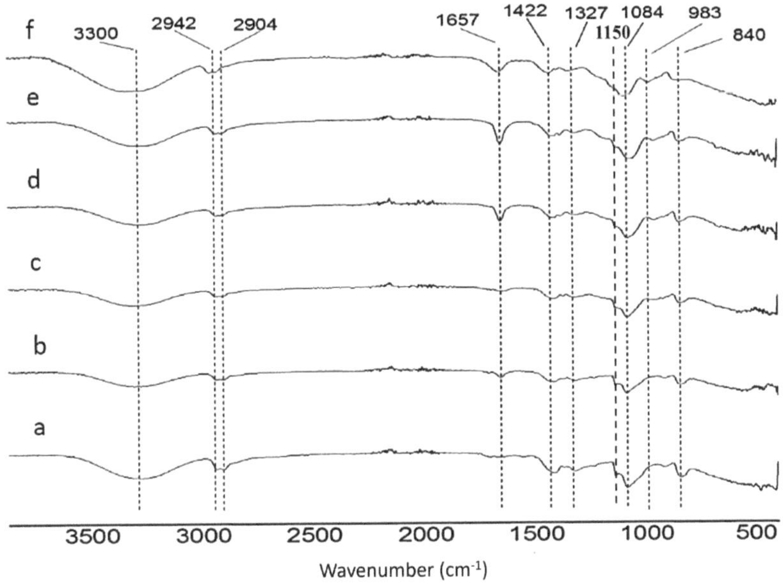

Figure 4. FTIR spectra of PVA scaffolds with different Bg concentrations: (**a**) 5% hybrid; (**b**) 10% hybrid; (**c**) 15% hybrid; (**d**) 20% hybrid; (**e**) 25% hybrid; and (**f**) 30% hybrid.

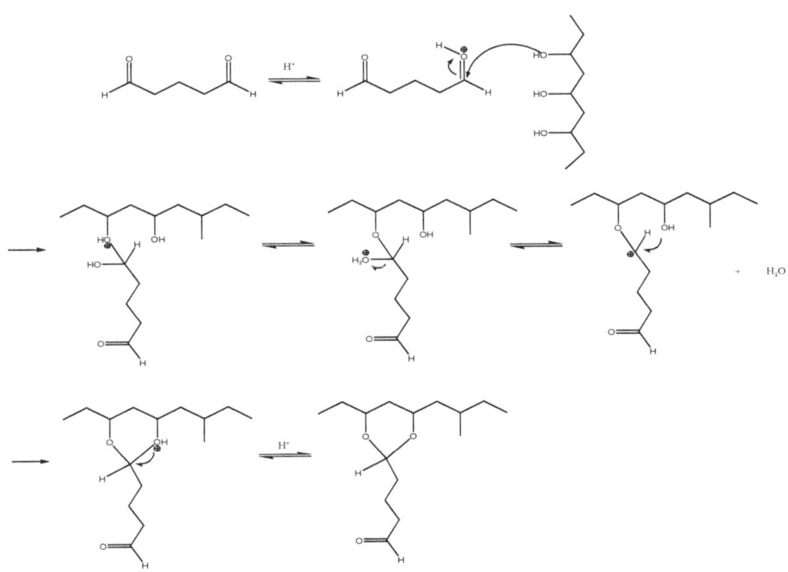

Figure 5. Schematic representation of the crosslinking reaction of PVA with GA.

A decrease in the signal intensity of the PVA OH^- group at 3300 cm^{-1} was observed after PVA crosslinking, which may be attributed to the formation of new bonds by glutaraldehyde interaction (Figure S4). No signals were observed from GA or toluene residues used for chemical crosslinking. It should be noted that the signals corresponding to Si-O-Si bonds were not perceptible at concentrations below 20% Bg in the PVA matrix.

3.1.4. Thermal Analysis by DSC and TGA

Differential Scanning Analysis, DSC

The T_g and T_m of the samples before and after PVA crosslinking are shown in Figure 6. The DSC thermogram of the non-crosslinked PVA/Bg samples showed an endothermic transition between 50 °C and 61 °C because of evaporation (T_{Ev}, evaporation temperature) of low-molecular-weight residual solvents. After evaporation of the residual solvents and cooling, the second heating cycle clearly showed that the T_g was in the range between 63 °C and 88 °C [42].

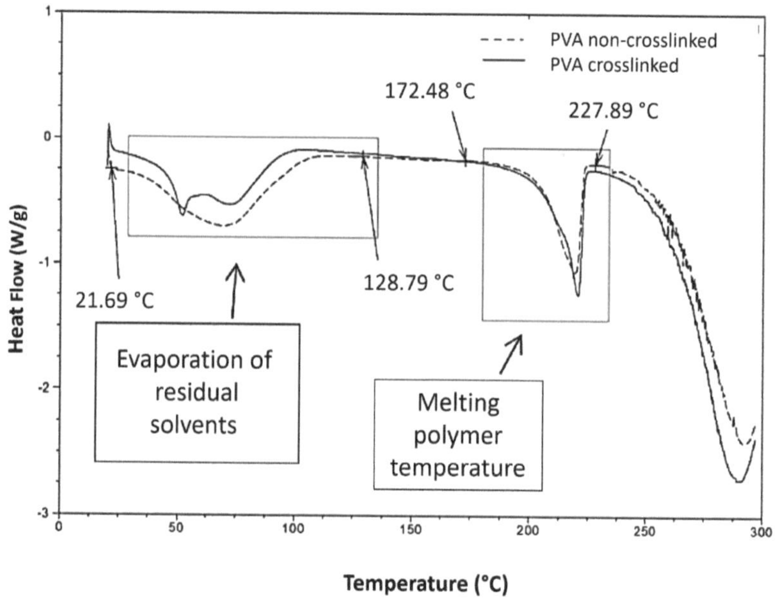

Figure 6. PVA DSC thermograms before and after crosslinking.

The glass transition (T_g) of PVA depends on the degree of hydrolysis. The T_g reported for PVA with a degree of hydrolysis below 98–99% is close to 85 °C [43–46]. The melting temperature, T_m, of PVA also depends on the degree of hydrolysis. The reported values range from 180 °C to 230 °C. The PVA used here exhibited a T_m of 218 °C. It was reported that the crystalline phase of PVA has an average T_m of 230 °C. For partially hydrolyzed PVA, some authors have reported that PVA has an average T_m between 180 °C and 190 °C, [5,47], while other authors have reported a T_m of 209 °C [48].

The T_m may also vary depending on the average molecular weight of the PVA. The T_m of the PVA/Bg hybrid scaffolds shift slightly to higher temperatures as the Bg concentration increases (Table 2). Bg is assumed to adsorb energy before transferring into the polymer, which is enhanced by increasing the Bg concentration in PVA. This effect is more pronounced in hybrid scaffolds where PVA is chemically crosslinked with GA. For PVA/Bg samples, the T_m was observed between 218 °C and 300 °C, Figure S5.

Table 2. Thermal transitions of PVA/Bg hybrids scaffolds.

Bg Content in PVA, (%)	Non-Crosslinked PVA			Crosslinked PVA	
	T_{Ev} (°C)	T_g (°C)	T_m (°C)	T_g (°C)	T_m (°C)
Pristine PVA	50	63	218	66	218
(a) 5%	-	63	219	66	219
(b) 10%	-	70	205	68	262
(c) 15%	-	73	260	69	300
(d) 20%	60	78	263	78	300
(e) 25%	61	88	254	92	300
(f) 30%	56	86	261	98	300
Bioglass	-	385	-	385	-

Thermogravimetric Analysis, TGA

The TGA thermograms of PVA/Bg scaffolds before and after crosslinking showed three clearly defined stages for both groups of scaffolds. The weight loss resulted from the Bg content ranging from 5 to 30%, Table 3. For the crosslinked scaffolds, all samples showed a weight loss of 3 to 16% in the first step, which ranged from 40 °C to 200 °C, due to the presence of residual solvents such as ethanol and water adsorbed on the scaffold. For the non-crosslinked samples, the weight loss ranged from 3 to 10%. In the second step, at 200–350 °C, all scaffolds mostly lost weight due to thermal degradation of PVA molecules. PVA gradually loses OH^- groups at temperatures above 200 °C, which decreases its degree of hydrolysis. As the temperature continues to rise, the weight loss is attributed to degradation products of the hydroxyl groups on the PVA backbone such as acetic acid. For non-crosslinked PVA/Bg fibers, the weight loss ranged from 50 to 66%, while for crosslinked samples, the weight loss ranged from 28 to 66%, with Bg ranging from 5 to 30%. As expected, crosslinking increased the thermal stability of the hybrid composites. The third range, from 350 to 600 °C, showed the mass loss of gases released by the complete combustion of PVA [48,49], and, in general, the crosslinked scaffolds were more stable in the previous two steps, showing a higher weight loss in the third step than the non-crosslinked scaffolds.

Table 3. Weight loss determined by TGA of PVA/Bg hybrid scaffolds before (left) and after (right) crosslinking.

Bg Concentration in PVA, (%)	%Weight Loss before Crosslinking			%Weight Loss after Crosslinking		
	Temperature Intervals (°C)			Temperature Intervals (°C)		
	40–200	200–350	350–600	40–200	200–350	350–600
5	5	66	25	4.3	63.8	21.6
10	10	50	40	3	45	30
15	5	55	18	13	60	25
20	3	45	10	16	40	20
25	3	50	10	6	48	18
30	10	50	10	16	28	15
PVA	6	70	24	10	48	42
Bg	13.56	11	-	13.56	11	-

The TGA thermograms of PVA/Bg hybrid scaffolds before and after crosslinking are shown in Figure S6, in addition to PVA and Bg thermograms, respectively (Figures S7 and S8).

3.1.5. Biological Interaction of Ca^{2+} in Hybrid Scaffolds with Cardiomyocytes

Cardiomyocytes were seeded on scaffolds and cultured in a medium providing appropriate ionic and pH conditions and necessary nutrients, including O_2 and CO_2. The electrolytic medium conditions were enriched with Ca^{2+} and Si^{2+} ions released from the PVA/Bg scaffolds. The effect of calcium on the contraction patterns of cardiomyocytes on the hybrid scaffolds was analyzed by fluorescence microscopy. Figure 7 shows representative images of the behavior of cardiomyocytes on crosslinked and non-crosslinked scaffolds. The living cell population could be identified by its contractile activity, which is seen in the green fluorescence color. It is due to intracellular calcium fluctuations that appear to be associated with increased Bg concentration releasing Ca^{2+} and Si^{2+} ions. That is when the cells are depolarized to contract, implying that their molecular system is functional, including proteins like actin, myosin, troponin, tropomyosin, ATP, Ca^{2+}, etc.

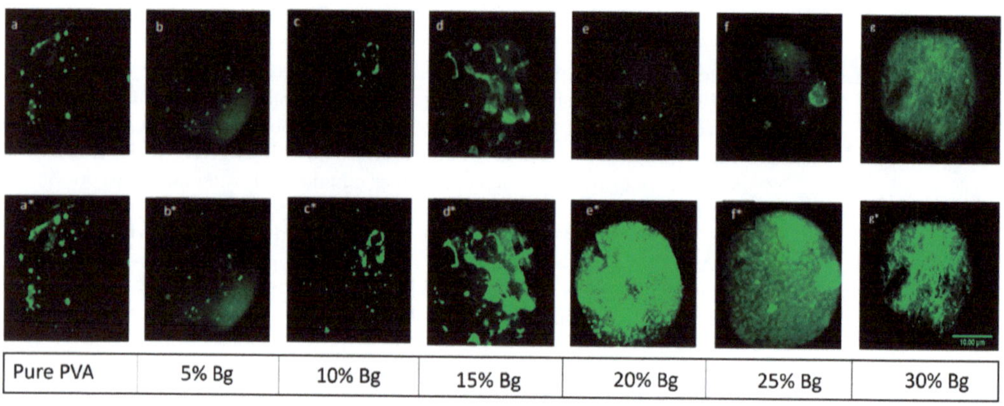

Figure 7. Fluorescence images of cardiomyocytes on hybrid PVA/Bg scaffolds, at different Bg concentrations: (**a**–**g**) at rest (top row) and (**a***–**g***) during contractile activity (bottom row). (4×) stereoscopic microscope.

The images correspond to a layer of thousands of cells, also called a cell monolayer. Therefore, it is not possible to distinguish individual cells with a Leica microscope at 4× magnification, which was used.

Images of PVA/Bg scaffolds at Bg concentrations below 15% in Figure 7a*–c* show independent cell aggregates (niches) with heterogeneous contractile activity. The formation of independent niches hinders the establishment of a wavefront that could reach distant cells and prevents synchronous periodic activity. On the scaffold with 15% Bg, a region of cell niches was observed on a thin layer of cells, Figure 7d*. This cell layer established an incipient but weak connection between cell aggregates that did not promote the spread of contractile activity over the entire scaffold surface. In contrast, the scaffolds with 20%, 25%, and 30% Bg, Figure 7e*–g*, allowed the formation of a strongly connected cell layer that covered the entire scaffold surface, also suggesting high cell adhesion. In this case, the contractile activity of the cells was able to spread along the scaffold, favoring a homogeneous, synchronous, and periodic activity.

Analysis of cardiomyocyte contractile activity shows that regions of interest (ROIs) exhibit homogeneous, synchronous, and rhythmic activity. Remarkably, inactivity is not observed in any ROIs throughout the time interval of the experiment (50 s). Figure 8 shows the sequence of contraction pulses in embryonic cardiomyocytes applied to crosslinked scaffolds, including the type and intensity of cellular electric pulses. Again, at Bg concentrations below 20%, cells developed in independent niches and did not exhibit homogeneous contraction pulses. At Bg concentrations of 20%, 25%, and 30%, the cells formed a uniform cell layer in the ROIs and showed a homogeneous sequence of contraction pulsations. Overall, it was observed that the electrical activity patterns in cardiomyocytes seeded on the electrospun scaffolds strongly depend on the Bg concentration, especially on the availability of Ca^{2+} ions.

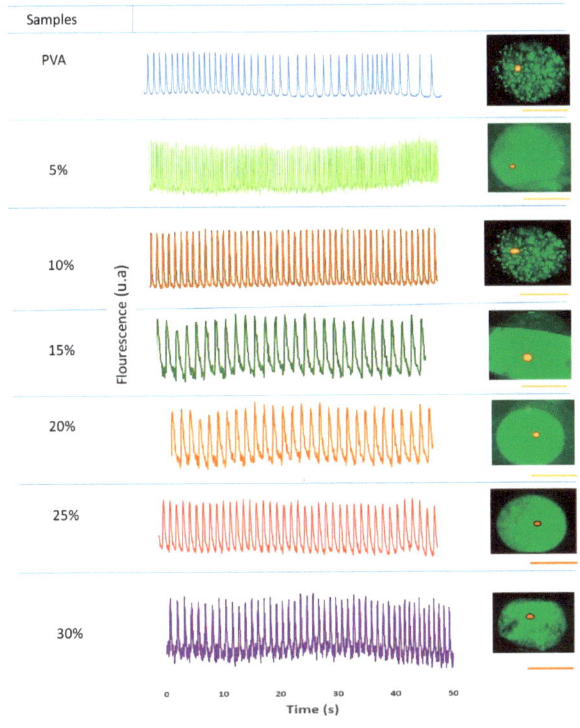

Figure 8. Contractile activity patterns revealed by fluorescence peaks attributable to intracellular ΔCa^{2+} are shown in the center panel; they were selected as regions of interest (ROIs) and are marked with a red dot on the corresponding image on the right. Right column: Fluorescence images of embryonic ventricular cardiomyocytes seeded on PVA/Bg-crosslinked scaffolds at six Bg concentrations as indicated in the left column. As can be appreciated, the different substrates do not inhibit contractile activity. However, the different Bg concentrations regulate adhesion and form separate cell aggregates (5, 10, and 15% Bg) or confluent layers (20, 25, and 30% Bg) with synchronized contractile activity in extended regions. Scale 100 μm.

3.1.6. Proposed Model of Ca^{2+} Ion Dissolution with Time for the Interaction with Cardiomyocytes

Next, a schematic showing the evolution of both the size of the radius of PVA fibers containing Bg and the concentration of Ca^{2+} ions near the fibers as a function of time when the scaffold is embedded in the body fluid is shown in Figure 9. As can be seen in the microscopic images of the Bg nanoparticles (Figure 2), the Ca^{2+} ions are uniformly distributed in the fibers, both on the surface and throughout their volume, as shown in the following schematic, Figure 9. Adenyika [50] showed that the dissolution of PVA in water is a function of concentration. At low concentrations (0.2% *w*/*v*), PVA particles dissolve in 30 to 70 min. Muriel [51] concluded that the rate of dissolution is a function of the pH of the medium: at a higher pH, dissolution is reduced to a few minutes. In studies performed with Bioglass 45S5 [52], it was found that the dissolution speed depends on the particle size. It was observed that particles from 2 µm to 16 µm dissolve in the first few seconds after being placed in a physiological medium, and after 20 min, the dissolution rate stabilizes. Based on these tests, the following qualitative model is proposed to explain the dissolution of Ca^{2+} ions in water as a function of time (Figure 9), assuming that all nanofibers decrease their radius uniformly with time.

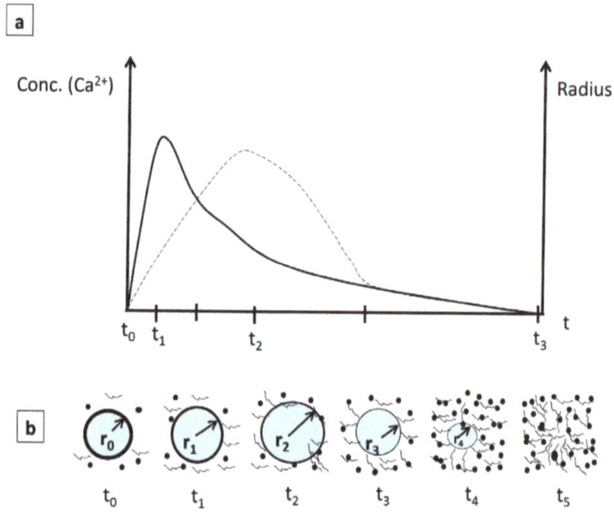

Figure 9. (**a**) Diagrams of dissolution of Ca^{2+} ions and fiber radius versus time: dissolution of Ca^{2+}, solid line, and fiber radius (r), dotted line. (**b**) Schematic of the change in fiber radius at different times.

Once the nanofibers come into contact with the aqueous environment, Si^{2+}, Ca^{2+}, and PO_4^{3-} ions are released. In particular, the release of Ca^{2+} ions tends to increase the pH of the solution. In contrast, the pH of the aqueous medium decreases with the onset of PVA dissolution and balances the pH of the medium [8]. This process ends when the PVA is completely dissolved.

During the first phase (t_o–t_1), the dissolution of Ca^{2+} ions from Bg increases continuously from zero to a maximum value. Thereafter, the dissolution of Ca^{2+} ions decreases while the radius (r) of the fibers continues to increase by swelling to a maximum (t_2). After t_2, both Ca^{2+} and the fiber radius decrease at different rates until t_3, where both rates become similar, followed by the complete dissolution of the fibers, Figure 9a.

In addition, there is water absorption and adsorption that occur in the fibers from t_o to t_1. The fiber radius increases from an initial value of r_0 to a higher value of r_1 at t_1 due to swelling ($r_1 > r_0$); here, the Bg concentration in the aqueous media reaches a maximum. After this point, the Ca^{2+} concentration continuously decreases to its minimum value; however, due to swelling, the fiber radius continues to increase to a maximum value (r_2) (Figure 9b). Thereafter, due to erosion of the PVA and dissolution in the aqueous medium, the fiber radius decreases to r_3 until the fibrillar structure is completely dissolved (t_5), Figure 9b. It is worth mentioning that the swelling phenomenon of the fibers is a consequence of the dissolution process of a polymer. This is described by a series of steps, including the diffusion of the solvent in the polymer matrix and the disentanglement of the polymer chains by a swelling process [53]. In the case of crosslinked polymers, the chains cannot slide between them due to the network structure formed by covalent bonds, and, therefore, infiltration by the solvent begins, causing swelling of the polymer structure.

Fluorescence images of seeded cardiomyocytes (Figure 10) show that the cells remain on the scaffold surface. There is no cell migration into the fiber interior because the pore size of the fibrillar structure is smaller than that of the cardiomyocytes. The mechanism by which ionic migration of Ca^{2+} to the interior of the cytoplasmic membrane occurs is shown schematically in Figure 10. After contact with the scaffold, cardiomyocytes adhere to the hydrophilic nanofiber surface; this adhesion is further enhanced by the release of Si^{2+}, Ca^{2+}, and PO_4^{3-} ions. When these ions dissolve in the culture medium, they interact with the cardiomyocytes across their cytoplasmic membrane.

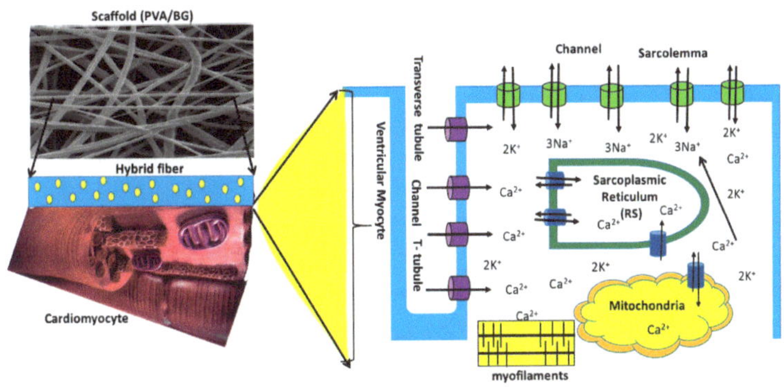

Figure 10. Schematization of cardiomyocytes across the PVA/Bg scaffold surface; cardiomyocytes adhere to the surface of nanofibers in the presence of a culture medium. Released Ca^{2+}, Si^{2+}, and PO_4^{3-} surrounding the cardiomyocyte surface and other ions from the culture medium.

The electrospun scaffolds act as a synthetic extracellular matrix (ECM) that stimulates cardiomyocytes to adhere and orient to the nanofibers, promoting the formation of epimysial and endomysial fibrils. These fibrils help cells form a coherent tissue that allows nutrient diffusion and gas exchange, including CO_2 and O_2 [54–57].

Some biopolymers loaded with metal ions generate electric current depending on the ion concentration [58]. As shown in Figure 8, Si^{2+}, Ca^{2+}, and PO_4^{3-} ions enhance electrical conduction in the case of PVA/Bg scaffolds in comparison with PVA pristine scaffold. After cardiomyocytes are seeded on the scaffold, the culture medium promotes cell interconnection through endomysium and epimysium fibrils, which form active cellular networks that provide mechanical support. Perimysium fibrils also facilitate mechanical work during the contraction and expansion cycles of these cell assemblies through Ca^{2+} exchange. All of these fibrils support action potential function at the cellular level [55].

Depending on Bg concentration, Ca^{2+} ions diffuse into cardiomyocytes and interact with the surface of the cytoplasmic membrane. The membrane senses the presence of these ions as they alter the electrical potential between intracellular and extracellular space by flowing through selective L-type Ca^{2+} channels (Cav1.2), [59,60]. Ca^{2+} ions are crucial intracellular messengers for cellular functions. An increase in concentration in the periphery of the cytoplasmic membrane leads to various cellular responses, including activation of the Ca^{2+} ion channel, ion current or flux, enzymatic activity, neurotransmitter and hormone secretion, and muscle contraction. In other cell types such as stem cells, proliferation, differentiation, and apoptosis are also influenced by Ca^{2+}. Once these channels are opened, Ca^{2+} flows into the cytosol and causes electrical excitation of the cell [58].

The discharge of intracellular deposits activates the influx of Ca^{2+} ions and is associated with membrane depolarization. There are two recognized Ca^{2+} ion channels: L-type and T-type. The first type is activated at a more negative potential difference, i.e., it operates with a low threshold, while the second type operates with a high threshold.

T-type Ca^{2+} channels are functionally expressed in embryonic hearts, whereas they are expressed at higher densities in adult cardiac tissue. They are found in pacemaker cells, so these channels play an important role in pacemaker synchrony. In most cases, these cells contribute to the automatic function of the heart [61].

Putney proposed that the emptying of cellular depots causes the entry of Ca^{2+} from the extracellular medium into the internal depots that form the sarcoplasmic reticulum and from these into the cytosol. During recovery, active pumps such as sarcoendoplasmic reticulum calcium ATPase (SERCA) fill the reservoirs with Ca^{2+} from the cytosol rather than from the extracellular environment. In the plasma membrane, Ca^{2+} channels allow Ca^{2+} from the extracellular environment (1 mM Ca^{2+}) to enter the cell interior (100 nM Ca^{2+}) in favor of a larger electrochemical gradient [54,62–66].

4. Conclusions

Electrospun scaffolds were successfully fabricated using hybrid blends containing PVA as an organic matrix with Bg nanoparticles in a typical composition of 58% SiO_2, 33% CaO and 9% P_2O_5. The sol-gel technique used here to synthesize Bg made it possible to obtain nanoparticles with diameters of 5–20 nm, which facilitated their uniform incorporation into fibers with diameters ranging from 130 nm to 340 nm. Chemical crosslinking of PVA and increasing the Bg concentration allowed the hybrid nanocomposite scaffolds to be thermally stabilized. The proposed model for the change in fiber radius and calcium concentration with time illustrates the process of nanofibers dissolution for better understanding when the scaffold is placed in a physiological medium.

Isolated embryonic ventricular cardiomyocytes seeded on the fibrous scaffolds were successfully examined in vitro for survival, surface adhesion, and spontaneous activity patterns. The second graphical model describes a mechanism by which Ca^{2+} diffuses from the ECM nanofiber interface to its entry through calcium ion channels into cardiomyocytes. Cell viability analyses demonstrated the influence of Ca^{2+} on the adhesion and contractility of cultured cardiomyocytes. Cardiomyocyte adhesion was enhanced by fiber morphology resembling an ECM and the availability of Ca^{2+} ions released by Bg. Calcium ions promote the modulation of spontaneous activity of cardiomyocytes in vitro and enhance the activation of myofilaments. If the cardiomyocytes do not adhere to the scaffold, they float in the saline solution and are easily washed out. The results show that the crosslinked PVA/Bg scaffolds at concentrations of 20%, 25%, and 30% provide the best physicochemical conditions for cardiomyocytes to spread evenly on the surface of the scaffolds and show contractile pulses of activity that spread homogeneously over the entire surface. These scaffolds have a better release of Ca^{2+}, which gradually dissolves from the scaffolds into the culture medium for the cardiomyocytes, inducing the generation of uniform electrical pulses that lead to the observed synchronous contraction patterns in vitro (video).

At least from the results obtained in this work, it can be confirmed that the morphological and chemical stability of the electrospun fibers that make up the scaffolds is stable over a period of 48 h, during which the fluorescence study with the cardiomyocytes was carried out. However, the stability of the scaffolds should be determined at different time intervals.

The results obtained suggest that these hybrid scaffolds are promising candidates for studying the regeneration of infarcted myocardium.

Supplementary Materials: The following supporting information can be downloaded at: https://www.mdpi.com/article/10.3390/nano14040372/s1, Supplementary Materials: Figure S1. Histograms of fiber diameter PVA scaffolds with different Bg concentrations before cross-linking, Figure S2. Histograms of fiber diameter PVA scaffolds with different Bg concentrations after cross-linking, Table S1. Functional groups observed by ATR-FTIR in PVA/Bg hybrid scaffolds, Figure S3. FTIR-ATR spectrum of PVA and 58S bioglass, Figure S4. FTIR-spectra of PVA before and after crosslinking, Figure S5. DSC thermogram of (I) non-crosslinked and crosslinked (II) hybrids scaffolds, Figure S6. TGA thermograms of PVA/Bg hybrid scaffolds. (I) before crosslinking and (II) after crosslinking reaction, Figure S7. PVA thermograms before and after crosslinking, Figure S8. Bioglass TGA thermogram, Determination of standard deviation in fiber diameter measurement in electrospun scaffolds, and Video S1: Cardiomyocytes video.

Author Contributions: Conceptualization, R.V.-G., A.M.-C. and G.H.G.-G.; methodology, F.R.-T., A.F.-N. and H.T.E.-P.; software, H.T.E.-P. and K.G.-L.; validation, F.R.-T., A.F.-N. and H.T.E.-P.; formal analysis, F.R.-T., G.H.G.-G., K.G.-L. and R.V.-G.; investigation, F.R.-T., A.M.-C. and H.T.E.-P.; resources, R.V.-G.; data curation, F.R.-T., A.M.-C., A.F.-N. and K.G.-L.; writing—original draft preparation, F.R.-T. and H.T.E.-P.; writing—review and editing, A.M.-C., G.H.G.-G., A.F.-N., K.G.-L. and R.V.-G.; visualization, F.R.-T.; supervision, A.M.-C., G.H.G.-G. and R.V.-G.; project administration, A.M.-C., G.H.G.-G. and R.V.-G.; funding acquisition, R.V.-G. All authors have read and agreed to the published version of the manuscript.

Funding: This research was funded by Dirección General de Asuntos del Personal Académico (DGAPA)-Universidad Nacional Autónoma de México (UNAM), PAPIIT grant number IN200423, and by Consejo Nacional de Humanidades, Ciencia y Tecnología (CONAHCYT) CF-19 grant 140617.

Data Availability Statement: Data are contained within the article and Supplementary Materials.

Acknowledgments: The authors acknowledge technical support from Adriana Tejeda Cruz for DRX, Josué Esaú Romero Ibarra for TEM, Omar Novelo Peralta and Lourdes Soledad Bazan Díaz for SEM, and Karla Eriseth Reyes Morales for DSC and TGA, as well as the helpful discussions with Gerardo Cedillo-Servin. This work was supported by Dirección General de Asuntos del Personal Académico (DGAPA)-Universidad Nacional Autónoma de México (UNAM), PAPIIT grant IN200423, and by Consejo Nacional de Humanidades, Ciencia y Tecnología (CONAHCYT) CF-19 grant 140617. Dra. Karla Gómez-Lizárraga would like to thank CONAHCYT for Catedra project CIR/0040/2022.

Conflicts of Interest: The authors declare no conflicts of interest.

References

1. Kargozar, S.; Hamzehlou, S.; Baino, F. Potential of Bioactive Glasses for Cardiac and Pulmonary Tissue Engineering. *Materials* **2017**, *10*, 1429. [CrossRef]
2. *Deaths from Cardiovascular Diseases Surges 60% Globaly on the Last 30 Years: Report*; World Heart Federation: Geneva, Switzerland, 2023.
3. Nystoriak, M.A.; Bhatnagar, A. Cardiovascular Effects and Benefits of Exercise. *Front. Cardiovasc. Med.* **2018**, *5*, 135. [CrossRef]
4. Freytes, D.O.; Santambrogio, L.; Vunjak-Novakovic, G. Optimizing Dynamic Interactions between a Cardiac Patch and Inflammatory Host Cells. *Cells Tissues Organs* **2011**, *195*, 171–182. [CrossRef]
5. Maria, T.M.C.; de Carvalho, R.A.; Sobral, P.J.A.; Habitante, A.M.B.Q.; Solorza-Feria, J. The Effect of the Degree of Hydrolysis of the PVA and the Plasticizer Concentration on the Color, Opacity, and Thermal and Mechanical Properties of Films Based on PVA and Gelatin Blends. *J. Food Eng.* **2008**, *87*, 191–199. [CrossRef]
6. Pogwizd, S.M.; Schlotthauer, K.; Li, L.; Yuan, W.; Bers, D.M. Arrhythmogenesis and Contractile Dysfunction in Heart Failure. *Circ. Res.* **2001**, *88*, 1159–1167. [CrossRef]
7. Dattola, E.; Parrotta, E.I.; Scalise, S.; Perozziello, G.; Limongi, T.; Candeloro, P.; Coluccio, M.L.; Maletta, C.; Bruno, L.; De Angelis, M.T.; et al. Development of 3D PVA Scaffolds for Cardiac Tissue Engineering and Cell Screening Applications. *RSC Adv.* **2019**, *9*, 4246–4257. [CrossRef]
8. Hirt, M.N.; Hansen, A.; Eschenhagen, T. Cardiac Tissue Engineering: State of the Art. *Circ. Res.* **2014**, *114*, 354–367. [CrossRef]
9. Kitsara, M.; Agbulut, O.; Kontziampasis, D.; Chen, Y.; Menasché, P. Fibers for Hearts: A Critical Review on Electrospinning for Cardiac Tissue Engineering. *Acta Biomater.* **2017**, *48*, 20–40. [CrossRef]
10. Cui, Z.; Yang, B.; Li, R.K. Application of Biomaterials in Cardiac Repair and Regeneration. *Engineering* **2016**, *2*, 141–148. [CrossRef]
11. Ruvinov, E.; Cohen, S. Alginate Biomaterial for the Treatment of Myocardial Infarction: Progress, Translational Strategies, and Clinical Outlook. From Ocean Algae to Patient Bedside. *Adv. Drug Deliv. Rev.* **2016**, *96*, 54–76. [CrossRef]
12. Liu, Z.; Wang, H.; Wang, Y.; Lin, Q.; Yao, A.; Cao, F.; Li, D.; Zhou, J.; Duan, C.; Du, Z.; et al. The Influence of Chitosan Hydrogel on Stem Cell Engraftment, Survival and Homing in the Ischemic Myocardial Microenvironment. *Biomaterials* **2012**, *33*, 3093–3106. [CrossRef]
13. Christman, K.L.; Fok, H.H.; Sievers, R.E.; Fang, Q.; Lee, R.J. Fibrin Glue Alone and Skeletal Myoblasts in a Fibrin Scaffold Preserve Cardiac Function after Myocardial Infarction. *Tissue Eng.* **2004**, *10*, 403–409. [CrossRef]
14. Lakshmanan, R.; Krishnan, U.M.; Sethuraman, S. Polymeric Scaffold Aided Stem Cell Therapeutics for Cardiac Muscle Repair and Regeneration. *Macromol. Biosci.* **2013**, *13*, 1119–1134. [CrossRef]
15. Chiono, V.; Mozetic, P.; Boffito, M.; Sartori, S.; Gioffredi, E.; Silvestri, A.; Rainer, A.; Giannitelli, S.M.; Trombetta, M.; Nurzynska, D.; et al. Polyurethane-Based Scaffolds for Myocardial Tissue Engineering. *Interface Focus* **2014**, *4*, 20130045. [CrossRef]
16. Plackett, D.; VÁzquez, A. Natural Polymer Sources. In *Green Composites: Polymer Composites and the Environment*; Woodhead Publishing: Sawston, UK, 2004; pp. 123–153. ISBN 9781855737396.
17. Chaudhuri, R.; Ramachandran, M.; Moharil, P.; Harumalani, M.; Jaiswal, A.K. Biomaterials and Cells for Cardiac Tissue Engineering: Current Choices. *Mater. Sci. Eng. C* **2017**, *79*, 950–957. [CrossRef]
18. Kamoun, E.A.; Chen, X.; Mohy Eldin, M.S.; Kenawy, E.R.S. Crosslinked Poly(Vinyl Alcohol) Hydrogels for Wound Dressing Applications: A Review of Remarkably Blended Polymers. *Arab. J. Chem.* **2015**, *8*, 1–14. [CrossRef]

19. Baker, M.I.; Walsh, S.P.; Schwartz, Z.; Boyan, B.D. A Review of Polyvinyl Alcohol and Its Uses in Cartilage and Orthopedic Applications. *J. Biomed. Mater. Res. B Appl. Biomater.* **2012**, *100 B*, 1451–1457. [CrossRef]
20. Nagarkar, R.; Patel, J. Polyvinyl Alcohol: A Comprehensive Study. *Acta Sci. Pharm. Sci.* **2019**, *3*, 34–44.
21. AM, A. A Mini-Review on the Bioactive Glass-Based Composites in Soft Tissue Repair. *Bioceram. Dev. Appl.* **2018**, *8*, 8–11. [CrossRef]
22. Bui, X.V.; Dang, T.H. Bioactive Glass 58S Prepared Using an Innovation Sol-Gel Process. *Process. Appl. Ceram.* **2019**, *13*, 98–103. [CrossRef]
23. Ding, Y.; Li, W.; Müller, T.; Schubert, D.W.; Boccaccini, A.R.; Yao, Q.; Roether, J.A. Electrospun Polyhydroxybutyrate/Poly(ε-Caprolactone)/58S Sol-Gel Bioactive Glass Hybrid Scaffolds with Highly Improved Osteogenic Potential for Bone Tissue Engineering. *ACS Appl. Mater. Interfaces* **2016**, *8*, 17098–17108. [CrossRef] [PubMed]
24. Bers, D.M. Calcium Fluxes Involved in Control of Cardiac Myocyte Contraction. *Circ. Res.* **2000**, *87*, 275–281. [CrossRef] [PubMed]
25. Sharma, D.; Ferguson, M.; Kamp, T.J.; Zhao, F. Constructing Biomimetic Cardiac Tissues: A Review of Scaffold Materials for Engineering Cardiac Patches. *Emergent Mater.* **2019**, *2*, 181–191. [CrossRef] [PubMed]
26. Kai, D.; Prabhakaran, M.P.; Jin, G.; Ramakrishna, S. Biocompatibility Evaluation of Electrically Conductive Nanofibrous Scaffolds for Cardiac Tissue Engineering. *J. Mater. Chem. B* **2013**, *1*, 2305–2314. [CrossRef] [PubMed]
27. Geng, X.; Liu, B.; Liu, J.; Liu, D.; Lu, Y.; Sun, X.; Liang, K.; Kong, B. Interfacial Tissue Engineering of Heart Regenerative Medicine Based on Soft Cell-Porous Scaffolds. *J. Thorac. Dis.* **2018**, *10*, S2333–S2345. [CrossRef] [PubMed]
28. Fakoya, A.O.J. New Delivery Systems of Stem Cells for Vascular Regeneration in Ischemia. *Front. Cardiovasc. Med.* **2017**, *4*, 7. [CrossRef] [PubMed]
29. Tallawi, M.; Rai, R.; Boccaccini, A.R.; Aifantis, K.E. Effect of Substrate Mechanics on Cardiomyocyte Maturation and Growth. *Tissue Eng. Part B Rev.* **2015**, *21*, 157–165. [CrossRef] [PubMed]
30. Zhang, W.; Kong, C.W.; Tong, M.H.; Chooi, W.H.; Huang, N.; Li, R.A.; Chan, B.P. Maturation of Human Embryonic Stem Cell-Derived Cardiomyocytes (hESC-CMs) in 3D Collagen Matrix: Effects of Niche Cell Supplementation and Mechanical Stimulation. *Acta Biomater.* **2017**, *49*, 204–217. [CrossRef]
31. Durán-Pastén, M.L.; González-Gómez, G.H. Recurrence Plots to Analyze the Dynamical Changes During. In *AIP Conference Proceedings*; AIP: Melville, NY, USA, 2019; Volume 2090, p. 050005.
32. Li, Y.; Yao, S. High Stability under Extreme Condition of the Poly(Vinyl Alcohol) Nanofibers Crosslinked by Glutaraldehyde in Organic Medium. *Polym. Degrad. Stab.* **2017**, *137*, 229–237. [CrossRef]
33. Durán-Pastén, M.L.; Cortes, D.; Valencia-Amaya, A.E.; King, S.; González-Gómez, G.H.; Hautefeuille, M. Cell Culture Platforms with Controllable Stiffness for Chick Embryonic Cardiomyocytes. *Biomimetics* **2019**, *4*, 33. [CrossRef]
34. Oropeza-Ramos, L.; Macias, A.; Juárez, S.; Falcón, A.; Torres, A.; Hautefeuille, M.; González-Gómez, G.H. Low cost micro-platform for culturing and stimulation of cardiomyocyte tissue. In Proceedings of the 2011 IEEE 24th International Conference on Micro Electro Mechanical Systems, Cancun, Mexico, 23–27 January 2011; pp. 912–915.
35. González, H.; Nagai, Y.; Bub, G.; Glass, L.; Shrier, A. Reentrant Waves in a Ring of Embryonic Chick Ventricular Cells Imaged with a Ca^{2+} Sensitive Dye. *BioSystems* **2003**, *71*, 71–80. [CrossRef]
36. González, H.; Nagai, Y.; Bub, G.; Glass, L.; Shrier, A. Resetting and Annihilating Reentrant Waves in a Ring of Cardiac Tissue: Theory and Experiment. *Prog. Theor. Phys. Suppl.* **2000**, *139*, 83–89. [CrossRef]
37. Bub, G.; Glass, L.; Publicover, N.G.; Shrier, A. Bursting Calcium Rotors in Cultured Cardiac Myocyte Monolayers. *Proc. Natl. Acad. Sci. USA* **1998**, *95*, 10283–10287. [CrossRef] [PubMed]
38. Miranda-Buendia, E.; González-Gómez, G.H.; Falcón-Neri, M.A.; Durán-Pastén, M.L.; Jiménez-Martínez, C.; Vera-Graziano, R.; Ospina-Orejarena, A.; Rivera-Torres, F.; Prado-Villegas, G.; Maciel-Cerda, A. Activity Patterns of Cardiomyocytes in Electrospun Scaffolds of Poly (ε-Caprolactone), Collagen, and Epicatechin. *Mater. Today Commun.* **2022**, *31*, 103405. [CrossRef]
39. Mansur, H.S.; Sadahira, C.M.; Souza, A.N.; Mansur, A.A.P. FTIR Spectroscopy Characterization of Poly (Vinyl Alcohol) Hydrogel with Different Hydrolysis Degree and Chemically Crosslinked with Glutaraldehyde. *Mater. Sci. Eng. C* **2008**, *28*, 539–548. [CrossRef]
40. Tang, C.; Saquing, C.D.; Harding, J.R.; Khan, S.A. In Situ Cross-Linking of Electrospun Poly(Vinyl Alcohol) Nanofibers. *Macromolecules* **2010**, *43*, 630–637. [CrossRef]
41. Figueiredo, K.C.S.; Alves, T.L.M.; Borges, C.P. Poly(Vinyl Alcohol) Films Crosslinked by Glutaraldehyde Under Mild Conditions. *J. Appl. Polym. Sci.* **2009**, *111*, 3074–3080. [CrossRef]
42. Guirguis, O.W.; Moselhey, M.T.H. Thermal and Structural Studies of Poly (Vinyl Alcohol) and Hydroxypropyl Cellulose Blends. *Nat. Sci.* **2012**, *4*, 57–67. [CrossRef]
43. Rault, J.; Gref, R.; Ping, Z.H.; Nguyen, Q.T.; Néel, J. Glass Transition Temperature Regulation Effect in a Poly(Vinyl Alcohol)-Water System. *Polymer* **1995**, *36*, 1655–1661. [CrossRef]
44. Taghizadeh, M.T.; Yeganeh, N.; Rezaei, M. The Investigation of Thermal Decomposition Pathway and Products of Poly(Vinyl Alcohol) by TG-FTIR. *J. Appl. Polym. Sci.* **2015**, *132*, 42117. [CrossRef]
45. Stavropoulou, A.; Papadokostaki, K.G.; Sanopoulou, M. Thermal Properties of Poly(Vinyl Alcohol)-Solute Blends Studied by TMDSC. *J. Appl. Polym. Sci.* **2004**, *93*, 1151–1156. [CrossRef]
46. Huang, K.; Cai, S.; Xu, G.; Ren, M.; Wang, X.; Zhang, R.; Niu, S.; Zhao, H. Sol-Gel Derived Mesoporous 58S Bioactive Glass Coatings on AZ31 Magnesium Alloy and in Vitro Degradation Behavior. *Surf. Coat. Technol.* **2014**, *240*, 137–144. [CrossRef]

47. Zanela, J.; Bilck, A.P.; Casagrande, M.; Grossmann, M.V.E.; Yamashita, F. Polyvinyl Alcohol (PVA) Molecular Weight and Extrusion Temperature in Starch/PVA Biodegradable Sheets. *Polimeros* **2018**, *28*, 256–265. [CrossRef]
48. Peng, Z.; Kong, L.X. A Thermal Degradation Mechanism of Polyvinyl Alcohol/Silica Nanocomposites. *Polym. Degrad. Stab.* **2007**, *92*, 1061–1071. [CrossRef]
49. Taghi Taghizadeh, M.; Sabouri, N. Thermal Degradation Behavior of Polyvinyl Alcohol/Starch/Carboxymethyl Cellulose/Clay Nanocomposites. *Univers. J. Chem.* **2013**, *1*, 21–29. [CrossRef]
50. Aina, A.; Morris, A.; Gupta, M.; Billa, N.; Madhvani, N. Dissolution Behavior of Poly Vinyl Alcohol in Water and Its Effect on the Physical Morphologies of PLGA Scaffolds. *Pharm. Biosci. J.* **2014**, *2*, 1–6. [CrossRef]
51. Józó, M.; Simon, N.; Yi, L.; Móczó, J.; Pukánszky, B. Improved Release of a Drug with Poor Water Solubility by Using Electrospun Water-Soluble Polymers as Carriers. *Pharmaceutics* **2022**, *14*, 34. [CrossRef] [PubMed]
52. Cerruti, M.G.; Greenspan, D.; Powers, K. An Analytical Model for the Dissolution of Different Particle Size Samples of Bioglass®in TRIS-Buffered Solution. *Biomaterials* **2005**, *26*, 4903–4911. [CrossRef] [PubMed]
53. Miller-Chou, B.A.; Koenig, J.L. A Review of Polymer Dissolution. *Prog. Polym. Sci.* **2023**, *28*, 1223–1270. [CrossRef]
54. Shapira, A.; Feiner, R.; Dvir, T. Composite Biomaterial Scaffolds for Cardiac Tissue Engineering. *Int. Mater. Rev.* **2016**, *61*, 1–19. [CrossRef]
55. Wu, Y.; Wang, L.; Guo, B.; Ma, P.X. Interwoven Aligned Conductive Nanofiber Yarn/Hydrogel Composite Scaffolds for Engineered 3D Cardiac Anisotropy. *ACS Nano* **2017**, *11*, 5646–5659. [CrossRef]
56. Jin, G.; He, R.; Sha, B.; Li, W.; Qing, H.; Teng, R.; Xu, F. Electrospun Three-Dimensional Aligned Nanofibrous Scaffolds for Tissue Engineering. *Mater. Sci. Eng. C* **2018**, *92*, 995–1005. [CrossRef]
57. Kim, H.N.; Hong, Y.; Kim, M.S.; Kim, S.M.; Suh, K.Y. Effect of Orientation and Density of Nanotopography in Dermal Wound Healing. *Biomaterials* **2012**, *33*, 8782–8792. [CrossRef]
58. Zhang, Y.; Jiang, X.; Snutch, T.P.; Tao, J. Modulation of Low-Voltage-Activated T-Type Ca^{2+} Channels. *Biochim. Biophys. Acta Biomembr.* **2013**, *1828*, 1550–1559. [CrossRef]
59. Vassort, G.; Talavera, K.; Alvarez, J.L. Role of T-Type Ca^{2+} Channels in the Heart. *Cell Calcium* **2006**, *40*, 205–220. [CrossRef]
60. Lieb, A.; Ortner, N.; Striessnig, J. C-Terminal Modulatory Domain Controls Coupling of Voltage-Sensing to Pore Opening in Cav1.3 L-Type Ca^{2+} Channels. *Biophys. J.* **2014**, *106*, 1467–1475. [CrossRef]
61. Ono, K.; Iijima, T. Cardiac T-Type Ca^{2+} Channels in the Heart. *J. Mol. Cell. Cardiol.* **2010**, *48*, 65–70. [CrossRef]
62. Putney, J.W. A Model for Receptor-Regulated Calcium Entry. *Cell Calcium* **1986**, *7*, 1–12. [CrossRef]
63. Putney, J.W. Capacitative Calcium Entry Revisited. Review Article. *Cell Calcium* **1990**, *11*, 611–624. [CrossRef]
64. Birkeland, J.A.; Sejersted, O.M.; Taraldsen, T.; Sjaastad, I. EC-Coupling in Normal and Failing Hearts. *Scand. Cardiovasc. J.* **2005**, *39*, 13–23. [CrossRef] [PubMed]
65. Eisner, D.; Bode, E.; Venetucci, L.; Trafford, A. Calcium Flux Balance in the Heart. *J. Mol. Cell Cardiol.* **2013**, *58*, 110–117. [CrossRef]
66. Campbell, D.L.; Strauss, H.C. Regulation of Calcium Channels in the Heart. *Adv. Second. Messenger Phosphoprot. Res.* **1995**, *30*, 25–88. [CrossRef]

Disclaimer/Publisher's Note: The statements, opinions and data contained in all publications are solely those of the individual author(s) and contributor(s) and not of MDPI and/or the editor(s). MDPI and/or the editor(s) disclaim responsibility for any injury to people or property resulting from any ideas, methods, instructions or products referred to in the content.

nanomaterials

MDPI

Article

The Effect of Substrate Properties on Cellular Behavior and Nanoparticle Uptake in Human Fibroblasts and Epithelial Cells

Mauro Sousa de Almeida [1], Aaron Lee [1,2], Fabian Itel [3], Katharina Maniura-Weber [4], Alke Petri-Fink [1,5] and Barbara Rothen-Rutishauser [1,*]

1 Adolphe Merkle Institute and National Center of Competence in Research Bio-Inspired Materials, University of Fribourg, Chemin des Verdiers 4, 1700 Fribourg, Switzerland; mauro.sousadealmeida@unifr.ch (M.S.d.A.) a.lee22@imperial.ac.uk (A.L.); alke.fink@unifr.ch (A.P.-F.)
2 Department of Bioengineering, Imperial College London, South Kensington, London SW7 2BP, UK
3 Empa, Swiss Federal Laboratories for Materials Science and Technology, Laboratory for Biomimetic Membranes and Textiles, Lerchenfeldstrasse 5, 9014 St. Gallen, Switzerland; fabian.itel@empa.ch
4 Empa, Swiss Federal Laboratories for Materials Science and Technology, Laboratory for Biointerfaces, Lerchenfeldstrasse 5, 9014 St. Gallen, Switzerland; katharina.maniura@empa.ch
5 Department of Chemistry, University of Fribourg, Chemin du Musée 9, 1700 Fribourg, Switzerland
* Correspondence: barbara.rothen@unifr.ch

Abstract: The delivery of nanomedicines into cells holds enormous therapeutic potential; however little is known regarding how the extracellular matrix (ECM) can influence cell–nanoparticle (NP) interactions. Changes in ECM organization and composition occur in several pathophysiological states, including fibrosis and tumorigenesis, and may contribute to disease progression. We show that the physical characteristics of cellular substrates, that more closely resemble the ECM in vivo, can influence cell behavior and the subsequent uptake of NPs. Electrospinning was used to create two different substrates made of soft polyurethane (PU) with aligned and non-aligned nanofibers to recapitulate the ECM in two different states. To investigate the impact of cell–substrate interaction, A549 lung epithelial cells and MRC-5 lung fibroblasts were cultured on soft PU membranes with different alignments and compared against stiff tissue culture plastic (TCP)/glass. Both cell types could attach and grow on both PU membranes with no signs of cytotoxicity but with increased cytokine release compared with cells on the TCP. The uptake of silica NPs increased more than three-fold in fibroblasts but not in epithelial cells cultured on both membranes. This study demonstrates that cell–matrix interaction is substrate and cell-type dependent and highlights the importance of considering the ECM and tissue mechanical properties when designing NPs for effective cell targeting and treatment.

Keywords: nanoparticle uptake; extracellular matrix; mechanobiology; nanofibers; stiffness; fibroblasts; epithelial cells

Citation: Sousa de Almeida, M.; Lee, A.; Itel, F.; Maniura-Weber, K.; Petri-Fink, A.; Rothen-Rutishauser, B. The Effect of Substrate Properties on Cellular Behavior and Nanoparticle Uptake in Human Fibroblasts and Epithelial Cells. *Nanomaterials* **2024**, *14*, 342. https://doi.org/10.3390/nano14040342

Academic Editors: Costantino Del Gaudio, Silvia Baiguera and Lucy Di Silvio

Received: 7 January 2024
Revised: 5 February 2024
Accepted: 7 February 2024
Published: 10 February 2024

Copyright: © 2024 by the authors. Licensee MDPI, Basel, Switzerland. This article is an open access article distributed under the terms and conditions of the Creative Commons Attribution (CC BY) license (https://creativecommons.org/licenses/by/4.0/).

1. Introduction

Intensive research in nanomedicine has been conducted to develop new systems such as nanoparticles (NPs) to deliver therapeutic agents, including drugs, to treat various diseases such as cancer [1,2]. The primary focus of researchers has been on engineering the physicochemical properties (e.g., size, shape, and surface chemistry) of these NPs to improve treatment efficacy [3,4]. However, more in-depth in vitro research into how cell NP uptake differs in healthy and diseased states is required. The efficient delivery of NPs to the target cells depends on cell mechanical properties and the physical and biochemical properties of the surrounding environment that directs cell behavior [5].

Tissue culture plastic (TCP) and track-etched porous membranes are the most commonly used reference substrate materials to grow cells in vitro for cell–NP interaction studies [6,7]. These membranes are usually incorporated into cell culture inserts and placed in well plates, separating the bottom (basal) and top (apical) compartments and facilitating the creation of physiologically representative barrier models [8,9]. These reference substrate materials are typically made of polystyrene, polyethylene terephthalate, and polyester and have mechanical behaviors significantly dissimilar to tissues in situ. Importantly, cells are influenced by substrate properties such as mechanical stiffness and consequently may account for differences in nanoparticle uptake behavior between cells cultured in vitro and cells cultured in vivo [10].

Biocompatible cell substrates such as scaffolds support cell attachment and growth. They are designed to mimic the extracellular matrix (ECM) properties of human tissues and are typically made up of fibrous components that act as attachment and growth cues for cells [11]. The ECM is a dynamic, acellular structure that, along with the surrounding cells, forms the tissue microenvironment [12]. The biomechanical properties of the ECM in human tissues can change in pathological conditions such as fibrosis and cancer [13]. Remodeling of the ECM in disease is typically associated with increased fiber deposition and overall stiffness [14,15]. Altered matrix stiffness and force signaling within the microenvironment can dramatically affect critical cell functions such as migration, proliferation, and differentiation [12,16]. Crucially, matrix stiffness can drive cell mechanotransduction pathways and promote the fibroblast-to-myofibroblast transition and the differentiation of mesenchymal stem cells [17,18]. Furthermore, changes in the ECM reorganization and topography, including fiber alignment, have been shown to limit tumor invasion in breast cancer [19]. This knowledge should be considered when designing and developing new therapeutics.

Investigation of cell behavior on cell substrates that more closely resemble the tissue microenvironment is required to support and facilitate data translation from in vitro culture to in vivo applications. We evaluate the role of substrate stiffness and topography in cell phenotype, inflammation, and the cellular uptake of NPs. Polyurethane (PU) has been used for various biomedical applications (e.g., scaffolds) as it is biocompatible and presents good mechanical properties (e.g., strength and elasticity) [20]. Electrospinning was used to create two soft and biocompatible PU membranes composed of non-aligned (randomly oriented) and aligned nanofibers. The influence of both soft membranes on the morphology of lung epithelial cells (A549) and fibroblasts (MRC-5) was investigated and compared with the stiff reference substrate TCP/glass. The biocompatibility of PU membranes and their effects on critical processes such as the inflammation and internalization of 80 nm silica (SiO_2) NPs were investigated.

2. Materials and Methods

2.1. Materials

Texin 985A is a polyether-based thermoplastic polyurethane (PU) obtained from Covestro, Leverkusen, Germany. N, N-dimethyl formamide (DMF, >99.8%), tetraethylorthosilicate (TEOS, >99%), (3-aminopropyl)triethoxysilane (APTES, 99%), dimethylsulphoxide (DMSO, >99%), hexamethyldisilazane (HMDS, 99.9%), bovine serum albumin (BSA; >98% heat shock fraction, protease-free, fatty acid-free, essentially globulin free, pH 7), sodium azide (99%), ethylenediaminetetraacetic acid (EDTA), paraformaldehyde (PFA), Triton-X, and 4′,6-diamidino-2-phenylindole (DAPI) were obtained from Sigma Aldrich, Darmstadt, Germany. Sulfo-Cyanine 5 (Cy5) NHS ester was bought from Lumiprobe, Germany. Absolute ethanol (>99.8%) was purchased from VWR, Dietikon, Switzerland, and ammonia solution (25%) was purchased from Merck, Buchs, Switzerland. Phalloidin-AlexaFluor 488 (Cat #A12379) and secondary antibody goat anti-mouse AF647 (Cat #A21235) were bought from Invitrogen, Thermo Fisher Scientific, Basel, Switzerland. Transforming growth factor beta (TGF-β, Cat #11343160) and tumor necrosis factor alpha (TNF-α, Cat #11343015) were acquired from Immunotools, Friesoythe, Germany. All cell culture materials were purchased from Gibco, Thermo Fisher Scientific, Basel, Switzerland, unless otherwise specified. This study made use of only Milli-Q water.

2.2. Electrospinning and Characterization of Polyurethane Fibers

An in-house electrospinning setup was used to fabricate PU electrospun fiber membranes. A 12.5% *w/v* PU solution in DMF was prepared by dissolving 2.5 g of Texin 985A in 20 mL of DMF for 3 days under magnetic stirring at RT. To fully dissolve the polymer, the temperature was raised to 60 °C for 2 h. The solution was transferred into a 3 mL syringe with a 21G blunt end needle and placed on a syringe pump (AL-1000, Aladdin, Sarasota FL, World Precision Instrument, USA). The pump flow rate was set to 400 μL/h, and random electrospun fibers were collected on a rectangular plate covered with aluminum foil placed 20 cm from the needle. The electrospinning process was run for 120 min at a constant voltage of 9 kV. A rotating drum spun at 1500 rpm was used to generate aligned fiber membranes. Before performing additional experiments, the samples were allowed to dry for a week at room temperature. The morphological properties of the electrospun fibers were evaluated using an S-4800 scanning electron microscope (SEM, Hitachi High-Technologies, Schaumburg, IL, USA) after sputter coating the samples with 8 nm Au/Pd (EM ACE600, Leica Microsystems, Opfikon, Switzerland). ImageJ (National Institutes of Health, Bethesda, MD, USA) and the OrientationJ plugin were used to determine the average diameter and orientation of the fibers. Tensile tests were conducted on a TA DMA Q800 dynamic mechanical analyzer. The samples were cut into rectangular shapes with a width of 5.27 mm and a gauge length of f ~15 mm. Tests were carried out at RT with a strain rate of 100% min^{-1}. The elastic region of the stress–strain curve was used to calculate the Young's modulus. Tensile data are reported as the mean of three independent measurements, and all errors are reported as the standard deviation. Micrometer measurements (Caliper, Mitutoyo MDH) were performed to determine the thickness of the membranes. Fourier-transform infrared (FTIR) spectroscopy of random and aligned fibrous membranes was performed using a Perkin Elmer (USA) Spectrum 65 FTIR spectrometer from 600 to 4000 cm^{-1} with 25 accumulations per sample at a resolution of 4 cm^{-1}.

2.3. Synthesis of SiO_2-Cy5 Nanoparticles

SiO_2 NPs were synthesized via a modified Stöber method, as previously reported [21–23]. To obtain SiO_2 NPs with sizes of ~80 nm, 104 mL of ethanol and 3.9 mL of ammonia solution were added into a round-bottom flask and magnetically stirred for 30 min at 60 °C. Approximately 11 mL of TEOS was added into the solution and stirred for 2 min, followed by the addition of 100 μL of a mixture containing 1 mg of sulfo-Cy5 NHS ester and 1.5 μL of APTES in DMSO. The reaction was allowed to run for 4 h, after which the obtained NPs were purified by means of dialysis for two weeks, with the first day in ethanol and the following days in ultrapure water (18.2 MΩ). The final solution containing NPs was kept in the dark at 4 °C. The physical properties, including size and shape, of the SiO_2-Cy5 NPs were evaluated using a transmission electron microscope (TEM, FEI Tecnai G2 Filter, Hillsboro, OR, US). ImageJ's particle size analysis was used to determine the particle size distribution [24]. The colloidal stability of NPs in water and cell culture medium was determined at 24 h through dynamic light scattering (DLS) at 90° with a commercial goniometer instrument (3D LS Spectrometer, LS Instruments AG, Fribourg, Switzerland). At least 10 independent measurements were performed. The hydrodynamic diameter and zeta potential of Cy5-labeled particles dispersed in water were determined with a 90Plus particle size analyzer and ZetaPALS (Brookhaven Instruments Corp., Holtsville, NY, USA).

2.4. Cell Culture

Human alveolar epithelial type II cells (A549) and human lung fibroblasts (MRC-5) from the American Tissue Type Culture Collection (ATCC) were used in this study. The cells were grown at 37 °C in 5% CO_2 and 95% humidity and passaged twice a week at 80–90% cell confluence. A549 cells were kept in Roswell Park Memorial Institute (RPMI)-1640 cell culture media supplemented with 10% fetal bovine serum (FBS), 2 mM L-Glutamine, 100 U/mL penicillin, and 100 μg/mL streptomycin. The supplemented solution of RPMI-1640 is further mentioned as complete RPMI (cRPMI). Human lung fibroblasts were grown in minimum essential medium (MEM) GlutaMAX supplemented with 100 U/mL penicillin, 100 µg/mL streptomycin, and 1 × non-essential amino acids.

2.5. Cell Experiments

In this study, the effect of fibrous membranes on cell–particle interactions was evaluated. Electrospun membranes were cut into discs with a scalpel and separated from the aluminum foil with the help of tweezers. Glass coverslips of 13 and 22 mm were used as support for the membranes. The extra edges were folded into the bottom of the coverslip to keep the membranes stable and flat. The coverslips were placed into 24- and 12-well tissue culture plates. Sterilization was performed by exposing the material to UV light for 20 min and adding 70 vol.% ethanol for 30 min. The membranes were washed three times with PBS and left in the incubator overnight before cell seeding. A549 (passage range 4–20) and MRC-5 cells (passage range 4–15) were seeded at 53,000 cells/cm^2 for 72 h and 48 h, respectively, before exposure to 20 µg/mL of NPs for 24 h. TGF-β (10 ng/mL for A549 and 5 ng/mL for MRC-5) and TNF-α (1 µg/mL for A549 and 10 ng/mL for MRC-5) were added for 24 h. Metabolic activity, fluorescence imaging, and SEM assays were performed in 24-well plates, whereas 12-well plates were used for cytotoxicity, inflammatory response, and flow cytometry assays.

2.6. Cytotoxicity

According to the manufacturer's protocol, a lactate dehydrogenase (LDH) assay was used to determine cytotoxicity caused by either the fibrous membranes or the NPs (Roche, Mannheim, Germany). Approximately 100 µL of cell supernatant and 100 µL of LDH reagent were pipetted onto a 96-well plate for each measurement, which were carried out in duplicate. 0.2 vol.% Triton X-100 was used as a positive control. A spectrophotometer (Benchmark microplate reader, BioRad, Cressier, Switzerland) was used to detect LDH release in the form of formazan at 490 nm with a reference wavelength of 630 nm.

2.7. Metabolic Activity

A WST-1 assay was used to determine the effects of the fibrous material on the metabolic activity of A549 and MRC-5 cells. After growing on the fibrous membranes, the cells were washed twice with PBS before being exposed to 100 µL of WST-1 solution diluted at 1:10 in supplemented cell culture medium for at least 30 min. The formed formazan dye was measured spectrophotometrically at 450 nm. Measurements were performed in triplicate.

2.8. Inflammatory Response

The release of pro-inflammatory cytokines (Interleukin (IL)-6 and IL-8) in the cell culture medium was determined using a DuoSet Enzyme-Linked Immunosorbent Assay Development Kit (ELISA, R&D Systems, Zug, Switzerland) and the manufacturer's protocol. TNF-α was used as a positive control. Standards and samples were measured in triplicates and duplicates. The concentration of both inflammatory mediators was determined based on the standard curves. The final values were normalized against the total protein content. Calculations were performed by fitting the values in a four-parameter logistic (4PL) model using GraphPad Prism 9 software (GraphPad Software Inc., San Diego, CA, USA).

2.9. Flow Cytometry

The cells grown in the 12-well plates and exposed to particles were washed three times with PBS and detached with 100 µL of trypsin–ethylenediaminetetraacetic acid (EDTA) for 10 min. Approximately 900 µL of supplemented cell culture medium was added, followed by centrifugation for 5 min at 300× *g* at 4 °C. The cells were then stained for 5 min in 200 µL of PBS containing 0.2 µg/mL of DAPI. The cells were washed by centrifugation and resuspended in 300 µL of flow cytometry buffer (PBS with 1 *w/v*.% BSA, 0.1 vol.% sodium azide, and 1 mM EDTA at pH 7.4). Data were collected using a BD LSR FORTESSA (BD

Biosciences, San Jose, CA, USA) equipped with a violet laser and a red laser and 450/50 and 670/30 bandpass filters. Data were analyzed with FlowJo v10.7.1.

2.10. Fluorescence Imaging

Cells cultured on the fibrous substrates were washed three times. A solution of 4% PFA in PBS was used to fix the cells for 15 min. Permeabilization with 0.2 vol.% Triton X-100 in PBS for 10 min was performed, followed by blocking with 1.0% *w/v* BSA and 0.1 vol.% Triton X-100 in PBS. Primary antibodies, 4 μg/mL of mouse anti-E-cadherin and 5 μg/mL of mouse anti-α-SMA, in blocking buffer were added for 2 h, followed by the addition of secondary antibody goat anti-mouse AF647 at 1:400 for 1 h. Cell nuclei and F-actin staining with 1 μg/mL DAPI and 13.2 μM Alexa Fluor 488® phalloidin were performed in PBS for 30 min. All steps were performed at room temperature, with three rounds of PBS washing in between. Coverslips were mounted using Kaiser's glycerol gelatin and stored at 4 °C in the dark. Images were captured using an inverted confocal laser scanning microscopy system (CLSM, Leica, Stellaris 5, Heerbrugg, Switzerland) outfitted with Power HyD S detectors, a Plan-Apochromat 63x/1.4 Oil CS2 objective (Leica, Heerbrugg, Switzerland), and LAS X software version 3.0 (Leica). Image acquisition was performed sequentially with a field of view of 184.70 μm × 184.70 μm and a pixel density of 1024 × 1024. Three different laser excitation wavelengths were used: 405 nm (DAPI), 488 nm (Alexa Fluor 488), and 633 nm (Alexa Fluor 647).

2.11. Scanning Electron Microscopy Analysis of Cells

The cellular interaction with the fibrous membranes was visualized using scanning electron microscopy (SEM) (TESCAN Mira 3 LM field emission, Czech Republic). The samples were fixed using a PBS solution containing 2.5 vol.% glutaraldehyde and 2 vol.% PFA for 1 h. Following three washes with water, the samples were dehydrated using a series of alcohol washes (20, 30, 50, 70, 80, and 100 vol.% ethanol) for 10 min each before being treated with a solution of HMDS in ethanol (1:1 volume ratio). Before imaging, the samples were left to dry for 3 days and sputter-coated with 3 nm of gold (Sputter Coater 108 Auto, Cressington Scientific Instruments, Watford, UK).

2.12. Statistical Analyses

The Shapiro–Wilk normality test was used to determine whether the data distribution was normal (GraphPad Prism). An unpaired *t*-test was used to compare two independent groups, while one-way ANOVA (with Dunnett's post hoc test for multiple comparisons) was used to compare more than two groups with one variable.

3. Results

3.1. Properties of Electrospun Polyurethane Fibers

Membranes of randomly oriented and aligned nanofibers were produced through electrospinning of a thermoplastic polyurethane solution (PU, Texin 985) using DMF as the solvent. Optimization of the electrospinning parameters, including polymer concentration, voltage, and flow rate, were tested. The chosen electrospinning conditions allowed for the production of homogeneous fibers in the nanoscale range (Figure 1). The random fibers presented a mean fiber diameter of 892 ± 157 nm. In contrast, the aligned fibers showed a mean fiber diameter of 882 ± 165 nm. The alignment was confirmed by determining the orientation of the fibers, with a clear direction observable for the aligned fibers (Figure 1B). The mechanical properties of the fibrous mats were investigated using a tensile test (Figure S1). The linear region of the stress-strain curve was used to calculate Young's moduli. Tensile tests were performed on membranes with random fibers and fibers aligned along (II) and perpendicular (T) to the strain direction. Membranes with fibers oriented parallel to the strain direction exhibited similar Young's moduli to those with randomly oriented fibers. Conversely, fibers oriented perpendicularly to the strain direction revealed a Young's modulus 4–5 times lower than that of the randomly oriented fibers. The thickness of both mats was close to 50 μm, with average thicknesses of 69 ± 23 and 48 ± 3 μm for randomly oriented and aligned nanofibers, respectively.

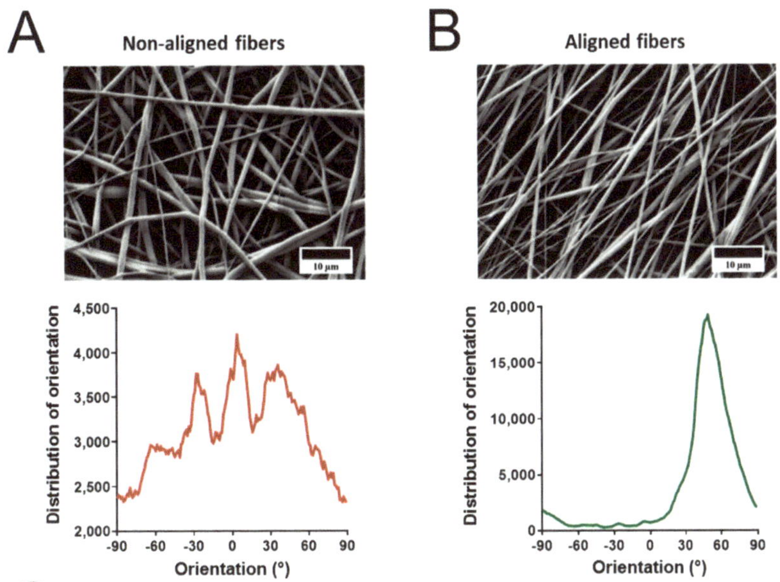

C

	Young's modulus (MPa)		*Fiber diameter (nm)*	*Membrane thickness (μm)*
Non-aligned	1.40 ± 0.20		892 ± 157	69 ± 23
Aligned	1.38 ± 0.40 (II)	0.28 ± 0.10 (T)	883 ± 165	48 ± 3

Figure 1. Characterization of non-aligned and aligned polyurethane electrospun fibrous membranes. Scanning electron microscopy images and fiber orientation of non-aligned (**A**) and aligned fibers (**B**). Tables presenting information regarding the Young's modulus, average fiber diameter, and membrane thickness (**C**). Data are presented as the mean ± standard error of the mean. II—fibers along the strain direction; T—fibers perpendicular to the strain direction.

FTIR spectroscopy was used to confirm the chemical properties of both materials after electrospinning (Figure S2). Both spectra are identical, with an absorption band at ~3300 cm^{-1} corresponding to NH stretching and two sharp peaks at ~2900 cm^{-1} associated with $-CH_2$ stretching [25]. The absorption band at ~1700 cm^{-1} is also linked to a C=O group, common in polyurethane.

3.2. Cell Interaction with Electrospun Membranes

The interaction between cells and the fibrous membranes, including cell attachment and morphology, was evaluated via CLSM (Figure 2). The results show that both cell types attach to the fiber membranes. A lower cell density can be observed for A549 cells on both membranes compared to the glass condition. A similar morphology for the A549 cells was observed between the different conditions. MRC-5 cells appear more elongated on the fibers, with more evidence on the aligned fibers. The same behavior can be observed in the SEM pictures (Figure 3). MRC-5 cells tend to attach and elongate in the direction of the fiber.

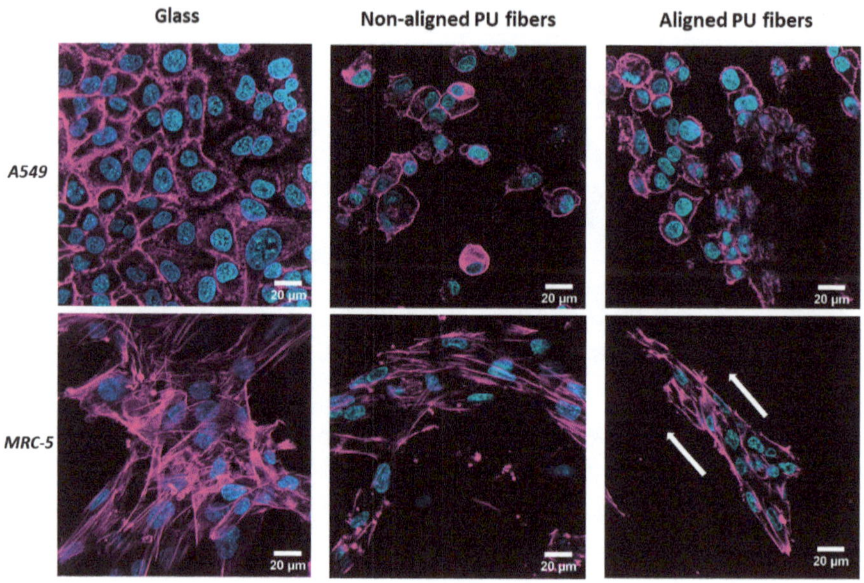

Figure 2. Cell interaction with different substrates. Confocal laser scanning microscopy (CLSM) images showing the morphology of the attached A549 lung epithelial cells (**top** row) and MRC-5 lung fibroblasts (**bottom** row) on glass, non-aligned polyurethane (PU) fibers, and aligned PU fibers. Nuclei are in cyan and f-actin is in magenta. Arrows indicate cell/fiber orientation.

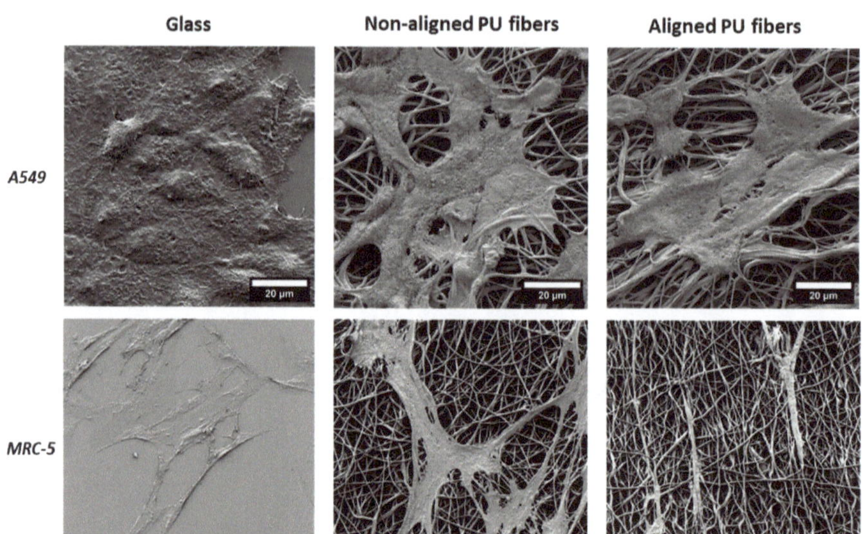

Figure 3. Scanning electron microscopy (SEM) images of lung epithelial cells (A549) and lung fibroblasts (MRC5) cultured on glass and aligned and non-aligned electrospun polyurethane (PU) membranes.

3.3. Cytotoxicity, Metabolic Activity, and Inflammatory Response

Cytotoxicity, metabolic activity, and inflammatory response were assayed to evaluate the biocompatibility of the produced mats. Cellular toxicity was assessed by measuring the amount of LDH released in the cell culture medium (Figure 4A,B). A compromised cell membrane leads to the release of intracellular LDH. There were no significant differences in LDH release between cells growing on the fibrous mats and tissue culture plastic (TCP). Similarly, no significant differences were observed in the metabolic activity of both cells (Figure 4C,D). Metabolically active cells convert the tetrazolium salt WST-1 into formazan. Cells growing on aligned and non-aligned PU fibers showed no significant difference in metabolic activity (WST-1 reduction) (Figure 4C,D).

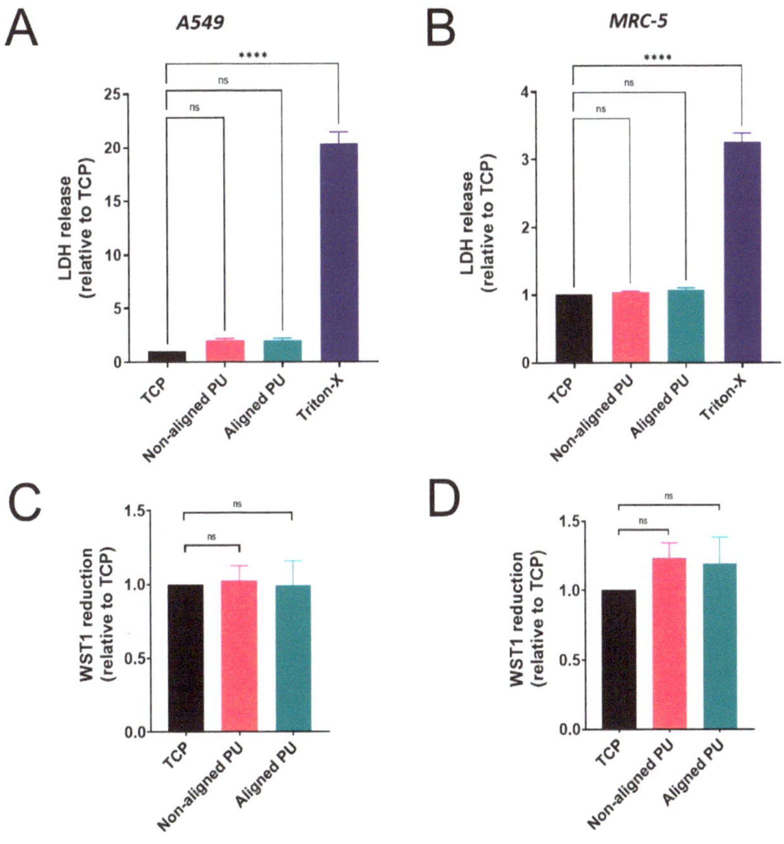

Figure 4. Impact of fibrous polyurethane (PU) membranes on the cytotoxicity and metabolic activity of lung epithelial cells (A549) and fibroblasts (MRC-5). Bar graphs, at the top, represent the release of lactate dehydrogenase relative to tissue culture plastic (TCP) from A549 (**A**) and MRC-5 (**B**) cells. Bar graphs, at the bottom, represent the reduction in WST-1 relative to TCP from A549 (**C**) and MRC-5 (**D**) cells. Absorbance values for the WST-1 assay were normalized against the total protein content. Data are presented as the mean ± standard error of the mean (n = 3). Statistical significance was determined via one-way ANOVA and Dunnett's post hoc test for multiple comparisons and is represented by **** $p \leq 0.0001$. ns: not significant.

In response to environmental stress, lung cells secrete inflammatory cytokines such as IL-6 and IL-8, which alert resident and circulating immune cells [26,27]. The release of both interleukins was measured using ELISA and normalized against the total protein content

(Figure 5 and Figure S3). TNF-α, at a non-toxic concentration, was used as a positive control for inducing cytokine release (Figure S4). Compared to TCP, IL-6 secretion increased about two-fold in epithelial cells on both fibrous membranes (Figure 5A), although the differences were not statistically significant. Contrastingly, IL-8 secretion increased significantly by more than two-fold on both substrates (Figure 5B). Significant differences were also found in fibroblasts, with IL-6 release increasing ~1.5 times for cells grown on both mats (Figure 5C). IL-8 release from fibroblasts, on the other hand, was significantly increased (two-fold) on aligned fibers but not on randomly aligned fibers (Figure 5D).

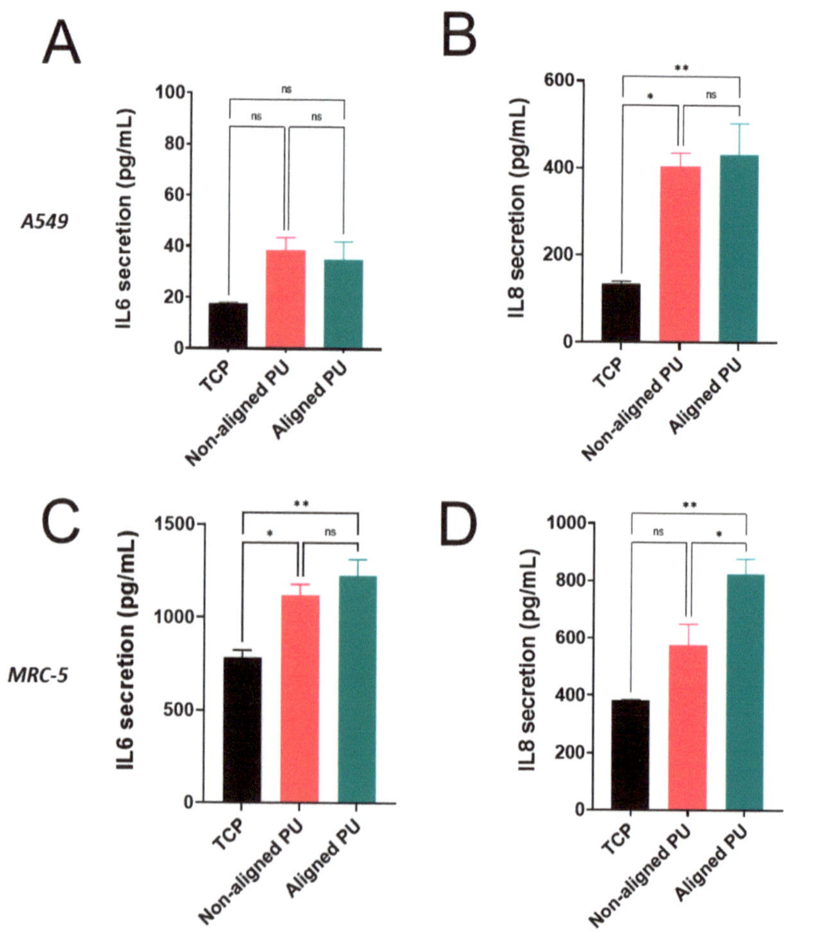

Figure 5. Impact of fibrous polyurethane (PU) membranes on the inflammatory response of lung epithelial cells (A549) and fibroblasts (MRC-5). Bar graphs represent the release of interleukin (IL)-6 and IL-8 from A549 (**A**,**B**) and MRC-5 (**C**,**D**) cultured on tissue culture plastic (TCP) and non-aligned and aligned PU. Values were normalized against the total protein content. Data are presented as the mean ± standard error of the mean ($n = 3$). Statistical significance was determined via one-way ANOVA and Tukey's post hoc test for multiple comparisons and is represented by * $p \leq 0.05$ and ** $p \leq 0.01$. ns: not significant.

3.4. *Formation of Stress Fibers Containing Alpha-Smooth Muscle Actin*

For MRC-5 cells, immunostaining of alpha-smooth muscle actin (α-SMA), a protein found in stress fibers, was performed (Figure 6). In healing tissues, fibroblasts develop a contractile phenotype defined by cytoskeleton reorganization and the de novo expression of α-SMA [28]. MRC-5 cells grown on glass revealed a weak signal for α-SMA. Nonetheless, the signal increased when TGF-β1 was added. Similar observations were denoted for fibroblasts growing on the different PU fibers, with cells displaying a high presence of α-SMA.

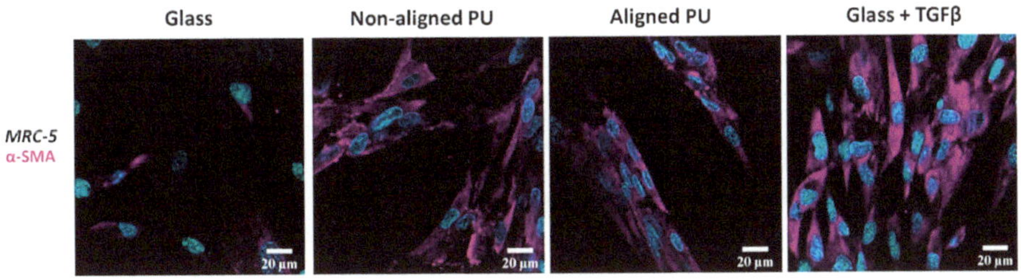

Figure 6. Immunofluorescence staining for α-smooth muscle actin (α-SMA). Confocal laser scanning microscopy (CLSM) images showing the immunostaining of MRC-5 fibroblasts with α-SMA cultured on glass, non-aligned and aligned polyurethane (PU) fibers, and glass + transforming growth factor beta (TGF-β). TGF-β was added for 24 h to MRC-5 cells at a concentration of 5 ng/mL. Nuclei are in cyan and α-SMA is in magenta.

3.5. Cellular Uptake of Nanoparticles

The cell type and the NP characteristics are critical factors in cell–NP interaction [23]. Other parameters, such as cell substrate, can influence cell behavior and, subsequently, the interaction and uptake of NPs. To investigate the impact of the produced fibrous membranes on NP uptake, Cy5-labeled SiO_2 NPs were synthesized. The main NP properties, such as morphology and size, were characterized via TEM (Figure S5). The produced SiO_2 NPs present a spherical shape and an average diameter of 83 ± 9 nm (Figure S5A,B). Hydrodynamic diameter and zeta potential measurements were performed in water (Figure S5C). The particles presented an average hydrodynamic diameter of 99 nm with a low polydispersity index (PDI) of 0.04. In addition, zeta potential measurements revealed strongly negative values (−46 ± 3 mV).

The colloidal stability of NPs in cell culture medium was assessed via DLS after 24 h (Figure S5D). The correlation function between particles in the cell culture medium and water did not change substantially. The absence of a significant increase in lag time and a flat baseline confirm the stability of NPs in the cell culture medium.

An LDH assay was performed to evaluate the cellular cytotoxicity of SiO_2 NPs on the various substrates (Figure S6C,D). There were no significant differences in cell membrane permeability, i.e., cytotoxicity, after NP exposure to A549 and MRC-5 cells.

Flow cytometry and CLSM imaging were used to examine the cellular uptake of SiO_2 NPs on different substrates (i.e., TCP vs. fibrous membranes) (Figure 7). Lung epithelial cells and fibroblasts were exposed to SiO_2 NPs for 24 h. There were no significant differences in the flow cytometry median fluorescence intensity (MFI) values for epithelial cells on the different substrates (Figure 7A). Internalization of NPs in epithelial cells was observed for all substrates (i.e., TCP and aligned and non-aligned fibers) to similar degrees (Figure 7B). In contrast, the MFI values for fibroblasts on both non-aligned and aligned fibers were significantly higher (three-fold increase) than on TCP (Figure 7C). Intracellular localization of NPs in fibroblasts upon uptake was confirmed in all substrates (Figure 7D).

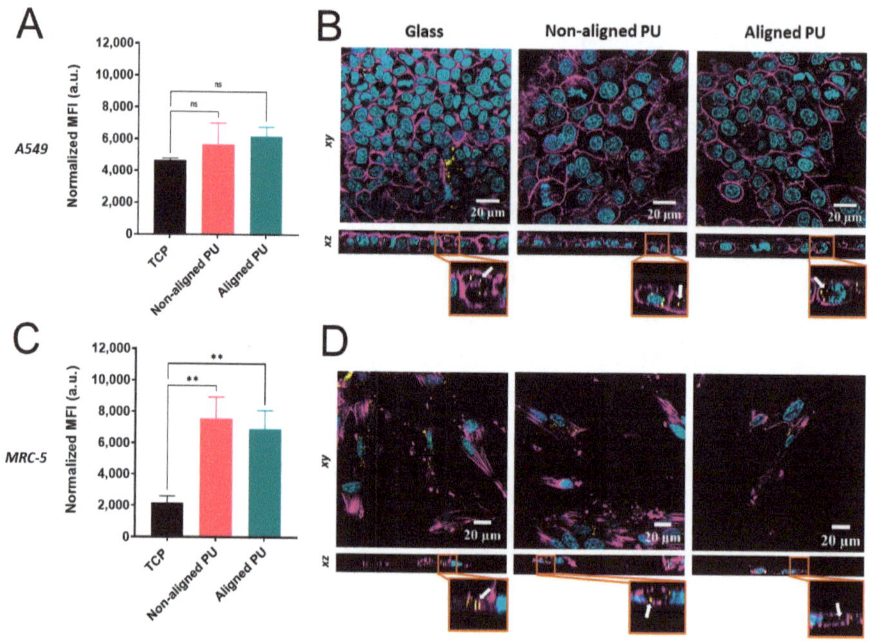

Figure 7. Impact of fibrous polyurethane (PU) membranes on SiO_2 nanoparticle (NP) uptake in lung epithelial cells (A549) and fibroblasts (MRC-5). Bar graphs show the median fluorescence intensity (MFI) values obtained from flow cytometry measurements upon 24 h of exposure of 20 μg/mL SiO_2 NPs to A549 (**A**) and MRC-5 cells (**C**) cultured on tissue culture plastic (TCP) and non-aligned and aligned PU fibers. Data are presented as the mean ± standard error of the mean (n = 3). Statistical significance was determined via one-way ANOVA and Dunnett's post hoc test for multiple comparisons and is represented by ** $p \leq 0.01$. ns: not significant. Confocal laser scanning microscopy (CLSM) images showing the internalization of SiO_2 NPs in A549 (**B**) and MRC-5 cells (**D**) cultured on glass and non-aligned and aligned PU fibers. Nuclei are in cyan, F-actin is in magenta, and SiO_2 NPs are in yellow. Arrows indicate the intracellular localization of NPs.

4. Discussion

The electrospinning of PU into fibrous membranes was used to create substrates that could mimic features of the ECM in lungs under healthy (random) and diseased (aligned) conditions. The ECM in the lungs is primarily composed of collagen with high tensile strength and elastin with low tensile strength [29]. Alveolar–capillary basement membrane (BM) is a specialized ECM responsible for maintaining the integrity of epithelial–endothelial cell layers [30]. In pathological conditions such as idiopathic pulmonary fibrosis (IPF), there is a loss of integrity of the alveolar epithelium and downregulation of cell–cell adhesion molecules, resulting in a compromised BM [31,32]. This is accompanied by ECM remodeling, typified by increased production of ECM proteins (e.g., collagen), fiber density, and alignment [33,34]. Consequently, ECM alignment and stiffness changes can change cell behavior and phenotype. Herein, the influence of fiber alignment on cell behavior was investigated. Mean fiber diameters of 892 nm and 882 nm were obtained for randomly oriented fibers and aligned fibers (fibers along the strain direction), respectively, which are values close to collagen and elastin fibers in human pulmonary alveolar walls (between 952 nm to 1270 nm) [35]. Additionally, a membrane stiffness of 1.4 MPa was obtained for both fibers, which is slightly lower but comparable to alveolar BM stiffness (between 2 and 3 MPa) [30].

Lung epithelial cells (A549) and fibroblasts (MRC-5) were cultured on soft PU membranes and stiff TCP/glass with stiffnesses in the GPa range. Both cell types could attach and grow on either PU membrane with no signs of cytotoxicity. However, compared to TCP/glass, a lower number of cells can be observed on PU membranes (Figure 2), confirmed by a two-fold decrease in total protein content (Figure S3). This decrease in cell number is most likely related to a reduction in cell proliferation. Cell morphology changes, such as elongated cells, were observed in fibroblasts growing in PU aligned fibers. Experiments were also carried out on spin-coated flat PU substrates, but cells were unable to attach, and these substrates were not used further.

Although no significant morphological changes were observed in epithelial cells, the biochemical response associated with cytokine release was altered. High levels of IL-6 and IL-8 are present in the early stages of IPF [36]. IL-6 is a crucial regulator of inflammation and tissue repair, and chronic activation of IL-6 signaling can contribute to the development of fibrosis [26]. IL-8 acts as a chemoattractant for immune cells, plays an important role in cell proliferation, invasion, and epithelial–mesenchymal transition, and is a driver of fibrotic progression [27,37]. Epithelial cells released higher levels of IL-8 when cultured on soft PU membranes than when cultured on TCP. Despite this, fiber alignment did not trigger significant differences in cytokine release. This demonstrates that cytokine release is triggered on soft PU membranes regardless of fiber alignment. In contrast, IL-6 release from fibroblasts increased in cells on both membranes compared to TCP, while fibroblasts cultured on aligned fibers elicited a stronger response. Significant differences in IL-8 release were observed for fibroblasts cultured on aligned fibers but not on non-aligned fibers. This outcome suggests that both cells can sense the physical changes in the environment (i.e., soft PU membranes) and convert them into biochemical signals (i.e., cytokine release) through mechanotransduction. Furthermore, the major differences in fibroblast cell morphology were related to greater cytokine release in cells on aligned fibrous membranes. Nevertheless, the differences in cytokine release in both cells cannot be directly attributed to one specific substrate feature (i.e., surface chemistry, topography, or stiffness). Even though we cannot isolate a specific substrate physical attribute, we may directly compare aligned and non-aligned fibers and observe a difference in IL-8 secretion in fibroblasts.

Lung diseases such as IPF are characterized by ECM remodeling and epithelial damage leading to several pathophysiological disorders, including fibroblast activation and differentiation to myofibroblasts [31,38]. TGF-β is a biomolecule capable of inducing fibroblast-to-myofibroblast transition (FMT) in vitro [39,40]. TGF-β causes disruptions in the homeostatic microenvironment by promoting cell activation, migration, or invasion in IPF [41]. This profibrotic cytokine promotes the production of proinflammatory and fibrogenic cytokines and collagen and the formation of α-SMA stress fibers [28,39,42]. Stress fibers containing α-SMA were observed for fibroblasts grown on both fibrous membrane types. This protein is a strong marker of myofibroblast differentiation, implying that PU membranes can induce fibroblast–myofibroblast transition (FMT) similar to that observed upon TGF-β stimulation of fibroblasts. However, to confirm a myofibroblast phenotype, the expression of additional FMT markers such as β-cadherins and mature focal adhesions should be assessed [40]. Similar observations were made for cardiac fibroblasts growing on randomly aligned and parallel aligned collagen fibers [43]. Fiber organization and orientation were important in determining the number of cardiac fibroblasts positive for α-SMA stress fibers cultured on both collagen fibers, which was higher than on a flat substrate.

Despite significant efforts to understand mechanotransduction in cells cultivated in various cell substrates and consequent cellular responses such as cell adhesion and inflammation, only a few studies have investigated NP cellular uptake in more relevant pathophysiological contexts. It has been shown that substrate properties such as stiffness and topography can influence the cellular uptake of NPs [44–46]. Here, we have demonstrated that the uptake of SiO_2 NPs by epithelial cells is not affected by the cell substrate (i.e., TCP or fibrous membranes). However, fibroblasts cultured on fibrous membranes rather than TCP could internalize more SiO_2 NPs, but fiber alignment did not affect particle uptake. According to Yhee et al., primary human fibroblasts cultured on TCP coated with a collagen matrix instead of uncoated TCP were able to internalize more glycol chitosan NPs [47]. They also concluded that macropinocytosis was the primary mechanism involved in the internalization of NPs. It has also been shown that cancer-associated fibroblasts, including myofibroblasts, present an increased rate of macropinocytosis in comparison with normal fibroblasts [48]. Lung fibroblasts cultured on PU membranes shown morphological changes and an increased quantity of α-SMA stress fibers. Increased NP uptake may be associated with increased macropinocytosis activity, but more research is needed to confirm this claim. This finding strongly demonstrates that the microenvironment surrounding lung fibroblasts influences NP cellular delivery.

In this study, we found that cell–substrate interaction influences the cellular delivery of SiO_2 NPs in a cell-type specific manner. Fiber orientation (aligned vs. non-aligned) did not affect particle internalization; however, as previously mentioned, it is unclear whether the increased uptake of SiO_2 NPs observed in fibroblasts is primarily influenced by substrate surface chemistry, stiffness, or topography (i.e., nanofibers vs. flat TCP/glass). Despite the limitations of the study, we were able to show that fibroblasts were more sensitive to changes in the extracellular environment due to differences in cell morphology, cytokine release, and NP uptake. Additionally, we believe that PU fibrous membranes are a better substrate for studying lung cell–NP interactions than TCP/glass substrates because they have better physicomechanical properties (i.e., topography and stiffness). Nevertheless, the fabrication of fibrous membranes made of PU blending with ECM proteins (e.g., collagen and elastin), which present mechanical properties that more closely resemble the ECM in healthy vs. diseased lungs, should be investigated.

5. Conclusions

This study suggests that cell–matrix interaction is substrate and cell-type dependent with consequences for cell behavior and NP internalization. PU fibrous membranes with different physical properties (i.e., topography and stiffness) compared to standard TCP increased cytokine release in epithelial cells and fibroblasts. Fiber alignment had little effect on epithelial cells but had an effect on fibroblast behavior by orienting cell alignment and increasing IL-8 secretion. The number of stress fibers containing α-SMA and the uptake of SiO_2 NPs increased in fibroblasts cultured on both membranes. However, the different substrates did not considerably affect the uptake of SiO_2 NPs in epithelial cells. This study emphasizes the significance of considering the ECM and tissue mechanical properties when designing and optimizing NPs for effective disease treatment. To progress in this direction, appropriate cell models, priming biological compounds, and more realistic cell substrates should be considered to more closely mimic and build a healthy/diseased model.

Supplementary Materials: The following supporting information can be downloaded at: https://www.mdpi.com/article/10.3390/nano14040342/s1. Figure S1—Tensile test measurements of both aligned and non-aligned polyurethane (PU) fibers. Figure S2—Fourier-transform infrared spectra of both aligned and non-aligned polyurethane fibers. Figure S3—Total protein content of A549 and MRC-5 cells after 72 and 48 h. Figure S4—Effect of the positive control tumor necrosis factor alpha on the secretion of inflammatory cytokines in lung epithelial cells and lung fibroblasts. Figure S5—Physicochemical properties of SiO_2 NPs and colloidal stability in cell culture medium. Figure S6—Cytotoxicity of positive controls and SiO_2 NPs in lung epithelial cells and fibroblasts.

Author Contributions: M.S.d.A. performed the experiments, analyzed the results, and wrote the manuscript; A.L. prepared the samples for SEM analysis and carried out the image acquisition; F.I. produced the polyurethane fibrous membranes. K.M.-W. reviewed the manuscript and supervised the project; A.P.-F. supervised the project; B.R.-R. thoroughly reviewed the manuscript and supervised the project; A.L., F.I., K.M.-W., A.P.-F. and B.R.-R. provided scientific advice. All authors have read and agreed to the published version of the manuscript.

Funding: This research was supported by the Swiss National Science Foundation through the National Center of Competence in Research Bio-Inspired Materials (grant number 51NF40-182881), the Partnerships for International Research and Education (PIRE) program under grant number IZPIP0_177995 and the Adolphe Merkle Foundation.

Data Availability Statement: The data presented in this study are openly available in the online repository Zenodo under the doi 10.5281/zenodo.10631246.

Acknowledgments: The authors are grateful to Laetitia Haeni and Liliane Ackerman Hirschi for their kind support with the biological assays and particle synthesis. The authors would like to acknowledge Jessica Caldwell, Chris Rader, and Sandor Balog for their assistance with the FTIR, tensile test measurements, and DLS data analysis, respectively.

Conflicts of Interest: The authors declare no competing financial interests.

References

1. Sousa de Almeida, M.; Susnik, E.; Drasler, B.; Taladriz-Blanco, P.; Petri-Fink, A.; Rothen-Rutishauser, B. Understanding nanoparticle endocytosis to improve targeting strategies in nanomedicine. *Chem. Soc. Rev.* **2021**, *50*, 5397–5434. [CrossRef] [PubMed]
2. Doane, T.L.; Burda, C. The unique role of nanoparticles in nanomedicine: Imaging, drug delivery and therapy. *Chem. Soc. Rev.* **2012**, *41*, 2885. [CrossRef] [PubMed]
3. Kinnear, C.; Moore, T.L.; Rodriguez-Lorenzo, L.; Rothen-Rutishauser, B.; Petri-Fink, A. Form Follows Function: Nanoparticle Shape and Its Implications for Nanomedicine. *Chem. Rev.* **2017**, *117*, 11476–11521. [CrossRef] [PubMed]
4. Zhao, J.; Stenzel, M.H. Entry of nanoparticles into cells: The importance of nanoparticle properties. *Polym. Chem.* **2018**, *9*, 259–272. [CrossRef]
5. Engin, A.B.; Nikitovic, D.; Neagu, M.; Henrich-Noack, P.; Docea, A.O.; Shtilman, M.I.; Golokhvast, K.; Tsatsakis, A.M. Mechanistic understanding of nanoparticles' interactions with extracellular matrix: The cell and immune system. *Part. Fibre Toxicol.* **2017**, *14*, 22. [CrossRef]
6. Katz, R.; Spohr, R. Ion Tracks and Microtechnology: Principles and Applications. *Radiat. Res.* **1991**, *126*, 111. [CrossRef]
7. Baptista, D.; Teixeira, L.M.; Birgani, Z.T.; van Riet, S.; Pasman, T.; Poot, A.; Stamatialis, D.; Rottier, R.J.; Hiemstra, P.S.; Habibović, P.; et al. 3D alveolar in vitro model based on epithelialized biomimetically curved culture membranes. *Biomaterials* **2021**, *266*, 120436. [CrossRef] [PubMed]
8. Lehner, R.; Wohlleben, W.; Septiadi, D.; Landsiedel, R.; Petri-Fink, A.; Rothen-Rutishauser, B. A novel 3D intestine barrier model to study the immune response upon exposure to microplastics. *Arch. Toxicol.* **2020**, *94*, 2463–2479. [CrossRef]
9. Lehmann, A.D.; Brandenberger, C.; Blank, F.; Gehr, P.; Rothen-Rutishauser, B. A 3D Model of the Human Epithelial Airway Barrier. *Altern. Technol. Anim. Test.* **2010**, *14*, 239–260.
10. Wells, R.G. The role of matrix stiffness in regulating cell behavior. *Hepatology* **2008**, *47*, 1394–1400. [CrossRef]
11. Chan, B.P.; Leong, K.W. Scaffolding in tissue engineering: General approaches and tissue-specific considerations. *Eur. Spine J.* **2008**, *17*, 467–479. [CrossRef] [PubMed]
12. Frantz, C.; Stewart, K.M.; Weaver, V.M. The extracellular matrix at a glance. *J. Cell Sci.* **2010**, *123*, 4195–4200. [CrossRef] [PubMed]
13. Bonnans, C.; Chou, J.; Werb, Z. Remodelling the extracellular matrix in development and disease. *Nat. Rev. Mol. Cell Biol.* **2014**, *15*, 786–801. [CrossRef] [PubMed]
14. Leight, J.L.; Drain, A.P.; Weaver, V.M. Extracellular matrix remodeling and stiffening modulate tumor phenotype and treatment response. *Annu. Rev. Cancer Biol.* **2017**, *1*, 313–334. [CrossRef]
15. Lampi, M.C.; Reinhart-King, C.A. Targeting extracellular matrix stiffness to attenuate disease: From molecular mechanisms to clinical trials. *Sci. Transl. Med.* **2018**, *10*, eaao0475. [CrossRef] [PubMed]
16. Zhao, D.; Xue, C.; Li, Q.; Liu, M.; Ma, W.; Zhou, T.; Lin, Y. Substrate stiffness regulated migration and angiogenesis potential of A549 cells and HUVECs. *J. Cell. Physiol.* **2018**, *233*, 3407–3417. [CrossRef] [PubMed]
17. Huang, X.; Yang, N.; Fiore, V.F.; Barker, T.H.; Sun, Y.; Morris, S.W.; Ding, Q.; Thannickal, V.J.; Zhou, Y. Matrix Stiffness–Induced Myofibroblast Differentiation Is Mediated by Intrinsic Mechanotransduction. *Am. J. Respir. Cell Mol. Biol.* **2012**, *47*, 340–348. [CrossRef] [PubMed]
18. Park, J.S.; Chu, J.S.; Tsou, A.D.; Diop, R.; Tang, Z.; Wang, A.; Li, S. The effect of matrix stiffness on the differentiation of mesenchymal stem cells in response to TGF-β. *Biomaterials* **2011**, *32*, 3921–3930. [CrossRef]
19. Provenzano, P.P.; Eliceiri, K.W.; Campbell, J.M.; Inman, D.R.; White, J.G.; Keely, P.J. Collagen reorganization at the tumor-stromal interface facilitates local invasion. *BMC Med.* **2006**, *4*, 38. [CrossRef]
20. Zdrahala, R.J.; Zdrahala, I.J. Biomedical Applications of Polyurethanes: A Review of Past Promises, Present Realities, and a Vibrant Future. *J. Biomater. Appl.* **1999**, *14*, 67–90. [CrossRef]
21. Stöber, W.; Fink, A.; Bohn, E. Controlled growth of monodisperse silica spheres in the micron size range. *J. Colloid Interface Sci.* **1968**, *26*, 62–69. [CrossRef]

22. Larson, D.R.; Ow, H.; Vishwasrao, H.D.; Heikal, A.A.; Wiesner, U.; Webb, W.W. Silica Nanoparticle Architecture Determines Radiative Properties of Encapsulated Fluorophores. *Chem. Mater.* **2008**, *20*, 2677–2684. [CrossRef]
23. Sousa de Almeida, M.; Taladriz-Blanco, P.; Drasler, B.; Balog, S.; Yajan, P.; Petri-Fink, A.; Rothen-Rutishauser, B. Cellular Uptake of Silica and Gold Nanoparticles Induces Early Activation of Nuclear Receptor NR4A1. *Nanomaterials* **2022**, *12*, 690. [CrossRef] [PubMed]
24. Zhang, S.; Wang, C. Precise Analysis of Nanoparticle Size Distribution in TEM Image. *Methods Protoc.* **2023**, *6*, 63. [CrossRef] [PubMed]
25. Trovati, G.; Sanches, E.A.; Neto, S.C.; Mascarenhas, Y.P.; Chierice, G.O. Characterization of polyurethane resins by FTIR, TGA, and XRD. *J. Appl. Polym. Sci.* **2010**, *115*, 263–268. [CrossRef]
26. Rincon, M.; Irvin, C.G. Role of IL-6 in Asthma and Other Inflammatory Pulmonary Diseases. *Int. J. Biol. Sci.* **2012**, *8*, 1281–1290. [CrossRef] [PubMed]
27. Mukaida, N. Pathophysiological roles of interleukin-8/CXCL8 in pulmonary diseases. *Am. J. Physiol. Cell. Mol. Physiol.* **2003**, *284*, L566–L577. [CrossRef] [PubMed]
28. Zhang, H.Y.; Gharaee-Kermani, M.; Zhang, K.; Karmiol, S.; Phan, S.H. Lung fibroblast alpha-smooth muscle actin expression and contractile phenotype in bleomycin-induced pulmonary fibrosis. *Am. J. Pathol.* **1996**, *148*, 527–537.
29. Burgstaller, G.; Oehrle, B.; Gerckens, M.; White, E.S.; Schiller, H.B.; Eickelberg, O. The instructive extracellular matrix of the lung: Basic composition and alterations in chronic lung disease. *Eur. Respir. J.* **2017**, *50*, 1601805. [CrossRef]
30. Welling, L.; Zupka, M.; Welling, D. Mechanical Properties of Basement Membrane. *Physiology* **1995**, *10*, 30–35. [CrossRef]
31. Camelo, A.; Dunmore, R.; Sleeman, M.A.; Clarke, D.L. The epithelium in idiopathic pulmonary fibrosis: Breaking the barrier. *Front. Pharmacol.* **2014**, *4*, 173. [CrossRef] [PubMed]
32. Janiszewska, M.; Primi, M.C.; Izard, T. Cell adhesion in cancer: Beyond the migration of single cells. *J. Biol. Chem.* **2020**, *295*, 2495–2505. [CrossRef] [PubMed]
33. Hu, M.; Ling, Z.; Ren, X. Extracellular matrix dynamics: Tracking in biological systems and their implications. *J. Biol. Eng.* **2022**, *16*, 13. [CrossRef] [PubMed]
34. Zhou, Y.; Horowitz, J.C.; Naba, A.; Ambalavanan, N.; Atabai, K.; Balestrini, J.; Bitterman, P.B.; Corley, R.A.; Ding, B.-S.; Engler, A.J.; et al. Extracellular matrix in lung development, homeostasis and disease. *Matrix Biol.* **2018**, *73*, 77–104. [CrossRef]
35. Sobin, S.S.; Fung, Y.C.; Tremer, H.M. Collagen and elastin fibers in human pulmonary alveolar walls. *J. Appl. Physiol.* **1988**, *64*, 1659–1675. [CrossRef]
36. Papiris, S.A.; Tomos, I.P.; Karakatsani, A.; Spathis, A.; Korbila, I.; Analitis, A.; Kolilekas, L.; Kagouridis, K.; Loukides, S.; Karakitsos, P.; et al. High levels of IL-6 and IL-8 characterize early-on idiopathic pulmonary fibrosis acute exacerbations. *Cytokine* **2018**, *102*, 168–172. [CrossRef]
37. Yang, L.; Herrera, J.; Gilbertsen, A.; Xia, H.; Smith, K.; Benyumov, A.; Bitterman, P.B.; Henke, C.A. IL-8 mediates idiopathic pulmonary fibrosis mesenchymal progenitor cell fibrogenicity. *Am. J. Physiol. Cell. Mol. Physiol.* **2018**, *314*, L127–L136. [CrossRef]
38. Di Gregorio, J.; Robuffo, I.; Spalletta, S.; Giambuzzi, G.; De Iuliis, V.; Toniato, E.; Martinotti, S.; Conti, P.; Flati, V. The Epithelial-to-Mesenchymal Transition as a Possible Therapeutic Target in Fibrotic Disorders. *Front. Cell Dev. Biol.* **2020**, *8*, 607483. [CrossRef]
39. Giacomelli, C.; Piccarducci, R.; Marchetti, L.; Romei, C.; Martini, C. Pulmonary fibrosis from molecular mechanisms to therapeutic interventions: Lessons from post-COVID-19 patients. *Biochem. Pharmacol.* **2021**, *193*, 114812. [CrossRef]
40. D'Urso, M.; Kurniawan, N.A. Mechanical and Physical Regulation of Fibroblast–Myofibroblast Transition: From Cellular Mechanoresponse to Tissue Pathology. *Front. Bioeng. Biotechnol.* **2020**, *8*, 609653. [CrossRef]
41. Fernandez, I.E.; Eickelberg, O. The Impact of TGF-β on Lung Fibrosis. *Proc. Am. Thorac. Soc.* **2012**, *9*, 111–116. [CrossRef] [PubMed]
42. Kim, K.K.; Kugler, M.C.; Wolters, P.J.; Robillard, L.; Galvez, M.G.; Brumwell, A.N.; Sheppard, D.; Chapman, H.A. Alveolar epithelial cell mesenchymal transition develops in vivo during pulmonary fibrosis and is regulated by the extracellular matrix. *Proc. Natl. Acad. Sci. USA* **2006**, *103*, 13180–13185. [CrossRef] [PubMed]
43. Bugg, D.; Bretherton, R.; Kim, P.; Olszewski, E.; Nagle, A.; Schumacher, A.E.; Chu, N.; Gunaje, J.; DeForest, C.A.; Stevens, K.; et al. Infarct Collagen Topography Regulates Fibroblast Fate via p38-Yes-Associated Protein Transcriptional Enhanced Associate Domain Signals. *Circ. Res.* **2020**, *127*, 1306–1322. [CrossRef] [PubMed]
44. Huang, C.; Butler, P.J.; Tong, S.; Muddana, H.S.; Bao, G.; Zhang, S. Substrate Stiffness Regulates Cellular Uptake of Nanoparticles. *Nano Lett.* **2013**, *13*, 1611–1615. [CrossRef] [PubMed]
45. Lee, A.; Sousa de Almeida, M.; Milinkovic, D.; Septiadi, D.; Taladriz-Blanco, P.; Loussert, C.; Balog, S.; Bazzoni, A.; Rothen-Rutishauser, B.; Fink, A.S. Substrate stiffness reduces particle uptake by epithelial cells and macrophages in a size-dependent manner through mechanoregulation. *Nanoscale* **2022**, *14*, 15141–15155. [CrossRef] [PubMed]
46. Huang, C.; Ozdemir, T.; Xu, L.-C.; Butler, P.J.; Siedlecki, C.A.; Brown, J.L.; Zhang, S. The role of substrate topography on the cellular uptake of nanoparticles. *J. Biomed. Mater. Res. Part B Appl. Biomater.* **2016**, *104*, 488–495. [CrossRef] [PubMed]

47. Yhee, J.Y.; Yoon, H.Y.; Kim, H.; Jeon, S.; Hergert, P.; Im, J.; Panyam, J.; Kim, K.; Nho, R. The effects of collagen-rich extracellular matrix on the intracellular delivery of glycol chitosan nanoparticles in human lung fibroblasts. *Int. J. Nanomed.* **2017**, *12*, 6089–6105. [CrossRef]
48. Zhang, Y.; Recouvreux, M.V.; Jung, M.; Galenkamp, K.M.O.; Li, Y.; Zagnitko, O.; Scott, D.A.; Lowy, A.M.; Commisso, C. Macropinocytosis in Cancer-Associated Fibroblasts Is Dependent on CaMKK2/ARHGEF2 Signaling and Functions to Support Tumor and Stromal Cell Fitness. *Cancer Discov.* **2021**, *11*, 1808–1825. [CrossRef]

Disclaimer/Publisher's Note: The statements, opinions and data contained in all publications are solely those of the individual author(s) and contributor(s) and not of MDPI and/or the editor(s). MDPI and/or the editor(s) disclaim responsibility for any injury to people or property resulting from any ideas, methods, instructions or products referred to in the content.

nanomaterials

MDPI

Article

Hyaluronic Acid/Collagen Nanofiber Tubular Scaffolds Support Endothelial Cell Proliferation, Phenotypic Shape and Endothelialization

Yuqing Niu [1,*] and Massimiliano Galluzzi [2]

1 Department of Pediatric Surgery, Guangdong Provincial Key Laboratory of Research in Structural Birth Defect Disease, Guangzhou Women and Children's Medical Center, Guangzhou Medical University, Guangzhou 510623, China
2 Materials Interfaces Center, Shenzhen Institutes of Advanced Technology, Chinese Academy of Sciences, Shenzhen 518055, China; galluzzi@siat.ac.cn
* Correspondence: niuyuqing@gwcmc.org

Abstract: In this study, we designed and synthetized artificial vascular scaffolds based on nanofibers of collagen functionalized with hyaluronic acid (HA) in order to direct the phenotypic shape, proliferation, and complete endothelization of mouse primary aortic endothelial cells (PAECs). Layered tubular HA/collagen nanofibers were prepared using electrospinning and crosslinking process. The obtained scaffold is composed of a thin inner layer and a thick outer layer that structurally mimic the layer the intima and media layers of the native blood vessels, respectively. Compared with the pure tubular collagen nanofibers, the surface of HA functionalized collagen nanofibers has higher anisotropic wettability and mechanical flexibility. HA/collagen nanofibers can significantly promote the elongation, proliferation and phenotypic shape expression of PAECs. In vitro co-culture of mouse PAECs and their corresponding smooth muscle cells (SMCs) showed that the luminal endothelialization governs the biophysical integrity of the newly formed extracellular matrix (e.g., collagen and elastin fibers) and structural remodeling of SMCs. Furthermore, in vitro hemocompatibility assays indicated that HA/collagen nanofibers have no detectable degree of hemolysis and coagulation, suggesting their promise as engineered vascular implants.

Keywords: collagen; hyaluronic acid; vascular endothelial cells; nanofibers; endothelization

Citation: Niu, Y.; Galluzzi, M. Hyaluronic Acid/Collagen Nanofiber Tubular Scaffolds Support Endothelial Cell Proliferation, Phenotypic Shape and Endothelialization. *Nanomaterials* **2021**, *11*, 2334. https://doi.org/10.3390/nano11092334

Academic Editors: Costantino Del Gaudio, Takuya Kitaoka, Silvia Baiguera and Lucy Di Silvio

Received: 28 June 2021
Accepted: 6 September 2021
Published: 8 September 2021
Corrected: 16 July 2024

Publisher's Note: MDPI stays neutral with regard to jurisdictional claims in published maps and institutional affiliations.

Copyright: © 2021 by the authors. Licensee MDPI, Basel, Switzerland. This article is an open access article distributed under the terms and conditions of the Creative Commons Attribution (CC BY) license (https://creativecommons.org/licenses/by/4.0/).

1. Introduction

Cardiovascular diseases are considered as the major cause of death worldwide [1]. Artery bypass grafting or replacement is an effective method for the treatment of degenerative arterial diseases, such as massive vascular occlusions caused by atherosclerosis. In recent years, various materials, such as polytetrafluoroethylene, polycaprolactone and decellularized extracellular matrix (ECM), have been used to prepare vascular grafts [2,3]. However, due to the incomplete coverage of endothelial cells (ECs) on the surface of vascular substitutes (causing intimal hyperplasia), tissue-engineered grafts for small vessels often cause occlusions (thrombosis) [4]. Complete endothelialization along the inner wall of a scaffold holds promises in solving this issue [5,6]. Using such a method, the in vitro expanded patient-derived ECs are firstly seeded on the luminal surface of the scaffold, and allowed to form a monolayer of evenly distributed ECs prior to transplantation.

The inner blood contact layer of native blood vessels is lined with a monolayer of ECs, which adhere to the basement membrane bed [7]. The ECM of the basement membrane consists mainly of elastin, collagen, laminin and heparan sulfate proteoglycan [8]. ECs and the correspondent ECM of basement membrane form the vascular endothelium, which can maintain antithrombotic properties and inhibit intimal hyperplasia.

From the perspective of vascular tissue engineering, biomaterial scaffolds have been regarded just as a structural framework, but now they are considered as important regulators of cell phenotype and function [9,10]. As the main component of natural ECM, collagen has been widely used as a scaffold material for tubular or hollow organs due to its wide availability, good biocompatibility and easy processability [11–13]. Collagen-based scaffolds fabricated by electrospinning technology can create a hierarchical nano-topography and microstructure similar to native ECM [14–16]. The greatest advantage of collagen as a scaffold material is to guide tissue regeneration by its temporary ECM function [17,18]. However, collagen scaffolds usually have limited anticoagulant activity and poor biological activity to regulate EC response [19,20]. Therefore, the endothelialization process in collagen-based scaffold grafts is relatively slow. Studies have shown that hyaluronic acid (HA) has double functionality as anticoagulant and promoter of endothelialization [21–23]. HA, as a unique glycosaminoglycan, widely exists in the ECM of various tissues [10,24,25]. It has been recognized that HA plays a key role in adjusting biological functions by binding to specific protein receptors (such as CD44) on different cell surfaces [26,27]. Therefore, it is a feasible strategy to integrate HA onto the surface of collagen scaffolds, regulating cell behavior and promoting blood compatibility and endothelialization of vascular transplantation.

Herein, tubular HA/collagen nanofibrous scaffolds were obtained by sequential electrospinning and crosslinking methods. The resultant scaffolds consisted of a thin inner layer and a thick outer layer, which, respectively, simulate the intima and middle layer of natural blood vessels. HA-functionalized collagen nanofiber on the inner wall of the tube was helpful to optimize the surface biochemistry, while the layered tubular bulk architecture provided mechanical support to cells. This allowed us to hypothesize that tubular HA/collagen nanofibrous scaffolds can have a favorable effect on the phenotypic shape, spatial distribution and behavior patterns of vascular ECs.

To verify the above hypothesis, we report the preparation, physicochemical properties and blood compatibility of HA/collagen layered nanofiber tubes. On this basis, the adhesion, elongation, proliferation and phenotypic shape of mouse primary aortic endothelial cells (PAECs) on HA/collagen nanofibers were studied. Finally, PAECs and their corresponding smooth muscle cells (SMCs) were sequentially seeded into the inner and outer layers of the layered nanofiber tubes. Histochemistry and immunofluorescence were employed to evaluate the morphology, phenotype and matrix secretion of the two types of primary cells in the layered nanofiber tubes.

2. Materials and Methods

2.1. Materials and Reagents

Hyaluronic acid (Mw ≈ ~800 kDa, F1177), collagen (11179179001), plasminogen (341578), Dulbecco's modified Eagle's medium (DMEM, D0819), 4′6-diamidino-2′-phenylindole dihydrochloride (DAPI, 10236276001), phosphate buffered saline (PBS, pH 7.4, P4417), bovine serum albumin (BSA, A3858), paraffin (327204), and dehydrated alcohol (1012772) were obtained from Sigma-Aldrich (St. Louis, MO, USA). Fetal bovine serum (FBS, 16140071), penicillin/streptomycin (P/S, 15070063), trypsin (0.25%, 15050065), primary antibody to CD31 (37-0700), Rhodamine Phalloidin (R415) were obtained from Gibco (Grand Island, NY, USA). Anti-alpha (α)-smooth muscle actin (α-SMA, ab5694) and Cell Counting Kit-8 (CCK-8, ab228554) were purchased from Abcam (Cambridge, UK). The 4% paraformaldehyde (C104190), Masson's trichrome (MTC), Verhoeff-Van Gieson (VVG, GP1035), Triton X-100 (WGT8200), hematoxylin eosin (H&E, G1004, G1002), secondary antibodies, Alexa Fluor-488, or Cy3-conjugated anti-mouse or anti-rabbit immunoglobulin-G (IgG) for fluorescence staining were obtained from Servicebio Science and Technology Co., Ltd (Wuhan, China). Hexafluoroisopropanol (HFIP, 920-66-1) was purchased from Aladdin (Shanghai, China). Mouse PAECs (SNP-M016) and SMCs (SNP-M015) were obtained from Sunncell Bioscience Inc. (Wuhan, China).

2.2. Preparation of HA/Collagen Nanofiber Scaffolds

Collagen (10%) and HA (2%) electrospinning solutions were prepared, respectively, by dissolving 1 g of collagen and 0.2 g of HA in 10 mL of HFIP. Firstly, 2% HA electrospinning solution was transferred into a 10 mL syringe and fixed on the precise injection pump of the electrospinning machine. The feeding rate of the solution was set at 0.36 mL/h. The voltage power supply was set at 11 kV, and the distance between the spinneret tip and the grounded drum collector was fixed at 20 cm. For comparison, collagen nanofiber films and tubular structures with similar parameters were prepared. The device collecting nanofiber in tubular structure was a stainless-steel tube with outer diameter of 3 mm, while thin films composed of nanofibers were collected on a stainless-steel rod with an outer diameter of 10 cm. Drum speed was set at 1200 rpm. The electrospinning environment was maintained at 18–25 °C and 40–45% (relative humidity). In order to obtain layered nanofiber tubes, the spinning time of electrospinning solution HA was 20 h, followed by collagen spinning solution for 40 h with the same electrospinning parameters. The spinning time of HA/collagen film was firstly set at 2 and 8 h, respectively.

2.3. *Crosslinking of Nanofibers*

Glutaraldehyde (0.1 mol/L) and hydrochloric acid (0.01 mol/L) were used to crosslink the collagen and HA/collagen nanofibers in acetone (80%) water mixture at 25 °C for 24 h. Before further use, all electrospun nanofibers were placed in a vacuum drying oven for 24 h at 25 °C to remove organic solvents.

2.4. *Characterizations*

The morphology of electrospun collagen and HA/collagen nanofibers before and after crosslinking treatment was observed by scanning electron microscopy (SEM, SU8010, Hitachi, Tokyo, Japan) with an accelerating voltage of 5 kV after sputter coating with gold [28]. Image analysis software ImageJ (http://imagej.net/citing) (accessed on 22 April 2017) was used to measure the average diameter and pore size distribution of electrospun nanofibers from SEM images. For each sample, 120 nanofibers or pores were randomly selected from 10000× magnification SEM images [29]. The fiber diameter and pore diameter were calculated by averaging 120 random nanofibers and pores, respectively.

The wettability of electrospun collagen and HA/collagen nanofibers was measured by water contact angle method (SL600, Solon Information Technology Co., Inc. (Solon, OH, USA). In short, after 2 s, 0.2 μL deionized water was dropped on the flat nanofiber membrane to take, then a static image of the water contact angle was measured perpendicularly to the nanofiber axis.

By measuring the resistance of nanofibers to plasminogen (human plasma fibrinolytic enzyme), the degradation properties of cross-linked electrospun nanofibers were measured. The initial weight (w_0) of electrospun nanofibers before treatment with plasminogen was recorded. Then, the samples were immersed in the digestive solution containing plasminogen (25 mg/mL) (pH 7.40) and incubated at 37 °C. The plasminogen solution was refreshed every three days. Samples were removed from the digestive solution and lyophilized at each time point (day 7, 14, 21 and 28). Finally, the residual mass (w_t) at each time point was recorded and the weight loss percentage (%) of the stent was calculated according to the following formula:

$$\text{Weight loss percentage (\%)} = (w_0 - w_t/w_0)100\%$$

Each nanofiber film was characterized by Fourier transform infrared spectroscopy (FTIR) in the range of 500–4000 cm^{-1} and attenuated total reflection (ATR) head (ATR-FTIR, thermo Nicolet, Waltham, MA, USA) with a scanning resolution of 2 cm^{-1}.

X-ray photoelectron spectroscopy (XPS) was performed using an ESCA-LAB 250Xi (Thermo Fisher, Waltham, MA, USA) to determine the surface chemistry of each nanofiber film, for which full (pass energy, 100 eV) and high-resolution spectra (pass energy, 20 eV) were recorded [30].

Wide-angle X-ray scattering (WAXS) measurements were performed using Cu k-alpha radiation (λ = 0.15418 nm) on an X-ray diffractometer (Bruker, D8 Advance, Billerica, MA, USA) equipped with an image plate detector and a graphite monochromator, and controlled by a fast/XRD diffractometer control software (RIGAKU America Corp. version 2.3.8, Tokyo, Japan). Fix the nanofiber sample vertically on the copper sample table. The diffraction data were collected for 20 min while the sample stage was rotated at a speed of 5° min^{-1}.

To ensure that the electrospun collagen and HA/collagen nanofiber scaffolds can withstand pressure changes in blood vessels, their ultimate tensile strength (UTS) and linear modulus (E) values were measured by a tensile testing instrument (Sans, Shenzhen, China). Rectangular specimens ($60 \times 12 \times 0.10$–0.15 mm) were held between the instrument grids with a 50 N load cell, and we conducted a fracture test at a strain rate of 2 mm/min with an initial holding distance of 60 mm. All measurements were performed at room temperature (RT). The tensile properties were obtained from the initial linear region of the stress–strain curve of the specimen [30,31]. Five samples were tested for each type of sample.

2.5. *In Vitro Hemocompatibility Evaluation*

2.5.1. Red Blood Cell (RBC) Morphology on Nanofibers

The animal-related experiment protocol was approved by the animal experimentation Ethics committee of Guangzhou Medical University. The collagen and HA/collagen nanofibers were cut into circular membranes with a diameter of 15 mm. The as-prepared nanofiber circular membranes were inserted in a 24-well plate, following addition of 70% ethanol solution to sterilize the nanofiber samples at room temperature (RT) for 2 h, and finally pre-warmed 1× PBS to rinse the sample 3 times for further use. A 4 mL blood quantity was sampled from healthy New Zealand white rabbits and trisodium citrate was employed for anticoagulation at a volume ratio of 9:1. After 1500 rpm centrifugation for 10 min at 4 °C, the supernatant was removed. The RBCs were washed with 1× PBS, and incubated with different nanofiber samples in 1× PBS for 60 min at RT. After that, the RBCs were fixed with 4% formaldehyde for 15 min, and dehydrated with ethanol gradient solutions (85%, 95% and 100%, *v*/*v*). After freeze-drying, the RBC nanofiber scaffolds were plated with gold and observed with SEM.

2.5.2. RBC Lysis

The hemolysis rate was evaluated by determining the hemoglobin concentration released by diluted red blood cells (RBC) exposed to the electrospun nanofiber material. The negative control consisted of adding 50 μL of RBC suspension (16% in 1× PBS, *v*/*v*) to a centrifuge tube in 2 mL 1× PBS (as a negative control). As a positive control, 2 mL deionized water was used in place of 1× PBS in order to induce maximum red blood cell lysis ($n = 4$). After incubating for a selected period of time, the RBC suspension was centrifuged at 2000 rpm for 5 min, and the supernatant was collected. Then, a microplate reader (Multiskan MK33, Thermo electron corporation, China) was used to measure the absorbance of hemoglobin (HB) released in the supernatant (200 μL) at 540 nm. By comparing the absorbance values of the tested supernatant and the positive control (i.e., 100% hemolysis), the hemolysis rate in the presence of different nanofibers was calculated.

2.5.3. Platelet Adhesion Test

Platelet-rich plasma (PRP) was obtained from the above-mentioned whole blood separated by centrifugation and used for the platelet adhesion test. A 100 microliters quantity of PRP was added to each well plate covered with nanofiber film samples, and then incubate for 2 h at 37 °C. Then, specimens were rinsed gently with pre-warmed 1× PBS twice, followed by subsequent fixation, gradient dehydration, freeze-drying, and SEM observation.

2.5.4. Activated Partial Thromboplastin Time (APTT) and Prothrombin Time (PT) Evaluation

Fresh anticoagulated whole blood was centrifuged at 1500 rpm/min for 10 min, and the supernatant (platelet-deficient plasma) was collected. The platelet-poor plasma (270 μL) was added to the nanofiber-covered well plate. According to the instructions of the reagent manufacturer, after adding the corresponding reagents, the APTT and PT of the samples were measured with an automatic coagulation analyzer (H1201, Jiansu Horner Medical Instrument Co., Ltd., Changshu, China).

2.6. In Vitro Cell Scaffold Interactions

Mouse PAECs (passage 3) were cultured in DMEM containing 10% FBS, 1% P/S (penicillin/streptomycin) in a 37 °C, 5% CO_2 humidified incubator. Cells were seeded on nanofibers at a concentration of 1×10^5 cells/cm^2, and the medium was changed every two days. Before cell seeding, each crosslinked nanofiber sample was sterilized with 70% ethanol for 2 h, and then washed with 1× PBS 5 times. The tissue culture plate (TCP in vitro conventional two-dimensional cell culture mode) was set as the control group.

At 12 h seeding, the sterilized nanofiber film samples (15 mm in diameter) were added in to each pore of the TCP and cultured for 24 and 48 h followed by staining using a live/dead assay kit (L3224, Invitrogen) using previously described methods [32,33]. In short, after sucking and discarding the culture medium, we washed the cells and nanofiber film samples in each well plate well with 1× PBS, then added 1 mL staining solution containing 0.4 μL Calcein AM and 2 μL ether-homodimer-1 to each well, and incubated in the dark for 30 min. A confocal laser scanning microscope (CLSM, TCS SP5, Leica, Wetzlar, Germany) observed cell staining, with the excitation/emission filter set at 488/530 nm to observe living cells (stained green), and detected dead cells (stained red) at 530/580 nm. Five images were randomly selected from each hole containing nanofiber film samples for shooting and analysis. Wells without nanofiber film samples were set as positive controls.

At 24 and 36 h seeding, the cell growth and morphology on different substrates were studied by SEM. Briefly, the cell scaffold samples were washed twice with 1× PBS to remove non-adherent cells, and fixed with 4% paraformaldehyde at RT for 15 min. Then, samples were rinsed with 1× PBS 3 times, followed by a series of ethanol solution gradient dehydration, freeze-drying, and SEM observation.

The proliferation of mouse PAECs on each nanofiber and TCP (positive control) was analyzed using a CCK-8 kit at time points 24, 48, 72, 96 and 120 h after seeding, as described previously [34].

2.7. Cell Phenotype Analysis

A total of 72 h after seeding, the cells' phenotype and morphology were studied by immunofluorescence. Briefly, the cell scaffold samples were fixed with 4% paraformaldehyde for 15 min at RT. Subsequently, cells were permeabilized with 0.1% Triton X-100 for 15 min, and blocked with 1% BSA for 30 min. The cell scaffold samples were treated with primary antibody CD31 diluted 1:150 at 4 °C overnight. Then, Alexa Fluor-488-conjugated anti-rabbit IgG diluted 1:400 was added and incubated at RT for 45 min. After rinsing the sample twice with 1× PBS, 1:400 dilution of Rhodamine Phalloidin was added to stain the cell actin cytoskeleton for 30 min, then a 1:2000 dilution of DAPI was used to counter-stain the nucleus. After rinsing 3 times with 1 × PBS, the samples were imaged with CLSM. The CD31 staining positive area was averaged from 6 random fields of each cell scaffold sample (n = 4 samples per group) with the software ImageJ.

2.8. Co-Culture In Vitro

To demonstrate that endothelialization is necessary for the growth and structural remodeling of vascular SMCs, we successively inoculated mouse PAECs and SMCs into the inner and outer layers of HA/collagen nanofiber scaffolds. In short, HA/collagen nanofiber tube scaffold (2.5 cm length, 3 mm inner diameter) was inserted in a 3.5 cm cell culture plate, 70% ethanol solution was added to sterilize the scaffold at RT for 2 h and finally, pre-warmed 1× PBS was added to wash the scaffold 3 times. The sterilized tube scaffold was closed at both ends by sterile 5–0 sutures (VCP359H, Johnson and Johnson, New Brunswick, NJ, USA) to prevent cells from leaking out. PAECs were injected into the lumen at a density of 1 × 10^6 cells/mL via a 25 G needle (SYRINGE, Jiangyin Fanmei Medical Device Co, Ltd., Jiangyin, China). The cell scaffold was transferred to a 15 mL test tube containing 10 mL of culture medium and rotated at a speed of 10 rpm in an incubator at 37 °C. After 24 h, the suture was removed; the cell scaffold was transferred to a 3.5 cm cell culture plate, and incubated at 37 °C and 5% CO_2 for 3 days. Then, the mouse aortic SMCs were seeded into the outer layers of the scaffolds at a density of 1 × 10^7 cells/cm^2 at 37 °C for 24 h. The cell scaffold was rotated 45 degrees to seed the SMCs into the outer layers of the scaffolds at the above-mentioned density at 37 °C for 24 h, continuously until the outer layer of the tubular scaffold was inoculated with SMCs. After 5 days of co-cultivation, the cell scaffold samples were collected and fixed in 4% paraformaldehyde for 15 min. Then, the samples were embedded with optimum cutting temperature compound (OTC, SAKURA, Torrance, CA, USA) and frozen sections cut at 15 μm. Cell scaffolds were permeabilized with 01% Triton X-100 in 1× PBS for 10 min and blocked with 1% BSA in 1× PBS for 30 min at RT.

The following antibodies, mouse anti-CD31 monoclonal antibody and rabbit anti-α-SMA polyclonal antibody, were used at a dilution of 1:150. Sections were then incubated for 45 min at RT with species-matched Cy3 or Alexa Fluor-488-conjugated secondary antibodies. Following specimen washing with 1× PBS, nuclei were counterstained with DAPI. Then, samples were imaged with CLSM. Histomorphometric evaluations (n = 3 samples per group) were performed on three independent microscopic fields (20× magnification) using ImageJ software (http://imagej.net/citing) (accessed on 22 April 2017) to quantify CD31 and α-SMA positivity.

The same cross-sections were stained with H&E, MTC and VVG using standard methods. Images of stained samples were performed with a Pannoramic DESK microscope. ImageJ software was employed to convert the blue colour intensity of the MTC stains, relative to collagen composition, in mean gray value with respect to tissue area. This protocol used hue (121–179), saturation (20–255) and brightness (10–255) to isolate collagen, also adaptable for elastin and muscle content. Both muscle content and elastin were identified using the YUV color space (muscle: Y = 0–145, elastin: Y = 0–120, both: U and V = 0–255).

2.9. Statistical Analysis

Where appropriate, a one-way or two-way analysis of variance (ANOVA) was performed using Graphpad Prism 7.0 software to determine the significant difference of the Tukey post hoc test ($p < 0.05$). Unless otherwise stated, data and error bars are reported as mean × standard deviation.

3. Results and Discussion

3.1. Preparation and Characterization of HA/Collagen Nanofibers

The tubular HA/collagen nanofibrous vessel scaffolds with hierarchical architecture were fabricated with HA and collagen by sequential electrospinning. The tube has an inner diameter of 3.0 mm and a wall thickness of ~0.62 mm (Figure 1A). The inner wall of the tube is a layer of pure HA nanofibers with a thickness of ~0.22 mm, and the outer wall is a layered pure collagen nanofibrous structure with a thickness of ~0.4 mm (Figure 1B). SEM micrographs showed that the inner wall surface of electrospun HA/collagen and collagen nanofiber tubes possesses a porous nanofiber network-like nano-topography (Figure 1B). The single nanofibers are in a continuous cylindrical shape. Quantitatively, the average nanofiber diameter of an HA/collagen nanofiber tube is (541.3 ± 19.3) nm, which is slightly smaller than that (580 ± 16) nm of pure collagen nanofiber, but not statistically significant (Figure 1C). The average pore diameter of the inner wall of the HA/collagen and pure collagen nanofiber is (2139 ± 81) and (2159 ± 77) nm, respectively (Figure 1D). HA was physically immobilized on the inner surface of the hollow tubular collagen scaffold, rather than covalently linked, which may change the activity of HA. Different sizes of vascular scaffolds were obtained by adjusting the mold (receiving device). As a proof of concept, this procedure can be applied for other hollow organs. The inner diameter of the scaffold is 3 mm to simulate the small diameter artery. By adjusting the mold of the receiving device, it can easily adapt to the size and diameter for specific patients.

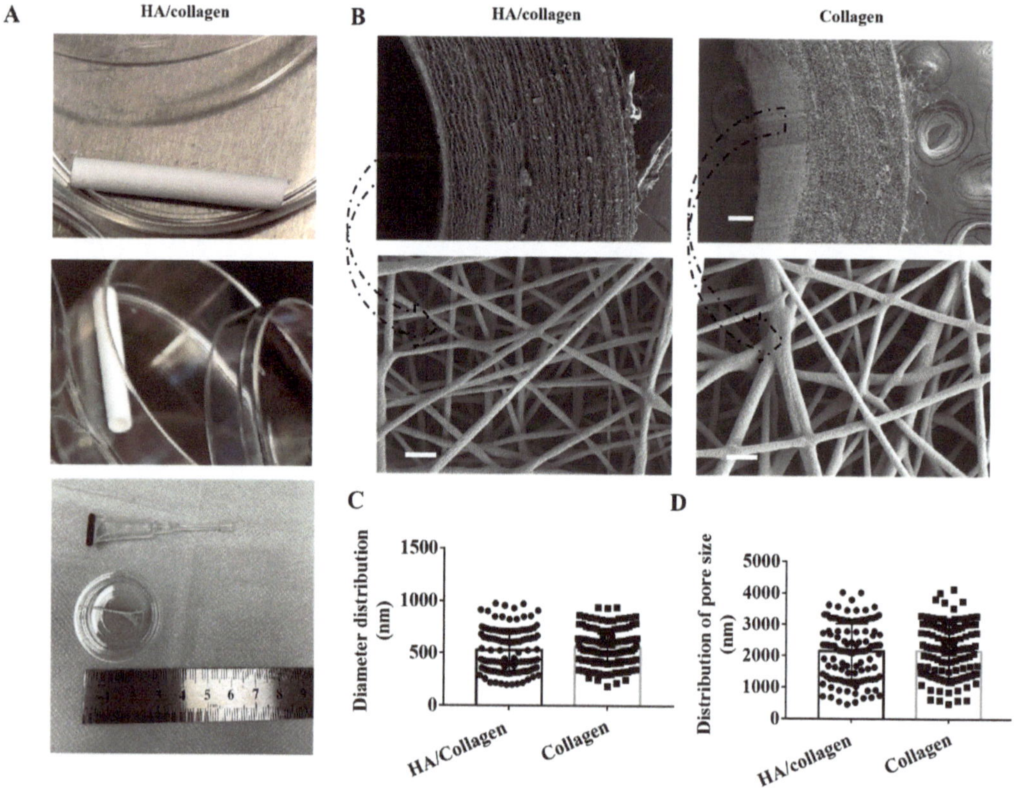

Figure 1. Physical characteristics of electrospun nanofibers. (**A**) The lateral (upper panel), cross-section (middle panel), and strip nanofiber concentric axis film (after cutting along the axis) in the image of an HA/collagen nanofiber tube. (**B**) SEM micrographs of the cross-section (upper panel) views and the inner wall surface (lower panel) of HA/collagen and collagen nanofibers before cross-linking. Scale bars: 100 μm (upper panel), 2 μm (lower panel). Statistical data of the diameter (**C**), pore size, and (**D**) distribution of various nanofibers.

HA is a hydrophilic bioactive macromolecule [19,34]. To prevent the HA nanofibers on the inner wall of the tube from swelling when the HA/collagen nanofiber tube was used under physiological conditions, we used glutaraldehyde/acetone treatment to induce cross-linking between molecules of HA and collagen. SEM morphology observation showed that electrospun HA/collagen and collagen nanofibers formed a tight interconnected nanofiber porous structure under the effect of glutaraldehyde/acetone crosslinking (Figure 2A). After cross-linking, the average nanofiber diameters of HA/collagen and collagen nanofibers were (527.3 $\pm$ 18) and (564 $\pm$ 16) nm, respectively (Figure 2B). The average pore size on the inner wall was (1699 $\pm$ 60) and (1411 $\pm$ 42) nm (Figure 2C). This indicates that crosslinking treatment has little effect on the average diameter of electrospun nanofibers, but has a significant effect on the average pore diameter of the electrospun nanofiber network structure.

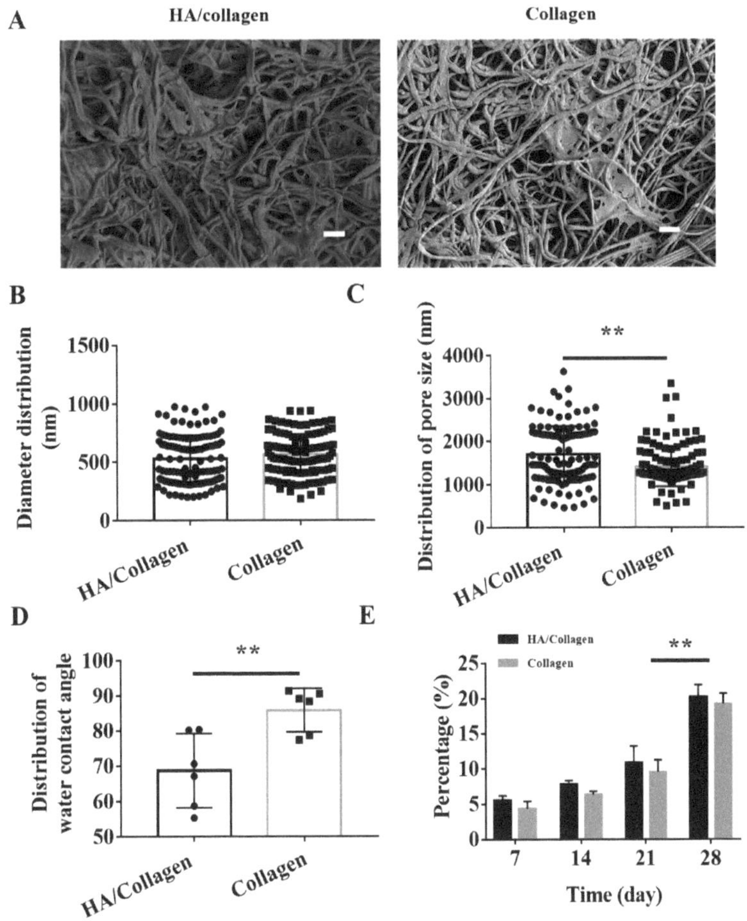

Figure 2. Morphology, hydrophilicity and degradability of cross-linked electrospun nanofibers in vitro. (**A**) SEM micrographs of the surface of cross-linked electrospun HA/collagen and collagen nanofibers. Scale bars: 4 μm. Statistical data of the diameter (**B**) and pore size (**C**) distribution of cross-linked nanofibers (n = 120). Average water contact angle (**D**) and degradability (**E**) of different nanofibers (n = 6). ** $p < 0.01$.

The averaged static water contact angle of the inner surface of the crosslinked HA/collagen nanofibers is $(68.7 \pm 4.3)^\circ$, which is significantly smaller than the $(85.9 \pm 2.5)^\circ$ of the pure collagen nanofiber tube (Figure 2D, data collected in Table S1). Due to the similar topography of electrospun HA/collagen and collagen nanofibers, the improved wettability of HA/collagen nanofibers can be attributed to the hydrophilic effect of HA.

The chemical properties of crosslinked collagen and HA/collagen nanofibers were detected by ATR-FTIR (Figure S1). The characteristic peaks of collagen nanofibers can be detected at 1656, 1556 and 1242 cm^{-1}, corresponding to amide I, amide II and amide III, respectively, which are consistent with a previous report [19]. The peaks show that crosslinking is not causing changes in the secondary structure of peptides and proteins. Compared with collagen nanofibers, the peaks of HA/collagen nanofibers in amide I, amide II and amide III absorption were 1658, 1552 and 1239 cm^{-1}, respectively. These changes may be due to the interaction between the aldehyde group of HA and the amine group of collagen. Figure S2 shows the XPS analysis of collagen and HA/collagen nanofibers detecting C, O, and N elements on their surface layers (Figure S2A), as expected from collagen-based on amino acids and HA, higher polysaccharide, composed of D-glucuronic acid and N-acetylglucosamine. In addition, a small amount of N1s (399.75 eV) signal appeared on the surface of collagen nanofibers, which was attributed to the amino group in collagen molecules (Figure S2B). After HA nanofibers covered the surface of collagen nanofibers, the content of N on the surface of HA/collagen nanofibers further increased (Figure S2B). The results show that HA nanofibers are successfully introduced into the surface of collagen nanofibers. The XRD patterns of HA/collagen and collagen nanofiber films are shown in Figure S3. The two nanofiber films show the characteristics of X-ray diffraction patterns of some crystalline materials. Collagen nanofiber film has a defined diffraction peak, and a wide peak at 2θ of 25.6°. This corresponds to the diameter of the triple helix collagen molecule. Compared with collagen nanofiber, the diffraction spectrum profile of the HA/collagen nanofiber film is similar, and the maximum peak intensity changes slightly, indicating that addition of HA has no effect on the crystallinity of collagen nanofiber film.

To determine the stability of the HA/collagen nanofiber tube structure, we tested the resistance of these electrospun fibers to plasminogen (an important enzyme in the blood that keep blood vessels and glandular ducts unobstructed). As shown in Figure 2E, by the seventh day, the weight loss rates of HA/collagen and collagen nanofiber tubes were (5.5 ± 0.7)% and (4.3 ± 1.1)%, respectively. By the 14th and 21st days, the weight loss rate of each sample increased slightly, but it was not statistically significant. As the experiment time was extended to the 28th day, the weight loss rates of HA/collagen and collagen nanofiber tubes were significantly greater than that on the 21st day. The weight loss rate of HA/collagen nanofiber tubes was 20.8%, slightly higher than 19.2% of collagen. These data indicate that HA/collagen nanofiber tubes have certain resistance to degradation in human plasma plasmin aqueous solution.

3.2. Mechanical Properties

To analyze the mechanical properties of the scaffold, each electrospun nanofiber film was tested in uniaxial tension mode to generate a stress–strain curve and derived tensile properties (Figure 3A). HA/collagen nanofiber films showed an ultimate tensile strength of (0.97 ± 0.1) MPa (Figure 3B), Young's modulus of (0.56 ± 0.02) MPa (Figure 3C) and elongation at break of (532.8 ± 18.62)% (Figure 3D), respectively collected in Table S1. Although the Young's modulus is equivalent to collagen, the ultimate stretch of collagen nanofibers is slightly lower than that of HA/collagen nanofibers, but not statistically significant. Both HA/collagen and collagen nanofibers have relatively high elongation at break (Figure 3A,D), indicating that both specimens are very soft and tough materials, exhibiting good ductility after stretching. These data suggest that the introduction of HA does not affect the mechanical properties of collagen. The mechanical properties of HA/collagen nanofiber tubes are derived from collagen nanofibers, while HA fixed on the inner wall surface of HA/collagen scaffold is helping to optimize the surface biochemistry. In previous studies we showed that mimicking porosity and mechanical properties of natural ECM promotes tissue regeneration [35].

3.3. Hemocompatibility of the HA/Collagen Nanofibers In Vitro

The morphology and lysis of RBCs after 1 h incubation on HA/collagen and collagen nanofibers were observed by SEM (Figure 4A). On different substrates, the shape of the RBC is a biconcave disk with a diameter of about 7–8 μm, with a concave central part (about 1 μm thick) and thick edges (about 1.9 μm thick). No morphological changes or cracks were observed. Furthermore, the hemoglobin released from the red blood cells was measured using a colorimetric method to determine the hemolysis rate. As shown in Figure 4B, according to the ISO 10993-4:2002 standard and ASTM F756-00 (2000), all nanofibers are non-hemolytic and will not cause any detectable hemolysis.

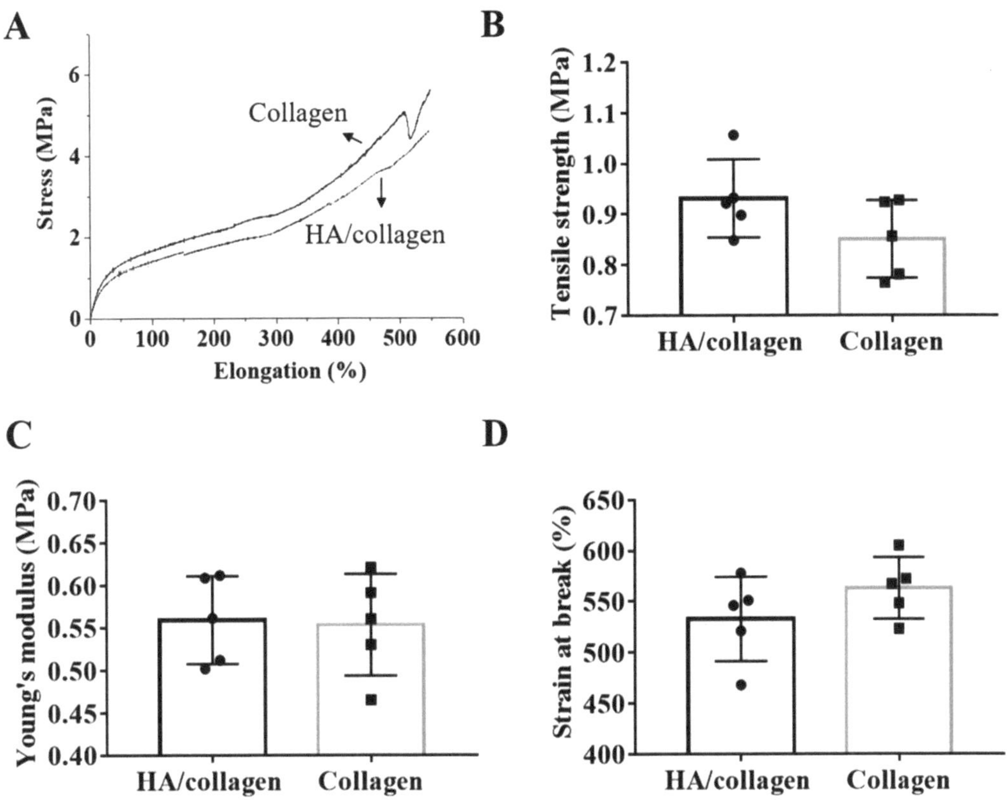

Figure 3. Mechanical properties. Relative to cross-linked electrospun HA/collagen and collagen nanofibers: (**A**) typical stress–strain curves, (**B**) ultimate tensile strength, (**C**) Young's modulus and (**D**) strain at break.

The antithrombotic properties of electrospun nanofiber scaffolds were evaluated by in vitro platelet adhesion measurement. The platelet adhesion of platelet-rich plasma (PRP) after 1 h incubation on different nanofiber substrates was observed by SEM (Figure 4C). A small amount of platelets was observed on the network structure of HA/collagen nanofibers, the shape was round, oval or irregular, and the diameter was about 1.5 μm. In contrast, on collagen nanofibers, some of the adherent platelets are activated, and the shape of the prosthetic foot is flat and extended.

Moreover, we evaluated the clotting time by calculating APTT to observe the influence of the material on the delay of clotting through the internal pathway [36]. The results showed that, compared with the control, the APTT value of HA/collagen nanofibers was (56.3 $\pm$ 2.2) s ($p < 0.05$), while the APTT value of collagen was (49.7 $\pm$ 1.7) s (Figure 4D, $p < 0.05$). In addition, we also tested the PT of each electrospun fiber to evaluate the material's effect on coagulation in the external pathway of the coagulation cascade [19]. Compared with the control group, the PT values of HA/collagen and collagen nanofibers increased significantly (Figure 4D, $p < 0.05$). These results indicate that the introduction of HA can improve the anticoagulant activity of the substrate.

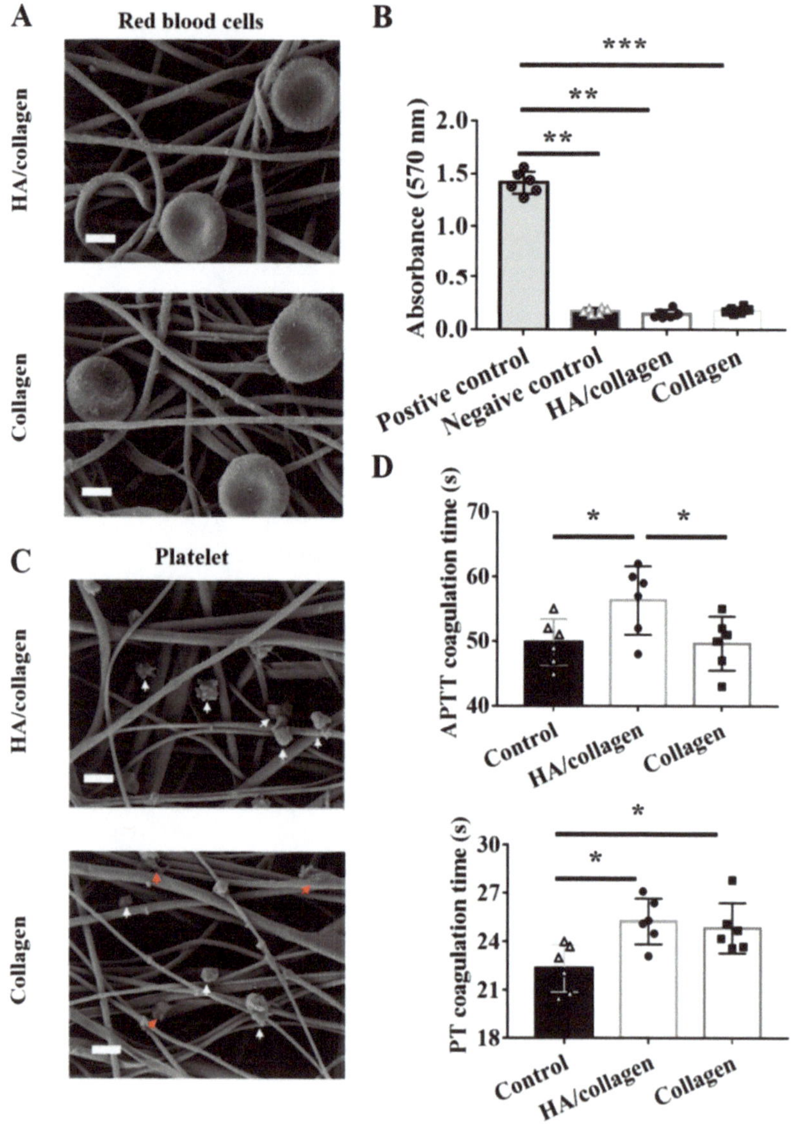

Figure 4. Hemolysis and anticoagulation tests. Representative SEM micrographs of red blood cells (**A**) and platelets (**C**) on HA/collagen and collagen nanofibers after 2 h seeding. White arrows indicate non-adherent platelets, red arrows indicate adherent platelets. Scale bars: 1 μm. Hemolysis rate (**B**) of different nanofibers. Statistical data of APTT and PT (**D**) of control native plasma and plasma incubated with different nanofibers. $n = 3$, * $p < 0.5$; ** $p < 0.01$; *** $p < 0.001$.

3.4. HA/Collagen Nanofibers Can Promote Complete Endothelialization of Vascular EC

Before measuring cell growth behavior, we evaluated the cytotoxicity of collagen and HA/collagen nanofibers by cell Live/dead staining. As shown in Figure 5, PAECs were co-cultured with collagen and HA/collagen nanofiber films for 48 h. The results showed that collagen and HA/collagen nanofiber films did not demonstrate obvious cytotoxicity to PAECs after 48 h of co-culture, and the cell viability was (96.4 ± 1.7)% and (98.2 ± 0.5)%, respectively. Since the measured cell viability is greater than 95%, it can be considered that cross-linked collagen and HA/collagen nanofibers can be used as vascular scaffolds [19,20].

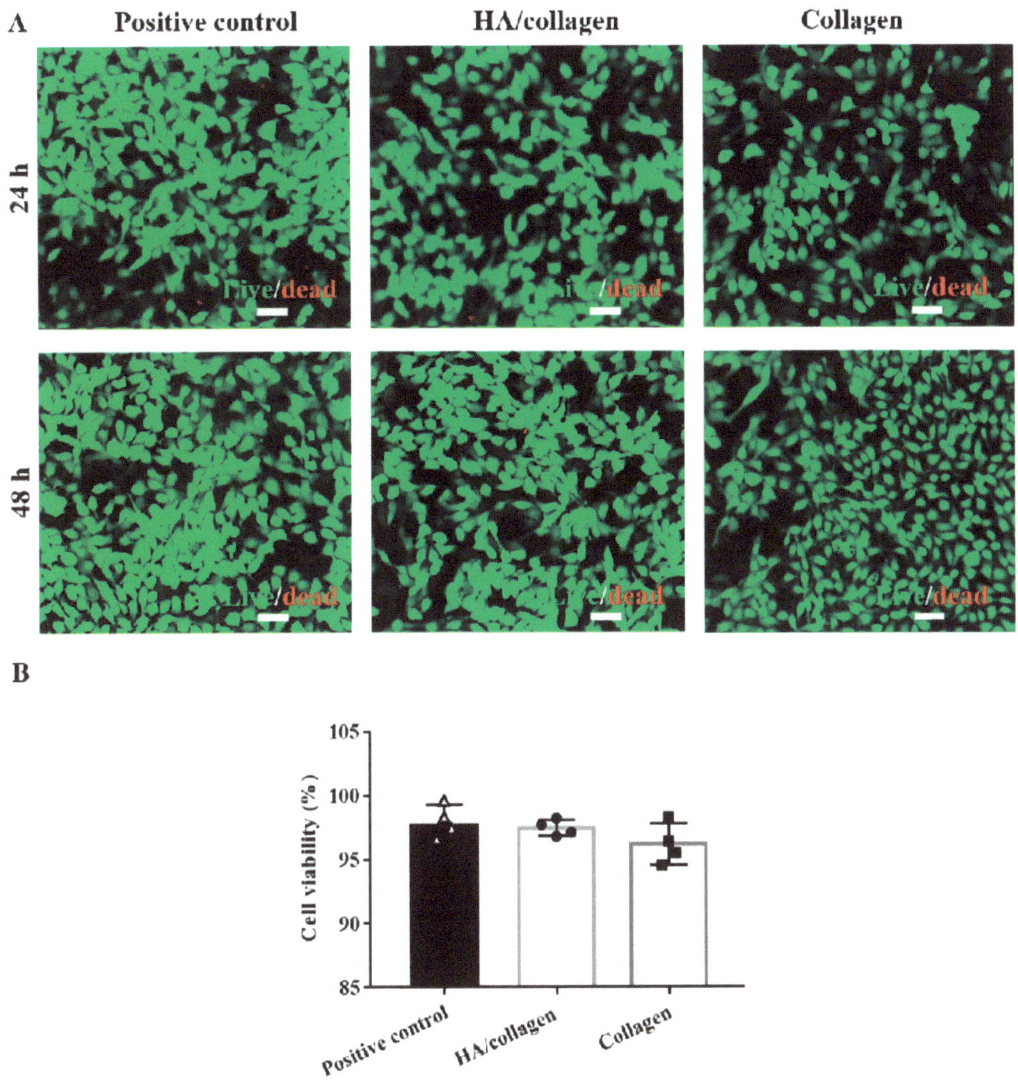

Figure 5. In vitro cytocompatibility analysis. (**A**) Fluorescence microscopy graphs of stained mouse PAECs with and without crosslinked nanofiber films after 24 and 48 h of culture. Scale bars, 50 μm. (**B**) The cell viability of mouse PAECs with and without crosslinked nanofiber films after 48 h of culture.

Next, we studied the growth behaviour of PAECs grown on HA/collagen nanofibers. As shown in Figure 6A, 72 h after inoculation, the cytoskeleton (red fluorescence) of mouse PAECs on the surface of HA/collagen and collagen nanofibers was completely extended, indicating that intercellular junctions were basically established. The staining of platelet EC adhesion molecule-1 (CD31), a specific marker for vascular ECs [37], showed that the signals of CD31-positive cells (green fluorescence) exist in vascular ECs growing on the surface of HA/collagen nanofibers, while the CD31-positive signal was weak in vascular EC growing on the surface of collagen nanofibers. Quantitatively, the density of CD31+ signals on HA/collagen nanofibers was higher than that of collagen nanofibers (Figure 6B, $p < 0.01$). It is reported that increased expression of CD31 contributes to endothelialization, while decreased expression of CD31 indicates cell dysfunction [19]. The surface biochemistry and biophysical characteristics of biomaterial scaffold are the key factors of cell adhesion, growth, proliferation and differentiation [3,17,23]. For example, ECM stiffness can regulate the integrity, permeability and leukocyte migration of vascular EC monolayers [12,23,25,38,39]. Moreover, controlling nanotopography and mechanical properties can be used to stimulate cells by mimicking their natural environment, triggering the correct functionalities [40,41]. The vascular EC growth on exogenous HA or its oligomer functionalized polymer scaffolds showed improved phenotypic shape and proliferation behaviour [30]. Scaffolds loaded with soluble biochemical factors (e.g., vascular endothelial growth factor, VEGF) can promote the adhesion, migration, proliferation and differentiation of vascular EC [42,43]. In this study, the surface of HA/collagen nanofibers may be an ideal matrix for the rapid and complete endothelialization of PAECs.

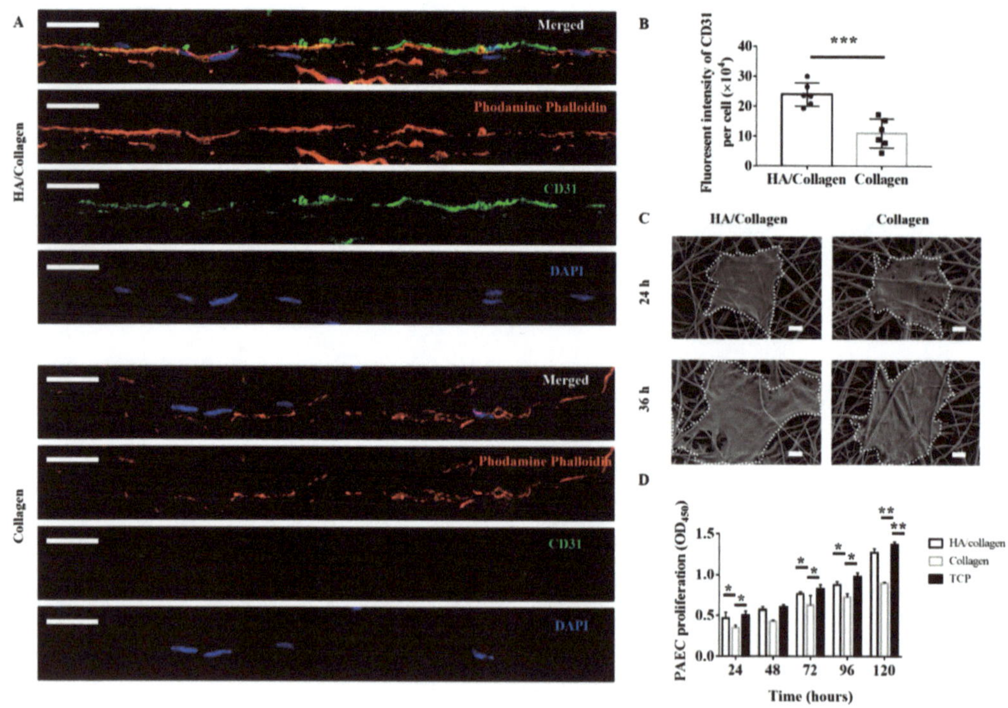

Figure 6. In vitro phenotypic shape and proliferation analysis of mouse PAECs on different substrates. Representative CLSM micrograph cross-section (**A**) of PAECs on different nanofibers after 72 h seeding. Nucleus (blue), cytoskeleton (red), CD31 (green). Scale bars: 20 µm. Statistical data of CD31 fluorescence (**B**) in ECs after 72 h seeding ($n = 5$). *** $p < 0.001$. Representative SEM micrographs (**C**) of PAECs on different nanofibers after 24 and 36 h seeding. Scale bars: 5 µm. The proliferation (**D**) of PAECs on different nanofibers and TCP (positive control) during a period of 5 days ($n = 3$). * $p < 0.5$; ** $p < 0.01$.

We further used SEM to observe the morphology of PAECs at the early stage (24–36 h after seeding) on these two electrospun nanofiber substrates (Figure 6C). After 24 h, PAECs adhered to the surface of all nanofibers with varying degrees of elongation. By 36 h, the PAECs on the surface of HA/collagen nanofibers showed a more significant elongation, suggesting that the HA/collagen substrate is more beneficial to mouse PAEC elongation.

To determine whether electrospun HA/collagen nanofibers affect the proliferation of vascular ECs, we used CCK-8 to analyse the growth of mouse PAECs on different nanofiber substrates. As shown in Figure 6D, mouse PAECs can proliferate on HA/collagen nanofibers for more than 5 days. It should be noted that on day 5, the optical density (OD) value of mouse PAECs cultured on HA/collagen nanofibers was significantly higher than that of collagen nanofibers, and was close to those of mouse PAECs cultured on TCP. As pure collagen nanofibers could not promote the proliferation and phenotypic expression of PAECs, they were not used in subsequent studies. These results are consistent with previous reports showing that scaffold surface biophysical and biochemical properties are crucial for vascular EC functions [19–23].

3.5. Complete Endothelialization of Tubular HA/Collagen Nanofibers Dictating Vascular SMCs Infiltration and Alignment

A healthy and confluent EC lining is essential for the structural reconstruction of vascular wall tissue damaged by vascular injury or disease [3,39,44–47]. To verify whether the endothelialization of the inner wall surface of the tubular scaffold plays a beneficial role in the subsequent spatial distribution of vascular SMCs, we first seeded PAECs on the inner surface of HA/collagen nanofibers to form an endothelialized ECs monolayer, and then SMCs were seeded on the outer surface of HA/collagen nanofibers. After 5 days of coculturing in vitro, the distribution of ECs and SMCs in the cellularized scaffolds was observed. As shown in Figure 7A, H&E staining showed that PAECs lined the inner wall luminal surface of the tube, forming a thin layer of flat epithelium, arranged closely. The nucleus is in the middle and slightly protruding. The basal surface of the cell adheres to the inner basal plate. The morphology of PAECs was similar to that of ECs in mouse aortic endothelium (Figure 7A). SMCs adhered to and grew into the wall under the PAEC layer. They were evenly distributed in layers and showed a wavy shape, similar to the shape of natural arteries. MTS staining on the same cross-section showed that PAECs and SMCs on the cross-section of HA/collagen scaffold could synthesize collagen (blue), and the collagen nanofibers were curly and wavy, and evenly distributed around PAECs and SMCs. At this time, a small number of smooth muscle bundles (red) were formed under the EC layer. VVG staining of the same cross-section showed that both PAECs and SMCs on the cross-section of HA/collagen scaffold could synthesize elastin (black) in a wavy shape. Quantitatively, the contents of collagen nanofibers, smooth muscle bundles and elastin in the tissue-engineered vascular grafts constructed by 5 days of culture in vitro were significantly lower than those in the native arteries (Figure 7B). These findings suggest that the bulk hierarchical architecture of HA/collagen scaffolds directs the spatial distribution of PAECs and SMCs as well as the synthesis of ECM.

In addition, we also studied the phenotypic shape of PAECs and SMCs on tubular HA/collagen nanofiber scaffolds. The PAECs and SMCs on the inner wall and outer wall surfaces were labelled with anti-mouse CD31-Cy3 and anti-rabbit α-SMA (a SMC specific marker [23])-Alexa staining, respectively (Figure 8A). The CLSM images show that a thin layer of CD31 positivity (red fluorescence) is evenly distributed on the inner wall of HA/collagen nanofiber tubes, and the contractile protein α-SMA positive (green fluorescence) of SMCs is restricted under the EC layer and evenly distributed. The characteristics of endothelium and smooth muscle are close to those of a native artery. Quantitatively, the content of CD31 positivity and α-SMA positivity was slightly lower than that of a native artery (Figure 8B). On the contrary, the α-SMA-positive (green fluorescence) signal in vascular SMCs grown on non endothelialized tubular HA/collagen nanofiber scaffolds was very weak, and vascular SMCs preferentially gathered on the inner surface of tubular HA/collagen scaffolds (Figure S4). Therefore, complete endothelialization may determine the phenotypic shape and growth pattern of vascular SMCs on tubular HA/collagen scaffolds.

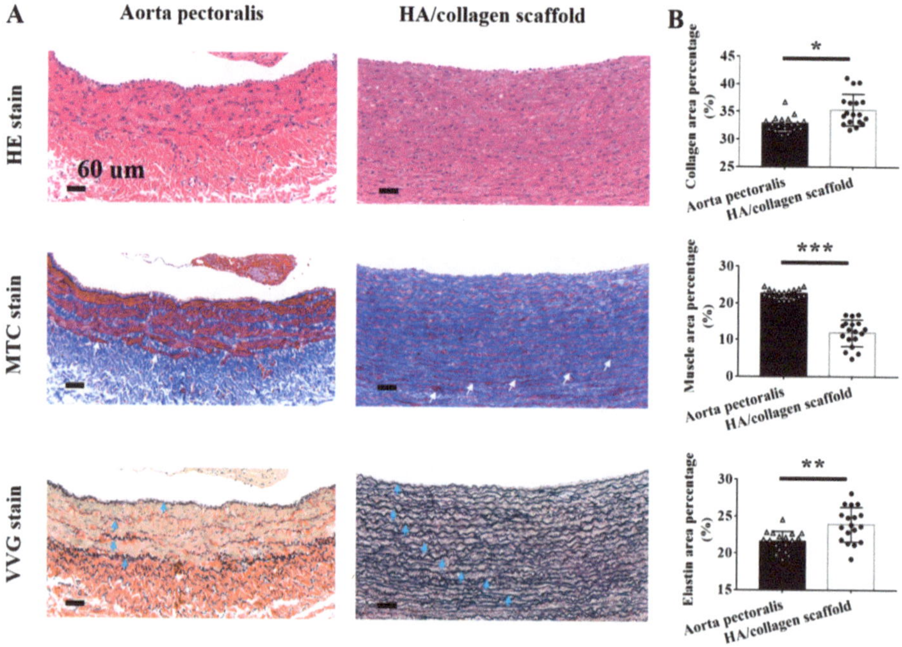

Figure 7. Histological evaluation. (**A**) Representative H&E, MTS, and VVG staining of rabbit native aorta pectoralis and cellularized tubular HA/collagen scaffold. Scale bars: 60 μm. (**B**) Statistical data of collagen, muscle bundles and elastin of rabbit native aorta pectoralis and cellularized tubular HA/collagen scaffolds (n = 3). * $p < 0.5$; ** $p < 0.01$; *** $p < 0.001$.

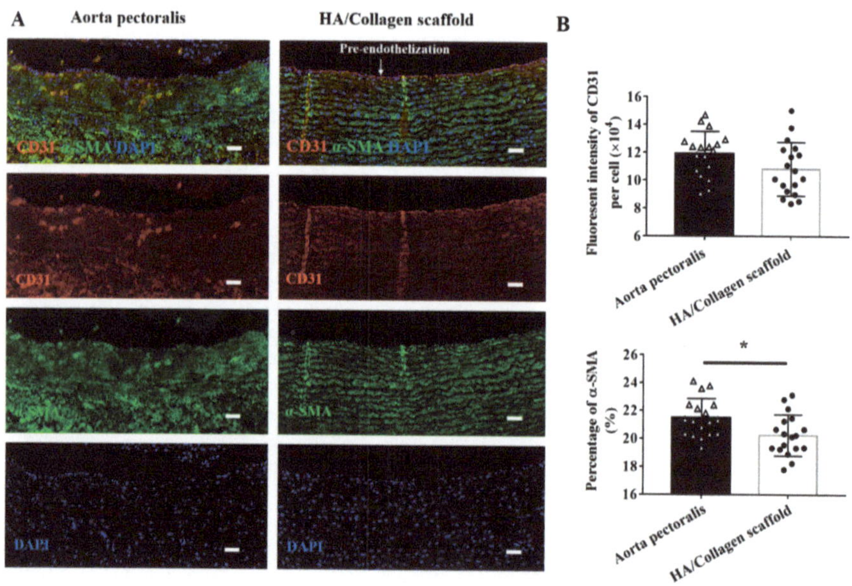

Figure 8. Fluorescence immunostaining. (**A**) Representative CLSM micrographs of rabbit native aorta and cellularized tubular HA/collagen scaffold. Nucleus (blue), CD31 in vascular EC (red), α-SMA in vascular SMC (green). Scale bars: 60 μm. (**B**) Statistical data of CD31 and α-SMA fluorescence of rabbit native aorta and cellularized tubular HA/collagen scaffolds. * $p < 0.5$.

4. Conclusions

Electrospinning technology was used to prepare a hierarchical architecture nanofiber scaffold based on HA and collagen, which can temporarily create an environment similar to native artery tissue. Thin HA nanofibers were fixed on the inner surface of the tubular nanofibers, so that the resultant HA/collagen nanofibers showed anisotropic wetting behaviour and mechanical compliance, as required for vascular scaffolds. Primary vascular ECs cultured on HA/collagen nanofibers showed the phenotypic shape and proliferation ability ideal for complete endothelialization, which may be directed by the synergistic effect of biochemical and biophysical cues of scaffolds. Most importantly, the complete endothclialization of vascular ECs along the inner wall of HA/collagen nanofibers also facilitate vascular SMC infiltration and alignment along their hierarchical architecture. In turn, SMCs facilitate the maintenance of their contractile phenotypic shape as well as establish natural growth behaviour patterns. In addition, HA/collagen nanofibers have good stability and blood compatibility, highlighting their suitability for tissue engineering vascular implants. Future perspectives of this work may involve in vivo studies to evaluate the performance of cellularized vascular stents. If this scaffold is durable enough after implantation in animals, cell scaffolds constructed from human cells may be used as living vascular prostheses for small caliber arteries. Our vascular stent is attractive for this application because it is lined with functional endothelium and can heal at the anastomosis, truly integrating with the host vascular system.

Supplementary Materials: The following are available online at https://www.mdpi.com/article/10.3390/nano11092334/s1, Figure S1: ATR-FTIR spectra collected for crosslinked collagen and HA/collagen nanofiber films, Figure S2: Surface chemical element distribution of collagen and HA/collagen nanofiber films after crosslinking treatment., Figure S3: XRD pattern of collagen and HA/collagen nanofiber film surfaces after crosslinking treatment. Figure S4: Fluorescence immunostaining of cross-section of cellularized scaffold, Table S1: Contact water angles and mechanical properties of HA/collagen and Collagen nanofibers.

Author Contributions: Y.N. prepared the material and performed the experiments. M.G. performed the characterizations. Y.N. acquired the funding. Y.N. and M.G. wrote the manuscript. Both authors have read and agreed to the published version of the manuscript.

Funding: This research was funded by National Natural Science Foundation of China (No. 82070695 and No. 81900619), 2013 National Clinical Key Specialist Construction Project (pediatric surgery) (No. 5001-2140019), and Guangdong Provincial Key Laboratory of Research in Structural Birth Defect Disease (No: 2019B030301004).

Institutional Review Board Statement: In this study, all animals were performed according to the guidelines approved by the Institutional Animal Care and Use Committee of Servicebio biotechnology Co., Ltd. (Servicebio, No.2020168).

Data Availability Statement: The data presented in this study are available on request from the corresponding authors.

Conflicts of Interest: The authors declare no conflict of interest.

References

1. Fereydooni, A.; Gorecka, J.; Xu, J.B.; Schindler, J.; Dardik, A. Carotid endarterectomy and carotid artery stenting for patients with crescendo transient ischemic attacks: A systematic review. *JAMA Surg.* **2019**, *154*, 1055–1063. [CrossRef] [PubMed]
2. Arsalan, M.; Mack, M.J. Coronary artery bypass grafting is currently underutilized. *Circulation* **2016**, *133*, 1036–1045. [CrossRef] [PubMed]
3. Akentjew, T.L.; Terraza, C.; Suazo, C.; Maksimcuka, J.; Wilkens, C.A.; Vargas, F.; Zavala, G.; Ocaña, M.; Enrione, J.; García-Herrera, C.M.; et al. Rapid fabrication of reinforced and cell-laden vascular grafts structurally inspired by human coronary arteries. *Nat. Commun.* **2019**, *10*, 3098.
4. McGuigan, A.P.; Sefton, M.V. The influence of biomaterials on endothelial cell thrombogenicity. *Biomaterials* **2007**, *28*, 2547–2571. [CrossRef] [PubMed]
5. Blit, P.H.; McClung, W.G.; Brash, J.L.; Woodhouse, K.A.; Santerre, J.P. Platelet inhibition and endothelial cell adhesion on elastin-like polypeptide surface modified materials. *Biomaterials* **2011**, *32*, 5790–5800. [CrossRef] [PubMed]

6. Gu, H.; Chen, X.S.; Yu, Q.; Liu, X.L.; Zhan, W.J.; Chen, H.; Brash, J.L. A multifunctional surface for blood contact with fibrinolytic activity, ability to promote endothelial cell and inhibit smooth muscle cell adhesion. *J. Mater. Chem. B* **2017**, *5*, 604–611. [CrossRef]
7. Wagensil, J.E.; Mecham, R.P. Vascular extracellular matrix and arterial mechanics. *Physiol. Rev.* **2009**, *89*, 957–989. [CrossRef] [PubMed]
8. Niklason, L.E.; Lawson, J.H. Bioengineered human blood vessels. *Science* **2020**, *370*, eaaw8682. [CrossRef]
9. Gaharwar, A.K.; Singh, I.; Khademhosseini, A. Engineered biomaterials for in situ tissue regeneration. *Nat. Rev. Mater.* **2020**, *5*, 686–705. [CrossRef]
10. Ibrahim, S.; Joddar, B.; Craps, M.; Ramamurthi, A. A surface-tethered model to assess size-specific effects of hyaluronan (HA) on endothelial cells. *Biomaterials* **2007**, *28*, 825–835. [CrossRef]
11. Versteegden, L.R.; van Kampen, K.A.; Janke, H.P.; Tiemessen, D.M.; Hoogenkamp, H.R.; Hafmans, T.G.; Roozen, E.A.; Lomme R.M.; van Goor, H.; Oosterwijk, E.; et al. Tubular collagen scaffolds with radial elasticity for hollow organ regeneration. *Acta Biomater.* **2017**, *52*, 1–8. [CrossRef]
12. Weinberg, C.B.; Bell, E. A blood vessel model constructed from collagen and cultured vascular cells. *Science* **1986**, *231*, 397–400 [CrossRef] [PubMed]
13. Pezzoli, D.; Paolo, J.D.; Kumra, H.; Fois, G.; Candiani, G.; Reinhardt, D.P.; Mantovani, D. Fibronectin promotes elastin deposition elasticity and mechanical strength in cellularised collagen-based scaffolds. *Biomaterials* **2018**, *180*, 130–142. [CrossRef] [PubMed]
14. Zhang, F.; Bambharoliya, T.; Xie, Y.; Liu, L.J.; Celik, H.; Wang, L.; Akkus, O.; King, M.W. A hybrid vascular graft harnessing the superior mechanical properties of synthetic fibers and the biological performance of collagen filaments. *Mater. Sci. Eng. C Mater. Biol. Appl.* **2021**, *118*, 111418. [CrossRef]
15. Pant, B.; Park, M.; Ojha, G.P.; Kim, D.; Kim, H.Y.; Park, S.J. Electrospun salicylic acid/polyurethane composite nanofibers for biomedical applications. *Int. J. Polym. Mater.* **2018**, *67*, 739–744. [CrossRef]
16. Wani, T.U.; Rather, A.H.; Khan, R.S.; Beigh, M.A.; Park, M.; Pant, B.; Sheikh, F.A. Strategies to use nanofiber scaffolds as enzyme-based biocatalysts in tissue engineering applications. *Catalysts* **2021**, *11*, 536. [CrossRef]
17. Camasão, D.B.; González-Pérez, M.; Palladino, S.; Alonso, M.; Rodríguez-Cabello, J.C.; Mantovani, D. Elastin-like recombinamers in collagen-based tubular gels improve cell-mediated remodeling and viscoelastic properties. *Biomater. Sci.* **2020**, *8*, 3536–3548 [CrossRef] [PubMed]
18. Wu, C.X.; Huang, J.T.; Chu, B.; Deng, J.P.; Zhang, Z.; Tang, S.Q.; Wang, X.Y.; Wang, Z.P.; Wang, Y.F. Dynamic and hierarchically structured networks with tissue-like mechanical behavior. *ACS Nano* **2019**, *13*, 10727–10736. [CrossRef]
19. Kang, L.Z.; Jia, W.B.; Li, M.; Wang, Q.; Wang, C.D.; Liu, Y.; Wang, X.P.; Jin, L.; Jiang, J.J.; Gu, G.F.; et al. Hyaluronic acid oligosaccharide-modified collagen nanofibers as vascular tissue-engineered scaffold for promoting endothelial cell proliferation. *Carbohydr. Polym.* **2019**, *223*, 115106. [CrossRef]
20. Hou, Y.C.; Li, J.A.; Zhu, S.J.; Cao, C.; Tang, J.N.; Zhang, J.Y.; Guan, S.K. Tailoring of cardiovascular stent material surface by immobilizing exosomes for better pro-endothelialization function. *Colloids Surf. B Biointerfaces* **2020**, *189*, 110831. [CrossRef]
21. Lyu, N.; Du, Z.Y.; Qiu, H.; Gao, P.; Yao, Q.; Xiong, K.Q.; Tu, Q.F.; Li, X.Y.; Chen, B.H.; Wang, M.; et al. Mimicking the nitric oxide-releasing and glycocalyx functions of endothelium on vascular stent surfaces. *Adv. Sci.* **2020**, *7*, 2002330. [CrossRef]
22. Li, J.A.; Chen, L.; Zhang, X.Q.; Guan, S.K. Enhancing biocompatibility and corrosion resistance of biodegradable Mg-Zn-Y-Nd alloy by preparing PDA/HA coating for potential application of cardiovascular biomaterials. *Mater. Sci. Eng. C Mater. Biol. Appl.* **2020**, *109*, 110607. [CrossRef]
23. Chang, H.; Zhang, H.; Hu, M.; Chen, J.Y.; Li, B.C.; Ren, K.F.; Martins, M.C.L.; Barbosa, M.A.; Ji, J. Stiffness of polyelectrolyte multilayer film influences endothelial function of endothelial cell monolayer. *Colloids Surf. B Biointerfaces* **2017**, *149*, 379–387. [CrossRef]
24. Fuenteslopez, C.V.; Ye, H. Electrospun fibres with hyaluronic acid-chitosan nanoparticles produced by a portable device. *Nanomaterials* **2020**, *10*, 2016. [CrossRef]
25. Han, C.Z.; Luo, X.; Zou, D.; Li, J.G.; Zhang, K.; Yang, P.; Huang, N. Nature-inspired extracellular matrix coating produced by micro-patterned smooth muscle and endothelial cells endows cardiovascular materials better biocompatibility. *Biomater. Sci.* **2019**, *7*, 2686–2701. [CrossRef] [PubMed]
26. Aruffo, A.; Stamenkovic, I.; Melnick, M.; Underhill, C.B.; Seed, B. CD44 is the principal cell surface receptor for hyaluronate. *Cell* **1990**, *61*, 1303–1313. [CrossRef]
27. Kim, S.M.; Park, K.-S.; Lih, E.; Hong, Y.J.; Kang, J.H.; Kim, I.H.; Jeong, M.H.; Joung, Y.K.; Han, D.K. Fabrication and characteristics of dual functionalized vascular stent by spatio-temporal coating. *Acta Biomater.* **2016**, *38*, 143–152. [CrossRef]
28. Cheng, F.; Cao, X.; Liu, H.B.; Xie, X.; Huang, D.; Maharjan, S.; Bei, H.P.; Gomez, A.; Li, J.; Zhan, H.Q.; et al. Generation of cost-effective paper-based tissue models through matrix-assisted sacrificial 3D printing. *Nano Lett.* **2019**, *19*, 3603–3611. [CrossRef] [PubMed]
29. Kang, Y.; Wang, C.L.; Qiao, Y.B.; Gu, J.W.; Zhang, H.; Peijs, T.; Kong, J.; Zhang, G.C.; Shi, X.T. Tissue-engineered trachea consisting of electrospun patterned sc-PLA/GO-g-IL fibrous membranes with antibacterial property and 3D-printed skeletons with elasticity. *Biomacromolecules* **2019**, *20*, 1765–1776. [CrossRef] [PubMed]
30. Ma, T.B.; Zhao, Y.S.; Ruan, K.P.; Liu, X.R.; Zhang, J.L.; Guo, Y.Q.; Yang, X.T.; Kong, J.; Gu, J.W. Highly thermal conductivities, excellent mechanical robustness and flexibility, and outstanding thermal stabilities of aramid nanofiber composite papers with nacre-mimetic layered structures. *ACS Appl. Mater. Interfaces* **2020**, *12*, 1677–1686. [CrossRef]

31. Niu, Y.Q.; Stadler, F.J.; Fang, J.H.; Galluzzi, M. Hyaluronic acid-functionalized poly-lactic acid (PLA) microfibers regulate vascular endothelial cell proliferation and phenotypic shape expression. *Colloids Surf. B Biointerfaces* **2021**, *206*, 111970. [CrossRef]
32. Liu, X.Y.; Niu, Y.Q.; Chen, K.C.; Chen, S.G. Rapid hemostatic and mild polyurethane-urea foam wound dressing for promoting wound healing. *Mater. Sci. Eng. C Mater. Biol. Appl.* **2017**, *71*, 289–297. [CrossRef]
33. Niu, Y.Q.; Zhang, B.K.; Galluzzi, M. An amphiphilic aggregate-induced emission polyurethane probe for in situ actin observation in living cells. *J. Colloid Interface Sci.* **2021**, *582*, 1191–1202. [CrossRef]
34. Chen, Q.Q.; Passos, A.; Balabani, A.; Chivu, A.; Zhao, S.D.; Azevedo, H.S.; Butler, P.; Song, W.H. Semi-interpenetrating network hyaluronic acid microgel delivery systems in micro-flow. *J. Colloid Interface Sci.* **2018**, *519*, 174–185. [CrossRef]
35. Niu, Y.; Galluzzi, M. A biodegradable block polyurethane nerve-guidance scaffold enhancing rapid vascularization and promoting reconstruction of transected sciatic nerve in Sprague-Dawley rats. *J. Mater. Chem. B* **2020**, *8*, 11063–11073. [CrossRef]
36. Ma, L.; Qin, H.; Cheng, C.; Xia, Y.; He, C.; Nie, C.X.; Wang, L.R.; Zhao, C.S. Mussel-inspired self-coating at macro-interface with improved biocompatibility and bioactivity via dopamine grafted heparin-like polymers and heparin. *J. Mater. Chem. B* **2014**, *2*, 363–375. [CrossRef]
37. Newman, P.J.; Berndt, M.C.; Gorski, J.; White, G.C.; Lyman, S.; Paddock, C.; Muller, W.A. PECAM-I (CD31) cloning and relation to adhesion molecules of the immunoglobulin gene superfamily. *Science* **1990**, *247*, 1219–1222. [CrossRef] [PubMed]
38. Amorim, S.; Reis, C.A.; Reis, R.L.; Pires, R.A. Extracellular matrix mimics using hyaluronan-based biomaterials. *Cell Press* **2021**, *39*, 90–104. [CrossRef] [PubMed]
39. Zhang, W.J.; Zhang, Y.S.; Bakht, S.M.; Aleman, J.; Shin, S.R.; Yue, K.; Sica, M.; Ribas, J.; Duchamp, M.; Ju, J.; et al. Elastomeric free-form blood vessels for interconnecting organs on chip systems. *Lab Chip* **2016**, *16*, 1579–1586. [CrossRef] [PubMed]
40. Zhou, G.; Zhang, B.; Tang, G.; Yu, X.-F.; Galluzzi, M. Cells nanomechanics by atomic force microscopy: Focus on interactions at nanoscale. *Adv. Phys. X* **2021**, *6*, 1866668.
41. Schulte, C.; Rodighiero, S.; Cappelluti, M.A.; Puricelli, L.; Maffioli, E.; Borghi, F.; Negri, A.; Sogne, E.; Galluzzi, M.; Piazzoni, C.; et al. Conversion of nanoscale topographical information of cluster-assembled zirconia surfaces into mechanotransductive events promotes neuronal differentiation. *J. Nanobiotechnol.* **2016**, *14*, 18. [CrossRef] [PubMed]
42. Lee, S.J.; Kim, M.E.; Nah, H.; Seok, J.M.; Jeong, M.H.; Park, K.; Kwon, I.K.; Lee, J.S.; Park, S.A. Vascular endothelial growth factor immobilized on mussel-inspired three-dimensional bilayered scaffold for artificial vascular graft application: In vitro and in vivo evaluations. *J. Colloid Interface Sci.* **2019**, *537*, 333–344. [CrossRef] [PubMed]
43. Du, F.; Wang, H.; Zhao, W.; Li, D.; Kong, D.; Yang, J.; Zhang, Y. Gradient nanofibrous chitosan/poly ε-caprolactone scaffolds as extracellular microenvironments for vascular tissue engineering. *Biomaterials* **2012**, *33*, 762–770. [CrossRef]
44. Smith, R.J.; Nasiri, B.; Kann, J.; Yergeau, D.; Bard, J.E.; Swartz, D.D.; Andreadis, S.T. Endothelialization of arterial vascular grafts by circulating monocytes. *Nat. Commun.* **2020**, *11*, 1622. [CrossRef]
45. Song, H.-H.G.; Rumma, R.T.; Ozaki, C.K.; Edelman, E.R.; Chen, C.S. Vascular Tissue Engineering: Progress, Challenges, and Clinical Promise. *Cell Stem Cell* **2018**, *22*, 608. [CrossRef]
46. Huang, F.; Hsieh, Y.F.; Qiu, X.F.; Patel, S.; Li, S. Engineering the composition of microfibers to enhance the temodeling of a cell-free vascular graft. *Nanomaterials* **2021**, *11*, 1613. [CrossRef] [PubMed]
47. Guo, X.R.; Zhu, J.J.; Zhang, H.M.; You, Z.W.; Morsi, Y.; Mo, X.M.; Zhu, T.H. Facile preparation of a controlled-release tubular scaffold for blood vessel implantation. *J. Colloid Interface Sci.* **2019**, *539*, 351–360. [CrossRef]

nanomaterials

MDPI

Article

A Perfusion Bioreactor for Longitudinal Monitoring of Bioengineered Liver Constructs

Lisa Sassi [1,2,†], Omolola Ajayi [1,2,†], Sara Campinoti [1,2,†], Dipa Natarajan [1,2], Claire McQuitty [1,2], Riccardo Rayan Siena [1,2], Sara Mantero [3], Paolo De Coppi [4,5], Alessandro F. Pellegata [3,4], Shilpa Chokshi [1,2] and Luca Urbani [1,2,*]

1 Institute of Hepatology, Foundation for Liver Research, London SE5 9NT, UK; lisa.sassi@kcl.ac.uk (L.S.); l.ajayi@researchinliver.org.uk (O.A.); s.campinoti@researchinliver.org.uk (S.C.); dipa.natarajan@researchinliver.org.uk (D.N.); claire.mcquitty@researchinliver.org.uk (C.M.); rayan.siena@outlook.it (R.R.S.); s.chokshi@researchinliver.org.uk (S.C.)

2 Faculty of Life Sciences & Medicine, King's College London, London WC2R 2LS, UK

3 Department of Chemistry, Materials and Chemical Engineering "Giulio Natta", Politecnico di Milan, 20133 Milan, Italy; sara.mantero@polimi.it (S.M.); a.pellegata@ucl.ac.uk (A.F.P.)

4 Stem Cells and Regenerative Medicine and Biomedical Research Center, NIHR, Great Ormond Street Institute of Child Health, UCL, London WC1N 1EH, UK; p.decoppi@ucl.ac.uk

5 Specialist Neonatal and Pediatric Surgery, Great Ormond Street Hospital, London WC1N 3JH, UK

* Correspondence: luca.urbani@researchinliver.org.uk

† These authors contributed equally.

Citation: Sassi, L.; Ajayi, O.; Campinoti, S.; Natarajan, D.; McQuitty, C.; Siena, R.R.; Mantero, S.; De Coppi, P.; Pellegata, A.F.; Chokshi, S.; et al. A Perfusion Bioreactor for Longitudinal Monitoring of Bioengineered Liver Constructs. *Nanomaterials* **2021**, *11*, 275. https://doi.org/10.3390/nano11020275

Academic Editor: Su Ryon Shin
Received: 30 November 2020
Accepted: 19 January 2021
Published: 21 January 2021

Publisher's Note: MDPI stays neutral with regard to jurisdictional claims in published maps and institutional affiliations.

Copyright: © 2021 by the authors. Licensee MDPI, Basel, Switzerland. This article is an open access article distributed under the terms and conditions of the Creative Commons Attribution (CC BY) license (https://creativecommons.org/licenses/by/4.0/).

Abstract: In the field of in vitro liver disease models, decellularised organ scaffolds maintain the original biomechanical and biological properties of the extracellular matrix and are established supports for in vitro cell culture. However, tissue engineering approaches based on whole organ decellularized scaffolds are hampered by the scarcity of appropriate bioreactors that provide controlled 3D culture conditions. Novel specific bioreactors are needed to support long-term culture of bioengineered constructs allowing non-invasive longitudinal monitoring. Here, we designed and validated a specific bioreactor for long-term 3D culture of whole liver constructs. Whole liver scaffolds were generated by perfusion decellularisation of rat livers. Scaffolds were seeded with Luc$^+$HepG2 and primary human hepatocytes and cultured in static or dynamic conditions using the custom-made bioreactor. The bioreactor included a syringe pump, for continuous unidirectional flow, and a circuit built to allow non-invasive monitoring of culture parameters and media sampling. The bioreactor allowed non-invasive analysis of cell viability, distribution, and function of Luc$^+$HepG2-bioengineered livers cultured for up to 11 days. Constructs cultured in dynamic conditions in the bioreactor showed significantly higher cell viability, measured with bioluminescence, distribution, and functionality (determined by albumin production and expression of CYP enzymes) in comparison to static culture conditions. Finally, our bioreactor supports primary human hepatocyte viability and function for up to 30 days, when seeded in the whole liver scaffolds. Overall, our novel bioreactor is capable of supporting cell survival and metabolism and is suitable for liver tissue engineering for the development of 3D liver disease models.

Keywords: bioreactor; tissue engineering; decellularization; liver; extracellular matrix; bioluminescence

1. Introduction

Liver tissue engineering is emerging as a suitable tool to facilitate the unmet need for in vitro liver models with physiological features of the native organ niche. Bioengineered liver constructs could form robust models to investigate cell metabolism, pathological mechanisms and perform drug screening and toxicity assay. Assays based on 2D cellular monolayers are not suitable to mimic the natural behaviours of hepatic cells in response to stimuli [1,2], since the 2D condition does not provide the hepatic architecture, biochemical gradients, cell-cell communication and cell-extracellular matrix (ECM) interaction. Mechanical stress generated by the stiffness of a petri dish affects the hepatic cell behaviour, by inducing phenotype trans-differentiation [3]. Moreover, consistent and reliable isolation and expansion of primary human hepatocytes still represents a challenge for therapeutic transplantation and laboratory research: in the absence of a 3D environment, hepatocytes rapidly dedifferentiate and down-regulate synthesis of metabolic enzymes within 24 h in culture [4]. Bioengineered liver tissue represents a valid strategy in recapitulating the hepatic microenvironment despite the intrinsic technical challenges in engineering such a complex organ. The hepatic architecture needs to be reproduced in vitro since it plays a crucial role in promoting cell communication and functions: metabolic activity of the hepatocytes, indeed, changes spatially along the sinusoids, depending on gradients of oxygen and ECM composition (liver zonation) [5,6]. Another challenge is the selection of appropriate biomaterials for cell scaffolding tailored to guarantee an appropriate 3D microenvironment. Decellularized scaffolds maintain biochemical and mechanical properties of the original tissue, guiding tissue regrowth according to the so-called contact-guidance theory, for which the cell behaviour is strongly influenced by the geometrical patterns, architecture and surface topography of the scaffold. Thickness of the construct can be an issue as scaffolds of clinically relevant size often results in the development of necrotic regions due to a lack of nutrient transport and oxygen diffusion [7]. Based on these complex requirements, bioreactors have the potential to revolutionize the standard culture procedure and represent a key tool in overcoming the challenges described in engineering liver tissue constructs. Bioreactors provide a suitable environment for the development of biological systems, under tightly controlled conditions and close monitoring of the variables which are well known to affect cell behavior [8]. There is a long history of bioreactor use in cartilage and bone engineering, but robust methods to develop and use bioreactors for liver tissue are lacking.

In this study we have developed a novel bioreactor-based technology, allowing long-term in vitro culture of liver scaffolds and providing dynamic medium and gas supply in a 3D perfusion system. The technology is based on a closed, sterile chamber connected to a programmable syringe pump, which provides a constant flow through the scaffold, optimizing the mass transport and exchange and the delivery of oxygen and nutrients. This bioreactor is also designed to allow non-invasive monitoring of cellular and perfusion parameters during culture which may be very relevant for disease modelling and pre-clinical toxicology studies.

2. Materials and Methods

2.1. Bioreactor Design

The bioreactor was designed by computer-aided software (CAD, SolidWorks 2020, Waltham, MA, USA) (Figure 1). The chamber was designed with a keyhole shape: a circular part (90 mm of diameter) for housing of the organ and a rectangular element (20 × 42 mm) for housing of cannulas and tubes. To facilitate the outflow and avoid fluid stagnation the internal flooring has been designed with an inclination of 3°. Sterility was ensured by a silicone double gasket. Nylon 6.6 (DirectPlastics, Sheffield, UK) was selected as suitable material, since it is autoclavable, biologically inert, shows chemical resistance and is available in black, making it compatible with bioluminescence non-invasive imaging. The chamber was manufactured using Computer Numerical Control machine (CNC, Roland, Shizuoka, Japan). The settings of the CNC machine are reported in Table S1.

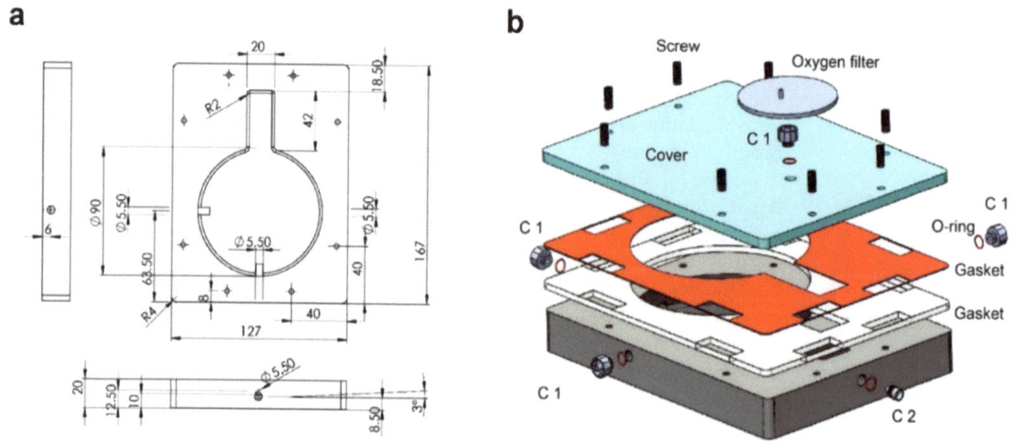

Figure 1. 2D (**a**) and 3D (**b**) drawing of the chamber designed by computer-aided software (CAD, SolidWorks 2020) and manufactured using Computer Numerical Control machine. C1 and C2: poly-propylene connectors.

2.2. Perfusion System

The system was based on a programmable syringe pump (World Precision Instruments R, Hitchin, UK) used as dispenser to perform a semi-continuous flow through the construct. To avoid presence of air, a Bubble Trap (Kinesis Scientific, Saint Neots, UK) linked to a Vacuum Assistance was implemented. The hydraulic circuit is characterized by an easy access three ways (Vygon, Swindon, UK) to perform media sampling. The system was assembled as shown in Figure 1b.

2.3. Organ Harvesting

All surgical procedures and animal husbandry were carried out in accordance with the recommendations in the Animal (Scientific Procedures) Act 1986 and the local ethics committee. Wild type healthy adult Sprague-Dawley rats of 280–300 g were euthanized by CO_2 inhalation, and the livers were isolated and harvested as previously described [9]. Briefly, the abdomen of the rat was sterilized with 70% Ethanol (EtOH; VWR, Leighton Buzzard, UK), the abdominal-pelvic cavity was exposed and the inferior vena cava (IVC) and portal vein (PV) were identified. The PV was cannulated with a 24G cannula (TERUMO, Fisher Scientific, Loughborough, UK) and the IVC was ligated with silk sutures (FST, Cambridge, UK). The whole liver was then released from the surrounding tissue. Sterile phosphate buffer saline (PBS) with 1 U/mL heparin (Sigma, Gillingham, Dorset, UK) was perfused to remove excess blood and check for leaks.

2.4. Decellularization of Rat Liver

Decellularization was performed through the vasculature network directly after rat liver harvesting. The cannulated PV was connected to a peristaltic pump (iPumps, Pamington, UK) to perfuse solutions. A bubble trap (Kinesis Scientific) was exploited to ensure that no bubbles were perfused into the vasculature of the liver. The liver was perfused with MilliQ water for 18 h at a flow rate of 4.5 mL/min at room temperature, followed by 4% sodium deoxycholate (SDC; Sigma) for 5 h at 6.5 mL/min. Next, the rat liver was perfused at 6.5 mL/min with PBS for 1 h, then 3 h with 25 mg/L DNAse-I (Sigma, UK) in saline solution (0.15 M NaCl/10 mM $CaCl_2$, Sigma), both pre-warmed and maintained at 37 °C. DNAse treatment was followed by perfusion of warm PBS for 1 h and finally PBS overnight at 1 mL/min at room temperature. Scaffolds were then sterilized by perfusion with 0.1% PAA/4% ethanol in milliQ water for 90 min, followed by 30 min of PBS with 1% penicillin-streptomycin (Sigma) and 50 ng/mL Primocin (Invitrogen, Waltham, MA, USA).

Decellularized livers were finally gamma irradiated and stored at 4 °C in sterile PBS with 1% penicillin-streptomycin (Sigma) and 50 ng/mL primocin (Invitrogen) until use.

2.5. HepG2 Cell Culture

HepG2 cells were cultured and transduced with pHIV-Luc ZSGreen based Lentivirus. The lentiviral transfer vector pHIV-LUC-ZsGreen was a gift from Dr. Bryan Welm [9] (Department of Surgery, University of Utah, purchased through Addgene, Inc., Watertown, MA, USA, plasmid #39196) (Addgene, Watertown, MA, USA). The ZSGreen and luciferase positive cells were then visualized with the substrate luciferin using an In Vivo Imaging System (IVIS). Lentivirus was added to cultures at multiplicities of infection in the range of 2–5, 100 μL for 10^5 cells per well, and left for 36–48 h ensuring that the cells were transduced and the lentivirus had inactivated. Following trypsinization, FACS analysis was performed in order to both quantify the percentage of transduced cells, and to select eGFP+ cells. Transduced HepG2 cells were cultured using Dulbecco's Modified Eagle Medium (DMEM) (Life Technologies, Renfrew, UK) supplemented of 10% fetal bovine serum (FBS; Life Technologies), 1% non-essential amino acids (NEAA; Sigma), 1% sodium pyruvate (100 mM; Life Technologies) and 1% L-glutamine (200 mM; Life Technologies).

2.6. HepG2 Seeding into Decellularized Rat Liver Scaffolds and Dynamic Perfusion Culture

The decellularized rat liver was primed with 9 mL of HepG2 media. A total of 50×10^6 HepG2 were seeded in the rat scaffold. The seeding was carried out though 4 perfusion steps, 12.5×10^6 HepG2 for each step with 30 min of rest between injections. Cells were delivered into the scaffold through the portal vein (PV) at 9 mL/min. After 24 h, the seeded construct was divided into two parts to obtain two scaffolds for either dynamic perfusion culture or static culture respectively, for 11 days. For dynamic perfusion 20 mL medium was pumped at 9 mL/min and subsequently withdrawn at the same speed using the automated syringe pump. The media were changed every 2 days and sampled every day from both 3D cultures.

2.7. Primary Human Hepatocytes Seeding

Primary human hepatocytes were kindly donated by Lonza (Morristown, NJ, USA) and stored in liquid nitrogen until the day of the seeding. A total of 5 vials of primary human hepatocytes were thawed in Hepatocytes Thawing medium (Lonza) following manufacturer's instructions. The average cell viability post thawing was 94.4%. Seeding was performed following the same protocol developed for HepG2 seeding in rat liver decellularized scaffold. A total of 50 million cells were seeded performing 5 perfusion steps through the portal vein (PV) at 4 mL/min in a volume of 6 mL (per perfusion step) of Hepatocytes Plating medium (Lonza). After 24 h, the seeded construct was divided into two parts to obtain two scaffolds for either dynamic perfusion culture (at 1 mL/min) or static culture respectively, for 30 days. Static and bioreactor 3D cultures were performed in Hepatocytes Incubation Medium composed of: William's E Medium (no phenol red, Sigma), 5% FBS (Life Technologies), 1% penicillin-streptomycin (Sigma) and 50 ng/mL primocin (Invitrogen), 1% glutamine (Invitrogen), 15 mM HEPES pH7.4 (Invitrogen), 0.1 μM dexamethasone (Sigma), 1X insulin-transferrin-selenium (IST-G, Life Technologies), 10 mM nicotinamide (Sigma) and 50 ng/mL human recombinant EGF (Peprotech, London, UK). The medium was changed every 4 days and sampled every other day from both 3D cultures.

2.8. Bioluminescence Analysis

For analysis at IVIS (Perkin Elmer, Waltham, MA, USA), the chamber was perfused with luciferin (10 μg/mL; Cayman Chemicals, Ann Arbor, MI, USA). Images were acquired using the Lumina III In Vivo Imaging System (IVIS) and the Living Image 4.4 Software. The black bioreactor chamber enhanced the bioluminescence visualization. The software generates pseudo-coloured, scaled images overlaid on grey scale images, providing 2-

dimensional localization of the source of light emission. All images were taken using stage E, with automated aperture setting, exposure time of 5 min and small binning (resolution) Regions of interest (ROIs) were selected using shape drawing tools and the light emission within the ROI was quantified as average radiance ([photons/sec/cm^2/steradiant]). ROI were kept constant between subjects within each experiment. Scaffolds were imaged every day or alternate days for bioluminescence.

2.9. Histology

2.9.1. Hematoxylin & Eosin

Tissue cryosections were prepared with a cryostat (Bright from Leica Biosystems, Wetzlar, Germany) for H&E staining. Sections of 10–12 µm were rehydrated in PBS at 37 °C and then left for 5 min in hematoxylin, 1 min in alcoholic acid and 5 min in eosin. Sections were then rehydrated (70%, 90% 100% EtOH, each phase for 1 min) and mounted with DPX after having immersed them in Histoclear (National Diagnostics, SLS, Nottingham, UK).

2.9.2. Immunofluorescence

Scaffold segments were fixed with 4% paraformaldehyde (PFA) for 1 h at RT. Tissue pieces were washed in PBS twice and left overnight at 4 °C in 30% sucrose (Sigma) in PBS. Subsequently, samples were embedded in 15% sucrose/7.5% gelatine in PBS, frozen in ice cold isopentane and sectioned at 10 µm with a cryostat (Bright). Sections were rehydrated in PBS at 37 °C and permeabilized in 0.1% triton in PBS (PBS-T) for 10 min followed by blocking using an appropriate blocking serum. Primary antibodies used, their concentration and blocking serum are listed in Table S2. Slides were incubated with primary antibodies overnight at 4 °C. Sections were then incubated with AlexaFluor-488 or AlexaFluor-568 conjugated secondary antibodies (dilution 1:500; Life Technologies) at RT for 1 h. Nuclei were counterstained with DAPI (dilution 1:1000, Life Technologies).

2.10. Albumin and Urea Quantification

To test the functionality of the hepatocytes and hepatoma cultures, aliquots of media was collected every 2 days from both static and dynamic perfusion cultures. Albumin content in the medium was measured using the Human Albumin ELISA kit (E88-129, Bethyl Laboratories, Inc. Montgomery, TX, USA) following manufacturer's instructions. Culture media was diluted 1:2 before assaying.

Urea content was measured using colorimetric Urea assay kit (ab83362, Abcam, Cambridge, UK) following manufacturer's instructions. Both static and dynamic perfusion culture media was diluted 1:32 before assaying.

2.11. RNA Isolation and RT-qPCR

2D cultured HepG2 and primary human hepatocytes and 3D seeded scaffolds were digested with Trizol TRI Reagent (Sigma-Aldrich, St Louis, MO, USA) and RNA was extracted following the manufacturer's instructions. Precipitated and dried RNA was re-suspended in nuclease free water (Qiagen, Hilden, Germany). RNA concentration was measured using Nanodrop1000 (ThermoScientific, Waltham, MA, USA). RNA was converted into cDNA using GoScript™ Reverse Transcriptase kit (Promega, Southampton, UK) according to the manufacturer's protocol. cDNA concentration was adjusted to 10 ng/µL. Quantitative (q)PCR was performed on 25 ng of cDNA using PCR master mix (PrecisionPLUS-R—Primerdesign Ltd. Chandler's Ford, UK) with low-ROX and Taqman qPCR probes (Integrated DNA Technology, Coralville, Ia, USA. List of probes in Supplementary Table S2) in MicroAmp Fast Optical 96 well Reaction Plates (Starlab, Milton Keynes, UK) using the ABI7500 Real-Time PCR System (Applied Biosystems, Foster City, CA, USA). Target genes are reported in Table S3.

2.12. DNA Quantification

To determine whether decellularisation had effectively removed native DNA, the total DNA was isolated from fresh liver and decellularised liver samples using DNeasy Blood and Tissue kit (Qiagen) as per the manufacturer's instructions. Purified DNA was quantified using Nanodrop (ThermoScientific, Waltham, MA, USA).

2.13. Mycoplasma Test

To test the sterility of the culture, mycoplasma contamination was periodically tested with a commercial detection kit, MycoProbe (R&D Systems, Minneapolis, MN, USA) following the manufacturer's instructions. Briefly, media or supernatants from cell cultures or scaffold storage cultures were incubated in specialized microplates coated with biotin-labelled capture oligonucleotide probes and digoxigenin-labelled detection probes. The presence of 16S ribosomal RNA of the eight most common mycoplasma contaminants in the samples would lead to a hybridization with the probes. The hybridization solutions were then transferred to a streptavidin-coated microplate, with any hybridized RNA being captured, the addition of an anti-digoxigenin alkaline phosphatase conjugate and a substrate solution would develop colour with absorbance at 490 nm spectrophotometrically. O.D. values were then compared to the positive control; samples were free of mycoplasma contamination when calculated O.D. value < 0.10.

2.14. Endotoxin Quantification

Endotoxin contamination was periodically tested using a Cloud-Clone Corp kit (SEB526Ge, Katy, TX, USA) ELISA kit for lipopolysaccharide (LPS). Cell culture supernatant was stored at −20 prior to testing and assayed neat.

2.15. Collagen Quantification

Total collagen content of fresh and decellularized rat liver samples was quantified with Collagen Assay Kit (QuickZyme, Biosciences, Leiden, The Netherlands) as per manufacturer's instructions. The concentration was evaluated via spectrophotometry measurements of hydrolysed samples at 555 nm via interpolation with a standard curve produced with known collagen concentrations.

2.16. Statistical Analysis

Statistical analysis was carried out using Graphpad Prism Software (Groningen, The Netherlands) and Student's t-test was used to compare between two groups. Two-tailed p-values of data were assessed using Student's t-test. In all figures, statistical significance is expressed as *** $p < 0.001$, ** $p < 0.01$ and * $p < 0.05$. Quantitative data was expressed as mean $\pm$ standard deviation (SD) and graphs as mean $\pm$ standard error (SEM).

3. Results

Decellularized rat livers were obtained using an established detergent-enzymatic treatment [10]. Perfusion of decellularization reagents was performed through the cannulated portal vein (PV). The process eliminated the cellular compartment preserving the ECM, resulting in a translucent scaffold with visible vasculature (Figure 2a). DNA quantification in tissue segments confirmed elimination of the cells upon decellularization (Figure 2b), while collagen, the main ECM protein in the liver, was preserved (Figure 2c). H&E staining highlighted absence of nuclei and cytoplasm in the decellularized scaffolds and preservation of the overall matrix structure (Figure 2d). Trypan blue dye perfused into the decellularized scaffolds through the PV allowed for clear visualization of the intact vascular network, showing no leakage of dye (Figure 2e).

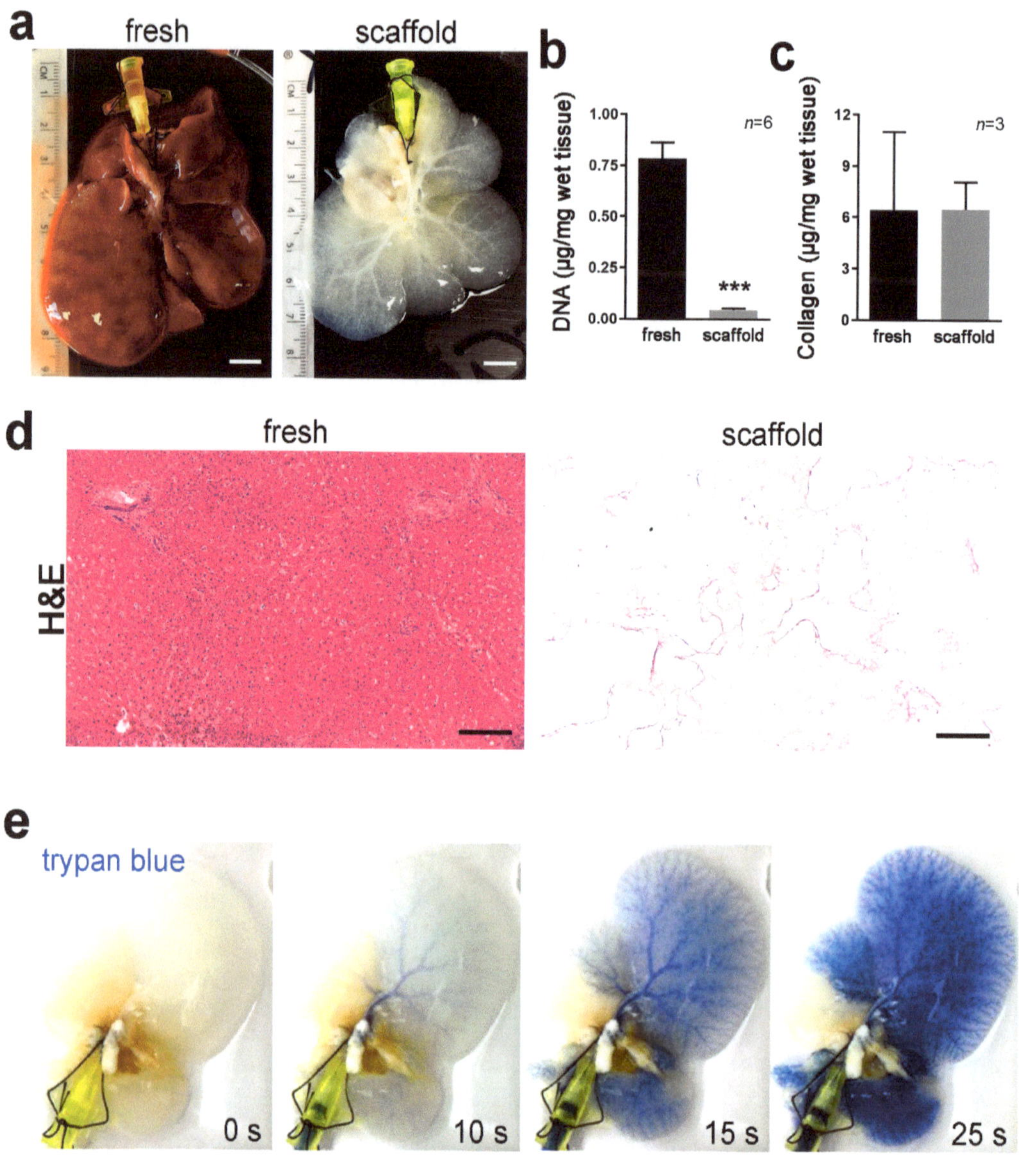

Figure 2. Decellularization of rat whole liver. (**a**). Rat liver with cannulated portal vein before (left) and after (right) detergent-enzymatic perfusion decellularization showing change in tissue colour during the process. Scale bar: 2 cm. (**b**). DNA quantification in segments of fresh liver tissue and decellularized liver scaffolds. *** = $p < 0.001$ *t*-test. (**c**). Collagen quantification in segments of fresh liver tissue and decellularized liver scaffolds. (**d**). H&E staining of fresh and decellularized rat liver tissue showing absence of nuclei and cytoplasm. Scale bar: 200 µm. (**e**). Snapshots recorded at 0, 10, 15 and 20 s of trypan blue dye perfusion through the portal vein of a decellularized scaffold to highlight intact vasculature tree.

HepG2 cells transduced with pHIV-Luc-ZSGreen based lentivirus (Luc$^+$HepG2) were perfused into the decellularized scaffolds through the PV using a syringe pump (Figure 3a), showing evident infiltration of cells in both median lobe (ML) and lateral left lobe (LLL) (Figure 3b). Cell retention into the scaffolds was evaluated by counting cells in the perfused media surrounding the liver scaffolds after seeding. Almost 100% of the cells perfused were retained inside the scaffolds (Figure 3c). The repopulated scaffolds were then separated placing the ML in static culture and the LLL into bioreactor perfusion culture, both cultures were incubated for up to 11 days (Figure 3a,d). Perfusion culture was obtained through the use of a closed-loop circuit, where the pump was connected to the chamber through two branches, the inlet branch and the outlet branch. When in "pumping" mode, the syringe pump pushed media through the inlet branch connected to the cannulated PV, diffusing through the vasculature network (Figure 3e). Media was then removed from the chamber back into the syringe once the pump was in "withdrawing" mode, via the outlet branch (Figure 3f). The presence of two check valves positioned at the two branches allowed correct directional functioning of the system: their role was to direct the fluid, allowing the pumping phase only through the inlet branch and the withdrawing phase through the outlet one.

Bioluminescence imaging was used for longitudinal assessment of cell distribution and viability by perfusing luciferin via the bioreactor or directly into the culture plate for static cultures. Bioluminescence clearly showed initial cell distribution in the proximal area of both static and perfused scaffolds, with comparable levels of bioluminescence detected in both conditions, indicating a similar number of viable cells in the two scaffolds (Figure 4a). A steady increase in cell viability and an overall homogenous distribution of cells was detected during culture, with significant higher cell viability in scaffolds cultured in perfusion conditions compared to static cultures, reported as average radiance (Figure 4b). Scaffolds repopulated with Luc$^+$HepG2 cells were embedded for cryosectioning or snap frozen at the end of the culture period to further assess cell density and distribution through DNA quantification and stainings. Although cells appeared to have reached different areas of the scaffolds under both culture conditions, a clear improvement in cell growth and distribution was shown in perfusion cultures compared to static conditions (Figure 4c–e). Cell density was significantly higher in bioreactor cultured scaffolds compared to static conditions as indicated by the higher number of cells per area (Figure 4d) and as appreciable by H&E staining (Figure 4e). Endotoxin levels and mycoplasma contamination were quantified in the circulating media or static media after 11 days of culture and both values resulted below the thresholds indicated by the kit's manufacturers and by regulatory agencies. No difference in endotoxin and mycoplasma levels was detected between bioreactor and static culture, demonstrating that the bioreactor maintained the same level of sterility in respect to conventional static cultures (Figure 4f).

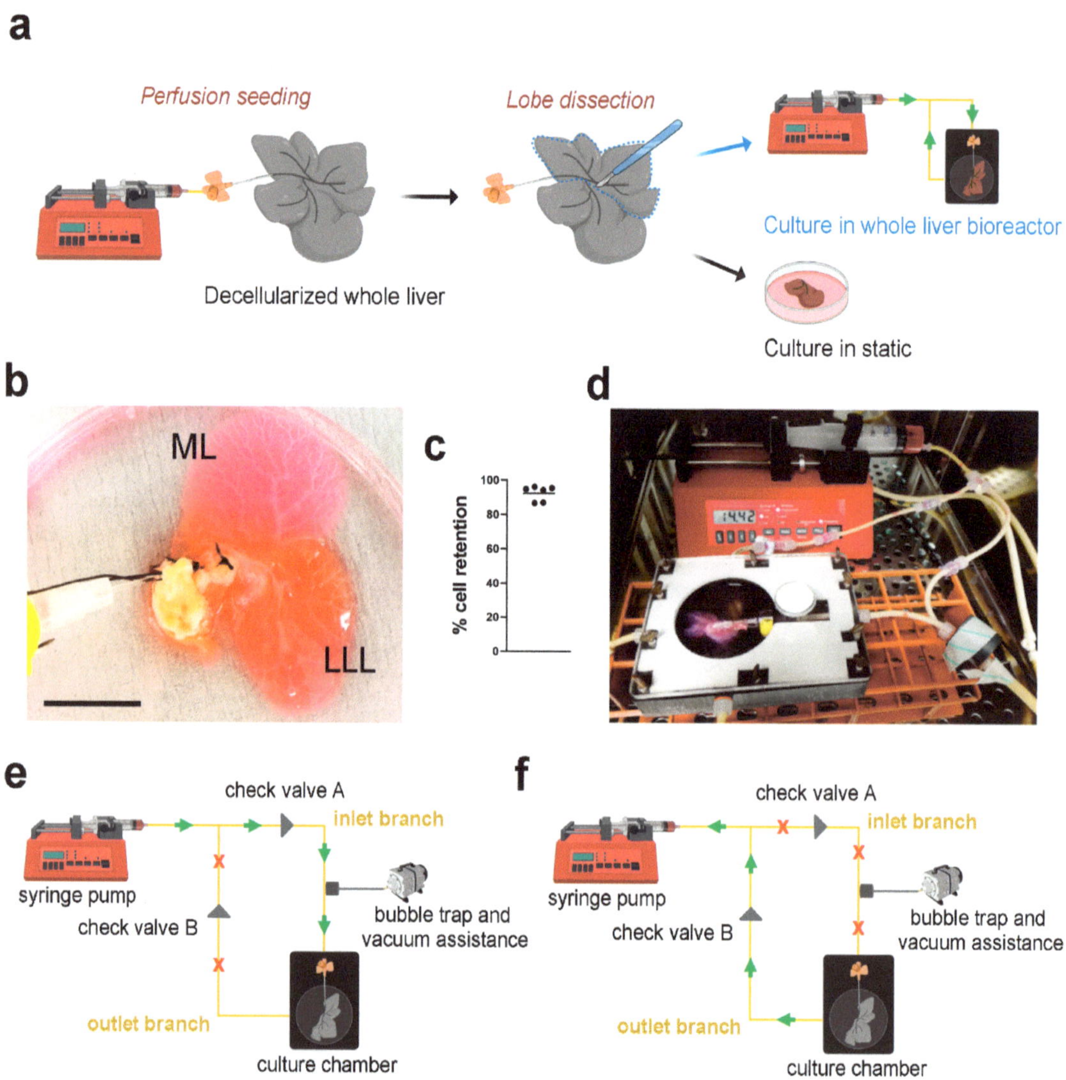

Figure 3. The perfusion system of the bioreactor. (**a**). Schematic of perfusion seeding into decellularized whole liver scaffolds, dissection and static or perfusion culture conditions. The scaffolds were seeded via the canula. Upon seeding, a lobe was dissected and cultured in static conditions while the other lobes were connected to the bioreactor circuit via the canula and cultured in dynamic perfusion condition. (**b**). Representative image of a decellularized scaffold seeded with Luc$^+$HepG2. Scale bar: 2 cm. (**c**). Percentage of cell retention in the scaffolds upon seeding. (**d**). Photo of the bioreactor system assembled with the chamber containing a decellularized scaffold primed with culture medium before seeding. The pump is connected to the chamber through two branches, the inlet branch and the outlet one. (**e**). Syringe pump set to "pumping" mode: medium is pushed through the inlet branch and diffused through the vasculature network. (**f**). Syringe pump set to "withdrawing" mode: medium is withdrawn through the outlet branch from the chamber, returning to the syringe. ML: median lobe; LLL: lateral left lobe.

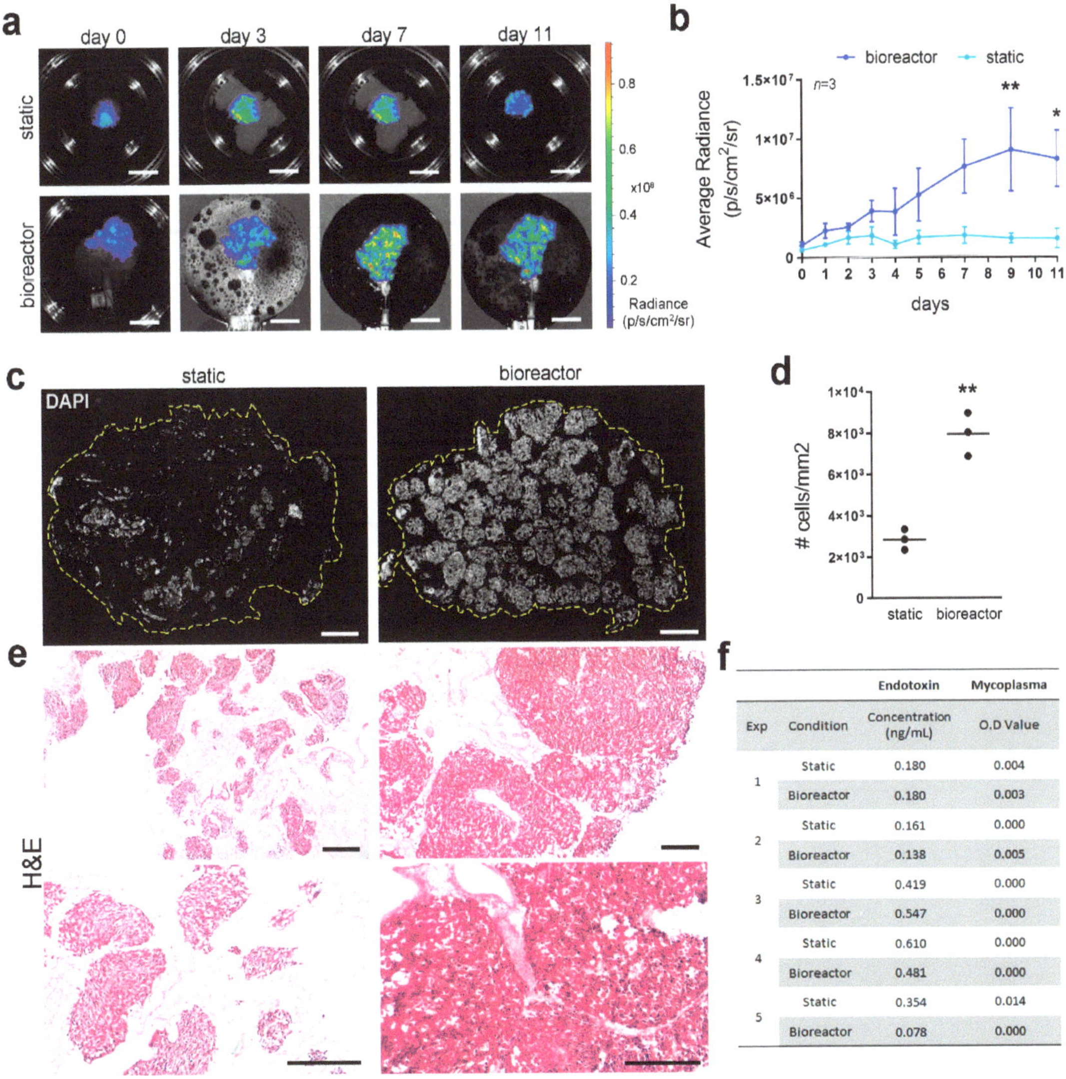

Exp	Condition	Endotoxin Concentration (ng/mL)	Mycoplasma O.D Value
1	Static	0.180	0.004
	Bioreactor	0.180	0.003
2	Static	0.161	0.000
	Bioreactor	0.138	0.005
3	Static	0.419	0.000
	Bioreactor	0.547	0.000
4	Static	0.610	0.000
	Bioreactor	0.481	0.000
5	Static	0.354	0.014
	Bioreactor	0.078	0.000

Figure 4. Cell viability, distribution, and density in 3D cultures. (**a**). Representative bioluminescence images at different time points of seeded ML and LLL from the same decellularized liver cultured in static and perfusion bioreactor conditions, respectively. Scale bar: 2 cm. (**b**). Bioluminescence readings up to 11 days of culture (n = 3). * = $p < 0.05$; ** = $p < 0.01$ 2-way ANOVA, Bonferroni's multiple comparison's test. (**c**). Representative images for staining with DAPI (grey) to show distribution of nuclei in cross-sections. Scale bar: 200 µm. (**d**). Number of cells per area determined in images from DAPI staining (**e**). Representative images of H&E staining of scaffolds cultured in static condition or in the bioreactor. Scale bar: 200 µm. (**f**). Mycoplasma and endotoxin concentration in the media at day 11 of static or bioreactor cultures in 5 different experiments.

Cell proliferation and apoptotic rate were assessed using immunofluorescence for Ki67 and caspase-3 on cryosections. Cell apoptosis and proliferation at day 11 seemed comparable between the two culture conditions with no significant difference in the percentage of caspase-3^{+} and Ki67^{+} cells (Figure 5a–d). Expression pattern of CK18 was also comparable between static and bioreactor perfusion culture (Figure 5c). Albumin was expressed by Luc⁺HepG2 cells cultivated in static and perfusion culture conditions as shown by immunofluorescence analysis, although some cells were negative for albumin expression in static cultures (Figure 5e). Difference in the metabolic activity of cells cultured in 3D scaffolds with or without the bioreactor was determined through albumin quantification and qPCR of hepatocyte specific factors. Bioreactor cultured constructs showed higher amount of albumin produced at day 11 of culture and a higher total cumulative albumin throughout the culture period compared to static culture (Figure 5f,g), suggesting enhanced hepatic cell metabolic function when cultured in the bioreactor. Gene expression profiling of 3D cultured Luc$^+$HepG2 cells was determined with qPCR. Transcription levels for hepatocyte nuclear factor 4-α (*HNF4α*), UDP-glucuronosyl- transferasy1-1 (*UGTA1*), *SERPINA1*, forkhead box A2 (*FOXA2*), cytochrome P450 family 1 subfamily A member 2 and family 3 subfamily A member 4 (*CYP1A2* and *CYP3A4*) and *MKI67* were analysed using *HPRT1* as reference gene and normalising gene expression on Luc$^+$HepG2 cells cultured in conventional 2D culture conditions (Figure 5h,i). The *SERPINA1* gene, which encodes for hepatocyte's serine protease inhibitor α-1-antitrypsin [11,12] resulted to be significantly upregulated in bioreactor cultures compared to static culture conditions and 2D cultured cells. Luc$^+$HepG2 3D cultured in the bioreactor showed upregulation of *HNF4α* in respect to conventional 2D cultures. HepG2 cultured in standard 2D conditions do not express *CYP3A4* at high levels [13,14], however, this gene was upregulated in 3D cultured Luc$^+$HepG2 cells, in particular in perfusion cultures. A similar upregulation was evident for other genes crucial for hepatocytes functions such as *FOXA2* and *CYP1A2*. As expected, *MKI67*, transcript of proliferation marker KI67, was downregulated in 3D conditions.

To test whether the bioreactor could support the long-term culture of human primary cells, we performed a proof of principle experiment seeding a whole rat liver scaffold with primary human hepatocytes (Figure 6a). Lobes were then separated to place the ML in static culture and the LLL in perfusion culture in the bioreactor for up to 30 days. H&E staining of scaffolds at 30 days of culture showed higher repopulation in scaffolds in the bioreactor compared to static culture conditions (Figure 6b). Immunofluorescence staining showed that hepatocytes in scaffolds retained their expression of CK18 and had comparable levels of albumin expression, while fewer cells were positive for apoptotic marker caspase-3 in scaffolds cultured in the bioreactor for 30 days (Figure 6c). Immunostaining for CYP3A4 evidenced similar distribution of positive cells between static and bioreactor cultured hepatocytes, with relative higher expression in perfusion cultured constructs (Figure 6c). The cellular metabolic activity in 3D cultures was determined through albumin and urea quantification (as a surrogate for hepatocyte mediated ammonia detoxification) and qPCR of hepatocyte specific markers. Over the 30 days of culture, primary human hepatocytes in bioreactor culture produced higher amount of albumin than cells in 3D static culture conditions (Figure 6d). Hepatocytes cultured in the bioreactor also produced more urea throughout the 30 days of culture (Figure 6e). qPCR analysis showed that hepatocytes cultured in perfusion conditions upregulated functional genes, such as *UGTA1*, *HNF4α* and *CYP1A2*. *CDH1* gene, encoding for E-cadherin adhesion molecule, was also upregulated in hepatocytes in the scaffold cultured in the bioreactor, indicating a better cell-cell and cell-ECM adhesion (Figure 6f). Albumin (*ALB*) transcripts levels were comparable between static and bioreactor culture conditions.

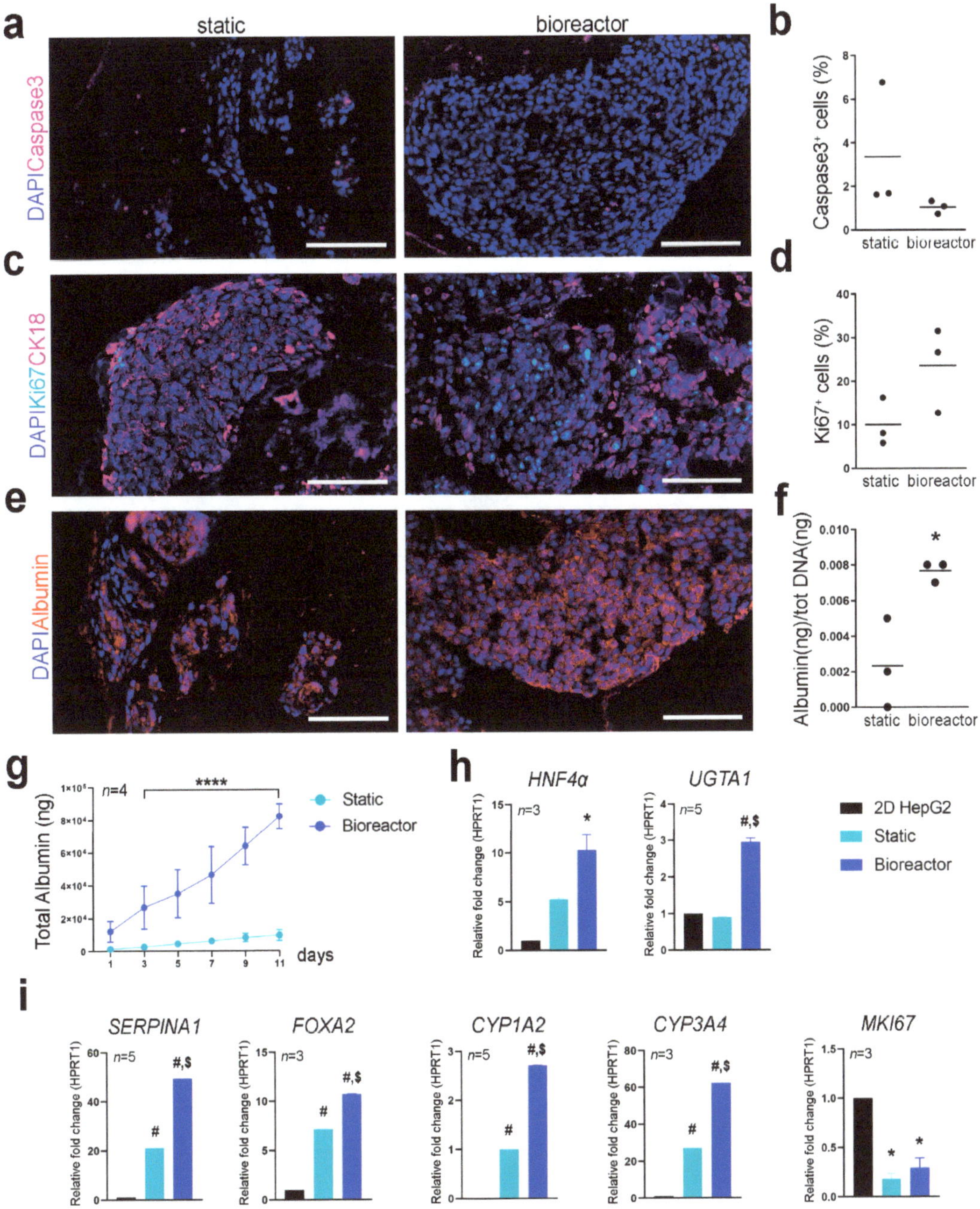

Figure 5. Characterization of scaffolds recellularized with Luc+HepG2 cells and cultured with or without bioreactor. (**a**,**c**,**e**). Immunofluorescent staining for caspase-3 (magenta) (**a**); Ki67 (cyan) and CK18 (magenta) (**c**); albumin (red) (**e**); nuclei were counterstained with DAPI (blue). (**b**,**d**) Quantification of caspase-3+ cells (**b**) and Ki67+ cells (**d**) over DAPI count per area, represented as mean ± SD. Scale bar: 100 µm. (**f**). Albumin quantification (in ng) in the media of static and bioreactor cultures at day 11. Nanograms of albumin were normalized against total DNA (ng) in scaffolds after static and bioreactor culture conditions respectively. (**g**). Albumin quantification represented as albumin production per day (in µg),

**** = $p < 0.0001$ Two-way ANOVA. (**h**,**i**). Gene expression profile of HepG2 cells in 2D conditions and in static and perfusion 3D cultures. qPCR was performed on 8 samples of scaffolds in static or dynamic perfusion culture for a total of 4 independent experiments. Data is shown as relative fold change in respect to the reference gene *HPRT1* as mean +/− SEM. Gene expression of HepG2 cells in 2D conditions was used as reference (1) for all genes with exception of CYP1A2, of which transcripts were not detected in HepG2 in 2D samples. * = $p < 0.05$; # = $p < 0.0001$ *t*-test vs. 2D cultures. $ = $p < 0.0001$ static vs. bioreactor culture.

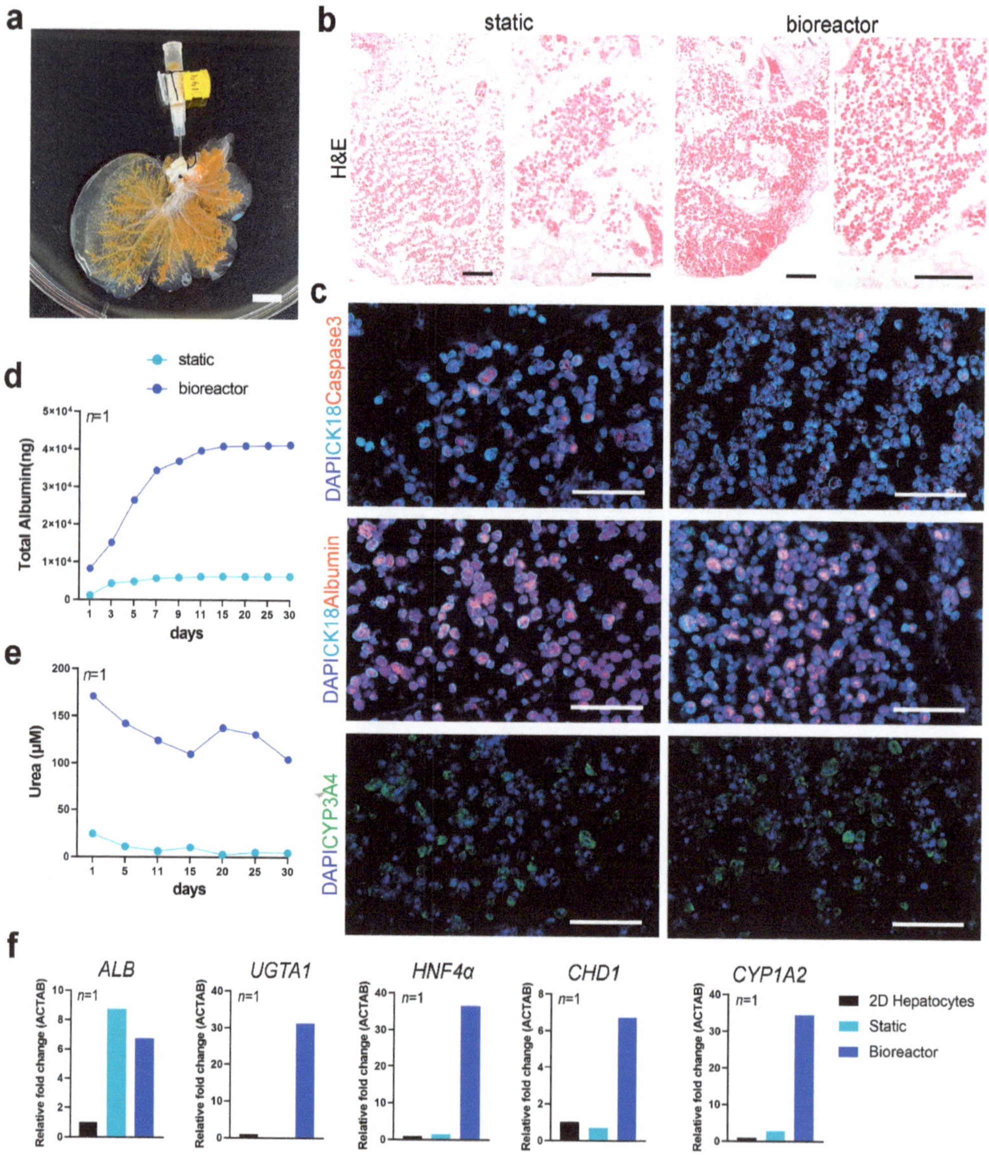

Figure 6. Characterization of scaffolds recellularized with primary human Hepatocytes and cultured with or without bioreactor for 30 days. (**a**). Representative image of a decellularized scaffold seeded with primary human Hepatocytes through the cannulated PV. Scale bar: 1 cm. (**b**). Representative images of H&E staining of scaffolds cultured in static

condition or in the bioreactor for 30 days. Scale bar: 200 μm (**c**). Immunofluorescent staining of scaffolds repopulated with primary hepatocytes and cultured in static conditions or in bioreactor for 30 days. Top: staining for caspase-3 (red) and CK18 (cyan); middle: staining for albumin (red) and CK18 (cyan). Bottom: staining for CYP3A4 (green). Nuclei were counterstained with DAPI (blue). Scale bar: 100 μm. (**d**). Albumin quantification (total albumin in the media, in ng) over the 30 days of culture of the scaffold in static condition or in the bioreactor. (**e**). Urea quantification (total urea in the media, in μM) over the 30 days of culture of the scaffold in static condition or in the bioreactor (**f**). Gene expression profile of primary hepatocytes in 2D conditions and in static and perfusion 3D cultures. Data is shown as relative fold change in respect to the reference gene *ACTAB*. Gene expression of hepatocytes cells in 2D conditions was used as reference (n = 1).

4. Discussion

In this study we have described a novel perfusion-based bioreactor technology to support the generation of whole-organ models. Our platform exploits the original hepatic vascular network as an effective route for seeding of hepatic cells, perfusion of nutrients and longitudinal monitoring of cell distribution, viability and function. The bioreactor fulfilled the most important ideal technical specifications, namely cytocompatibility, reliability, sterility, limited incumbrance, versatility, increase of nutrients and gas transport, biophysical stimulation, and automation.

The 3D ECM-scaffold was generated by established perfusion decellularization of rat livers. The maintained 3D native architecture provided an ideal platform for hepatic cultures. The preservation of vasculature networks was one of the key advantages of this model together with the custom design of the bioreactor allowing whole organ upscale, longitudinal non-invasive analysis and extended culture capability. The physiological environment constituted by the hepatic ECM is a tissue-specific architecture of both structural and functional proteins maintained by a precisely regulated equilibrium between synthesis and degradation [15]. The ECM harbors a range of growth factors and other matrix-associated molecules which influence cellular activity [16]. ECM also provides cells with signals for polarization, adhesion, migration, proliferation, survival and differentiation [17]. ECM obtained by decellularization provides a characteristically appropriate environment to support hepatic cell repopulation and tissue functionality, and its use was maximised in our perfusion culture system to extend and maintain long-term cultures.

Liver disease represents one of the most important global human health issues, which has driven a prominent strive forward in liver regenerative medicine. To investigate physiological and pathological hepatic mechanisms, conventional 2D cultures and animal models present significant drawbacks [18–21]. Recently, several reports have been published on the use of decellularized liver ECM for the development of models for liver function and diseases [22–25]. However, the development of an efficient bioreactor-based culture system for 3D hepatic structures represent one of the major constraints of the field. A limiting factor in whole-liver tissue engineering has been that bioreactors usually play a marginal role, confined to a temporary support for the generation of the construct [26–28].

The bioreactor in this study was developed in Nylon 6.6, which is FDA-approved and inert, with high chemical resistance and is not susceptible to corrosion or release of cytotoxic products, events we confirmed were absent in our long-term culture experiments. The efficiency of cell seeding as well as monitoring cell fate during long-term culture is currently determined retrospectively by histological analyses, DNA quantification assay and metabolic activity analyses. These techniques provide information at a fixed time point and are limited by the requirement of termination of the experiment for analysis. In the case of long-term cultures, a reliable tool to monitor and track cells at different time points is crucial. Nylon 6.6 is compatible with the use of bioluminescence imaging (BLI) [9], representing an innovative feature not yet exploited in the field of bioengineered livers. BLI is a powerful tool utilised in our system which overcomes many of these drawbacks. The procedure for BLI is non-invasive, allowing for longitudinal monitor of cells in the bioreactor at different time points. The bioreactor has been successfully designed with a black nylon chamber to ensure enhanced BLI visualisation upon direct injection of luciferin inside the scaffold through the PV, and concomitantly sampling of culture media whilst limiting exposure to contaminants, maintaining sterility. The entire system fits into a standard incubator, allowing easy access to the control panel during culture, without the need for expensive and difficult adjunct modalities. This allows for simple assembly and use, changing of culture media and addition of chemical/toxic compound during culture by replacement of the syringe or using the three-way access point. The bioreactor is scalable and easy to use and could be applied to other whole-liver cultures or other bioengineered whole-organs such as pancreas and kidney.

It is well-established that cells have a limited autonomy after a certain distance from a nutrient source, they can generally survive within an area of 1 mm away from a vessel [29]. This feature assumes paramount importance in the field of hepatic bioengineering because of the high rate of hepatocyte oxygen consumption; hepatic tissues should contain an extensive micro-vascular network to ensure constant nutrient supply and avoid any ischemic damage [30,31]. The perfusion within our system recreated a physiological-like delivery of nutrients, mimicking an environment where cells can reach optimal functionality, expansion and distribution. The flow rate selected for seeding and dynamic perfusion culture was 9 mL/min, the most conducive mechanical environment, closely emulating physiological blood flow through the PV of a rat liver [32]. Moreover, this flow rate has been previously reported to induce optimal HepG2 proliferation and tissue reorganization [26] The correlation between fluid dynamic stimulation and cell engraftment, proliferation and behaviour in engineered tissues has yet to be fully determined and further studies are required to investigate the role of liquid flow in these systems. Our bioreactor technology could represent a suitable device to explore the crosstalk between these parameters.

Hepatocyte or hepatoma cell lines represent a valid cell source in liver bioengineering. Due to their proliferation properties and metabolic prolife, HepG2 cells have been shown to provide a suitable alternative to primary hepatocytes, in pharmaceutical research, metabolism studies or sub-chronic to chronic hepatotoxicity studies [33], and in hepatic bioengineering with decellularized scaffolds [10,26,34]. Here we have reported a technology able to support culture and longitudinal analysis of HepG2-seeded scaffolds for up to 11 days, showing superior performances compared to static cultures. At 11 days, HepG2 cells were found repopulating almost all the scaffold area, with higher repopulation efficiency in dynamic perfusion condition in comparison to static. Furthermore, our system supported long-term survival and function of primary human hepatocytes for up to 30 days in proof-of-principle experiments. Cultivation of primary hepatocytes cells in liver scaffold employing a dynamic perfusion system has also been previously shown to improve cellular distribution within the scaffold and the oxygenation of the engineered construct [35]. Mazza et al. have shown increased functionality in 3D dynamic culture up to 10 days; here we support these data and we have implemented the analysis by cultivating primary human hepatocytes for a month; primary hepatocytes were found at high frequencies after prolonged in vitro culture, showing viability and expressing ontogenetic markers (i.e., CK18) and functional markers (i.e., albumin and CYP3A4). We showed that at gene and protein levels, both human primary hepatocytes and HepG2 cells were more functional in 3D bioreactor perfusion cultures than in static conditions, supporting the essential role of the bioreactor. Hepatic cell cultures in our perfusion bioreactor showed albumin secretion and urea production comparable to what has been previously shown by Robertson et al., 2018 in a similar rat liver model cultivated for 28 days with the use of a bioreactor system [24].

These results show that bioreactor-perfusion culture of complex bioengineered livers provides a more physiological environment that supports long-term culture of functional hepatocytes. This perfusion-based bioreactor could be easily used to culture scaffolds seeded with iPS-derived hepatocytes or hepato-biliary progenitors from liver organoids to evaluate whether a dynamic culture could support their survival, proliferation, and maturation. In this context, Wang and colleagues have shown advantages of ECM derived liver scaffold versus PLLA-collagen bioscaffolds in promoting specific hepatocyte marker expression and boosting the liver function of iPSCs [36]. Furthermore, in a recent report by de l'Hortet et al., the authors reported the biofabrication of genetically edited human liver tissue to mimic fatty liver disease starting from iPSC [37]. Liver organoids and iPSC are widening the possibilities to develop complex 3D models of disease and a perfusion bioreactor could extend/support these cultures, similarly to what recently shown in an in vitro whole-organ bioreactor grown artificial liver model (BALM), developed with iPSCs-derived hepatocyte-like cells [38]. Historically, models incorporating peristaltic pumps produce periodic compression of tubing, making the system unsuitable for the perfusion of circulating cells through the tissue or organ. Our system has the potential to incorporate perfusion of immune cells creating an immunocompetent liver model, highly sought in advanced liver disease modelling. This addition would permit intricate investigations of interactions between immune cells and hepatic cells in addition to hepatic cell-ECM interactions, recapitulating the complex liver microenvironment and inflammation-mediated pathology which is a central 'tenet' in the progression of chronic liver disease.

An important drawback of decellularisation is the loss of the organ endothelial layer. In the absence of such cells, coagulation can be easily triggered upon in vivo transplantation of the tissue engineered constructs, when blood is exposed to the ECM. Furthermore, the role of liver endothelial cells in physiological and pathological condition is critical and has to be taken into account in an appropriate liver disease model [39]. For this reason, it is essential to develop strategies that can allow hemocompatibility and re-endothelialisation of the scaffolds, similarly to what has been adopted for other organs [40,41]. Our perfusion seeding and culture system via canulation and the use of a syringe pump would also be appropriate for the reconstruction of the natural liver vascular tree and this is an area that warrants further investigation.

At present, bioreactors have been mainly exploited in clinical applications using human liver cells to support hepatic function in patients with acute liver failure [42,43]. Hollow-fibre bioreactors have been described as valuable tools to support the generation of small hepatic constructs as valid alternative for pharmacological studies [44,45]. These bioreactors are designed with the aim to maximise the delivery of nutrients and gas supply, but do not consider aspects crucial for liver function, for example the hepatic architecture. Bioartificial liver (BAL) support systems have the potential to provide temporary support to bridge patients waiting for liver transplant [46]. The development of BAL systems for short-term liver support must incorporate a functional cell source. As we were able to cultivate functional primary human hepatocytes for long term, our tissue engineering approach of culturing primary human cells within the native liver ECM could be adapted to further implement current liver support devices.

Finally, our hydraulic system could be easily upgraded into an automated circuit, as already described in other devices [47,48]. A reservoir connected to a pump controlled by a microcontroller-based unit, would provide automatic filling of the chamber with media followed by emptying and recirculation, and sampling of aliquots for analysis.

In conclusion, we have designed and validated a novel bioreactor for whole-liver bioengineering, showing stronger support of cell survival and metabolism compared to static cultures, longitudinal sampling and analysis of cell distribution and viability, maintenance of sterility and suitability for circulation of live cells for the development of complex 3D liver disease models.

Supplementary Materials: The following are available online at https://www.mdpi.com/2079-4991/11/2/275/s1, Table S1: CNC machine setting for the realization of the chambers; Table S2: Primary antibodies; Table S3: qPCR probes.

Author Contributions: Conceptualization: L.U., A.F.P., R.R.S.; acquisition and analysis of data: L.S., O.A., S.C. (Sara Campinoti), D.N., C.M., R.R.S.; interpretation of data: L.U., S.C. (Sara Campinoti), O.A.; writing of the manuscript: L.S., O.A., S.C. (Sara Campinoti), L.U.; revision of manuscript: P.D.C., S.C. (Shilpa Chokshi), S.M., A.F.P., L.U.; supervision: L.U., S.C. (Shilpa Chokshi). All authors have read and agreed to the published version of the manuscript.

Funding: P.D.C. is supported by National Institute for Health Research (NIHR-RP-2014-04-046), the UKRMP Universal cells to overcome HLA barriers in regenerative medicine (PI Qasim), the OAK Foundation Award W1095/OCAY-14-191; The GOSHCC the Horizon 2020 grant INTENS 668294.

Data Availability Statement: The data presented in this study are available on request from the corresponding author.

Acknowledgments: All research at Great Ormond Street Hospital NHS Foundation Trust and UCL Great Ormond Street Institute of Child Health is made possible by the NIHR Great Ormond Street Hospital Biomedical Research Centre.

Conflicts of Interest: The authors declare no conflict of interest.

References

1. Habibovic, P.; Woodfield, T.; de Groot, K.; van Blitterswijk, C. Predictive value of in vitro and in vivo assays in bone and cartilage repair–what do they really tell us about the clinical performance? *Adv. Exp. Med. Biol.* **2006**, *585*, 327–360. [CrossRef] [PubMed]
2. Astashkina, A.; Mann, B.; Grainger, D.W. A critical evaluation of in vitro cell culture models for high-throughput drug screening and toxicity. *Pharmacol. Ther.* **2012**, *134*, 82–106. [CrossRef] [PubMed]
3. Lachowski, D.; Cortes, E.; Pink, D.; Chronopoulos, A.; Karim, S.A.; Morton, J.P.; Del Rio Hernandez, A.E. Substrate Rigidity Controls Activation and Durotaxis in Pancreatic Stellate Cells. *Sci. Rep.* **2017**, *7*, 2506. [CrossRef] [PubMed]
4. Elaut, G.; Henkens, T.; Papeleu, P.; Snykers, S.; Vinken, M.; Vanhaecke, T.; Rogiers, V. Molecular mechanisms underlying the dedifferentiation process of isolated hepatocytes and their cultures. *Curr. Drug Metab.* **2006**, *7*, 629–660. [CrossRef]
5. Jungermann, K.; Thurman, R.G. Hepatocyte heterogeneity in the metabolism of carbohydrates. *Enzyme* **1992**, *46*, 33–58. [CrossRef]
6. Gebhardt, R. Metabolic zonation of the liver: Regulation and implications for liver function. *Pharmacol. Ther.* **1992**, *53*, 275–354. [CrossRef]
7. Chen, G.; Qi, Y.; Niu, L.; Di, T.; Zhong, J.; Fang, T.; Yan, W. Application of the cell sheet technique in tissue engineering. *Biomed. Rep.* **2015**, *3*, 749–757. [CrossRef]
8. Martin, I.; Wendt, D.; Heberer, M. The role of bioreactors in tissue engineering. *Trends Biotechnol.* **2004**, *22*, 80–86. [CrossRef]
9. Crowley, C.; Butler, C.R.; Camilli, C.; Hynds, R.E.; Kolluri, K.K.; Janes, S.M.; de Coppi, P.; Urbani, L. Non-Invasive Longitudinal Bioluminescence Imaging of Human Mesoangioblasts in Bioengineered Esophagi. *Tissue Eng. Part C Methods* **2019**, *25*, 103–113. [CrossRef]
10. Maghsoudlou, P.; Georgiades, F.; Smith, H.; Milan, A.; Shangaris, P.; Urbani, L.; Loukogeorgakis, S.P.; Lombardi, B.; Mazza, G.; Hagen, C.; et al. Optimization of Liver Decellularization Maintains Extracellular Matrix Micro-Architecture and Composition Predisposing to Effective Cell Seeding. *PLoS ONE* **2016**, *11*, e0155324. [CrossRef]
11. Matamala, N.; Lara, B.; Gomez-Mariano, G.; Martinez, S.; Retana, D.; Fernandez, T.; Silvestre, R.A.; Belmonte, I.; Rodriguez-Frias, F.; Vilar, M.; et al. Characterization of Novel Missense Variants of SERPINA1 Gene Causing Alpha-1 Antitrypsin Deficiency. *Am. J. Respir. Cell Mol. Biol.* **2018**, *58*, 706–716. [CrossRef] [PubMed]
12. Gomez-Mariano, G.; Matamala, N.; Martinez, S.; Justo, I.; Marcacuzco, A.; Jimenez, C.; Monzon, S.; Cuesta, I.; Garfia, C.; Martinez, M.T.; et al. Liver organoids reproduce alpha-1 antitrypsin deficiency-related liver disease. *Hepatol. Int.* **2020**, *14*, 127–137. [CrossRef] [PubMed]
13. Chiang, T.S.; Yang, K.C.; Chiou, L.L.; Huang, G.T.; Lee, H.S. Enhancement of CYP3A4 activity in Hep G2 cells by lentiviral transfection of hepatocyte nuclear factor-1 alpha. *PLoS ONE* **2014**, *9*, e94885. [CrossRef] [PubMed]
14. Berger, B.; Donzelli, M.; Maseneni, S.; Boess, F.; Roth, A.; Krahenbuhl, S.; Haschke, M. Comparison of Liver Cell Models Using the Basel Phenotyping Cocktail. *Front. Pharmacol.* **2016**, *7*, 443. [CrossRef]
15. Yarchoan, M.; Agarwal, P.; Villanueva, A.; Rao, S.; Dawson, L.A.; Llovet, J.M.; Finn, R.S.; Groopman, J.D.; El-Serag, H.B.; Monga, S.P.; et al. Recent Developments and Therapeutic Strategies against Hepatocellular Carcinoma. *Cancer Res.* **2019**, *79*, 4326–4330. [CrossRef]
16. McQuitty, C.E.; Chokshi, S.; Chokshi, S.; Urbani, L. Immunomodulatory Role of the Extracellular Matrix within the Liver Disease Microenvironment. *Front. Immunol.* **2020**. [CrossRef]
17. Urbani, L.; Camilli, C.; Phylactopoulos, D.E.; Crowley, C.; Natarajan, D.; Scottoni, F.; Maghsoudlou, P.; McCann, C.J.; Pellegata, A.F.; Urciuolo, A.; et al. Multi-stage bioengineering of a layered oesophagus with in vitro expanded muscle and epithelial adult progenitors. *Nat. Commun.* **2018**, *9*, 4286. [CrossRef]
18. Couzin-Frankel, J. When mice mislead. *Science* **2013**, *342*, 922–925. [CrossRef]
19. Ewald, D.A.; Noda, S.; Oliva, M.; Litman, T.; Nakajima, S.; Li, X.; Xu, H.; Workman, C.T.; Scheipers, P.; Svitacheva, N.; et al. Major differences between human atopic dermatitis and murine models, as determined by using global transcriptomic profiling. *J. Allergy Clin. Immunol.* **2017**, *139*, 562–571. [CrossRef]
20. Leist, M.; Hartung, T. Inflammatory findings on species extrapolations: Humans are definitely no 70-kg mice. *Arch. Toxicol.* **2013**, *87*, 563–567. [CrossRef]
21. Mazza, G.; Al-Akkad, W.; Rombouts, K. Engineering in vitro models of hepatofibrogenesis. *Adv. Drug Deliv. Rev.* **2017**, *121*, 147–157. [CrossRef] [PubMed]

22. Uygun, B.E.; Soto-Gutierrez, A.; Yagi, H.; Izamis, M.L.; Guzzardi, M.A.; Shulman, C.; Milwid, J.; Kobayashi, N.; Tilles, A.; Berthiaume, F.; et al. Organ reengineering through development of a transplantable recellularized liver graft using decellularized liver matrix. *Nat. Med.* **2010**, *16*, 814–820. [CrossRef] [PubMed]
23. Yagi, H.; Fukumitsu, K.; Fukuda, K.; Kitago, M.; Shinoda, M.; Obara, H.; Itano, O.; Kawachi, S.; Tanabe, M.; Coudriet, G.M.; et al. Human-scale whole-organ bioengineering for liver transplantation: A regenerative medicine approach. *Cell Transpl.* **2013**, *22*, 231–242. [CrossRef] [PubMed]
24. Robertson, M.J.; Soibam, B.; O'Leary, J.G.; Sampaio, L.C.; Taylor, D.A. Recellularization of rat liver: An in vitro model for assessing human drug metabolism and liver biology. *PLoS ONE* **2018**, *13*, e0191892. [CrossRef]
25. Debnath, T.; Mallarpu, C.S.; Chelluri, L.K. Development of Bioengineered Organ Using Biological Acellular Rat Liver Scaffold and Hepatocytes. *Organogenesis* **2020**, *16*, 61–72. [CrossRef]
26. Baptista, P.M.; Moran, E.C.; Vyas, D.; Ribeiro, M.H.; Atala, A.; Sparks, J.L.; Soker, S. Fluid Flow Regulation of Revascularization and Cellular Organization in a Bioengineered Liver Platform. *Tissue Eng. Part C Methods* **2016**, *22*, 199–207. [CrossRef]
27. Kojima, H.; Yasuchika, K.; Fukumitsu, K.; Ishii, T.; Ogiso, S.; Miyauchi, Y.; Yamaoka, R.; Kawai, T.; Katayama, H.; Yoshitoshi-Uebayashi, E.Y.; et al. Establishment of practical recellularized liver graft for blood perfusion using primary rat hepatocytes and liver sinusoidal endothelial cells. *Am. J. Transpl.* **2018**, *18*, 1351–1359. [CrossRef]
28. Ogiso, S.; Yasuchika, K.; Fukumitsu, K.; Ishii, T.; Kojima, H.; Miyauchi, Y.; Yamaoka, R.; Komori, J.; Katayama, H.; Kawai, T.; et al. Efficient recellularisation of decellularised whole-liver grafts using biliary tree and foetal hepatocytes. *Sci. Rep.* **2016**, *6*, 35887. [CrossRef]
29. Baptista, P.M.; Siddiqui, M.M.; Lozier, G.; Rodriguez, S.R.; Atala, A.; Soker, S. The use of whole organ decellularization for the generation of a vascularized liver organoid. *Hepatology* **2011**, *53*, 604–617. [CrossRef]
30. Kulig, K.M.; Vacanti, J.P. Hepatic tissue engineering. *Transpl. Immunol.* **2004**, *12*, 303–310. [CrossRef]
31. Nahmias, Y.; Schwartz, R.E.; Hu, W.S.; Verfaillie, C.M.; Odde, D.J. Endothelium-mediated hepatocyte recruitment in the establishment of liver-like tissue in vitro. *Tissue Eng.* **2006**, *12*, 1627–1638. [CrossRef] [PubMed]
32. Sakaeda, T.; Fukumura, K.; Takahashi, K.; Matsumura, S.; Matsuura, E.; Hirano, K. Blood flow rate in normal and tumor-bearing rats in conscious state, under urethane anesthesia, and during systemic hypothermia. *J. Drug Target* **1998**, *6*, 261–272. [CrossRef] [PubMed]
33. Lubberstedt, M.; Muller-Vieira, U.; Mayer, M.; Biemel, K.M.; Knospel, F.; Knobeloch, D.; Nussler, A.K.; Gerlach, J.C.; Zeilinger, K. HepaRG human hepatic cell line utility as a surrogate for primary human hepatocytes in drug metabolism assessment in vitro. *J. Pharmacol. Toxicol. Methods* **2011**, *63*, 59–68. [CrossRef]
34. Hussein, K.H.; Park, K.M.; Ghim, J.H.; Yang, S.R.; Woo, H.M. Three dimensional culture of HepG2 liver cells on a rat decellularized liver matrix for pharmacological studies. *J. Biomed. Mater. Res. B Appl. Biomater.* **2016**, *104*, 263–273. [CrossRef] [PubMed]
35. Mazza, G.; Al-Akkad, W.; Telese, A.; Longato, L.; Urbani, L.; Robinson, B.; Hall, A.; Kong, K.; Frenguelli, L.; Marrone, G.; et al. Rapid production of human liver scaffolds for functional tissue engineering by high shear stress oscillation-decellularization. *Sci. Rep.* **2017**, *7*, 5534. [CrossRef]
36. Wang, B.; Jakus, A.E.; Baptista, P.M.; Soker, S.; Soto-Gutierrez, A.; Abecassis, M.M.; Shah, R.N.; Wertheim, J.A. Functional Maturation of Induced Pluripotent Stem Cell Hepatocytes in Extracellular Matrix-A Comparative Analysis of Bioartificial Liver Microenvironments. *Stem Cells Transl. Med.* **2016**, *5*, 1257–1267. [CrossRef]
37. de l'Hortet, A.C.; Takeishi, K.; Guzman-Lepe, J.; Morita, K.; Achreja, A.; Popovic, B.; Wang, Y.; Handa, K.; Mittal, A.; Meurs, N.; et al. Generation of Human Fatty Livers Using Custom-Engineered Induced Pluripotent Stem Cells with Modifiable SIRT1 Metabolism. *Cell Metab.* **2019**, *30*, 385–401.E9. [CrossRef]
38. Lorvellec, M.; Pellegata, A.F.; Maestri, A.; Turchetta, C.; Alvarez Mediavilla, E.; Shibuya, S.; Jones, B.; Scottoni, F.; Perocheau, D.P.; Cozmescu, A.C.; et al. An In Vitro Whole-Organ Liver Engineering for Testing of Genetic Therapies. *iScience* **2020**. [CrossRef]
39. Hussein, K.H.; Park, K.M.; Kang, K.S.; Woo, H.M. Heparin-gelatin mixture improves vascular reconstruction efficiency and hepatic function in bioengineered livers. *Acta Biomater.* **2016**, *38*, 82–93. [CrossRef]
40. Pellegata, A.F.; Tedeschi, A.M.; de Coppi, P. Whole Organ Tissue Vascularization: Engineering the Tree to Develop the Fruits. *Front. Bioeng. Biotechnol.* **2018**, *6*, 56. [CrossRef]
41. Palikuqi, B.; Nguyen, D.T.; Li, G.; Schreiner, R.; Pellegata, A.F.; Liu, Y.; Redmond, D.; Geng, F.; Lin, Y.; Gomez-Salinero, J.M.; et al. Adaptable haemodynamic endothelial cells for organogenesis and tumorigenesis. *Nature* **2020**, *585*, 426–432. [CrossRef] [PubMed]
42. Coward, S.M.; Legallais, C.; David, B.; Thomas, M.; Foo, Y.; Mavri-Damelin, D.; Hodgson, H.J.; Selden, C. Alginate-encapsulated HepG2 cells in a fluidized bed bioreactor maintain function in human liver failure plasma. *Artif. Organs* **2009**, *33*, 1117–1126. [CrossRef] [PubMed]
43. Damania, A.; Kumar, A.; Teotia, A.K.; Kimura, H.; Kamihira, M.; Ijima, H.; Sarin, S.K.; Kumar, A. Decellularized Liver Matrix-Modified Cryogel Scaffolds as Potential Hepatocyte Carriers in Bioartificial Liver Support Systems and Implantable Liver Constructs. *ACS Appl. Mater. Interfaces* **2018**, *10*, 114–126. [CrossRef] [PubMed]
44. Hoffmann, S.A.; Muller-Vieira, U.; Biemel, K.; Knobeloch, D.; Heydel, S.; Lubberstedt, M.; Nussler, A.K.; Andersson, T.B.; Gerlach, J.C.; Zeilinger, K. Analysis of drug metabolism activities in a miniaturized liver cell bioreactor for use in pharmacological studies. *Biotechnol. Bioeng.* **2012**, *109*, 3172–3181. [CrossRef] [PubMed]

45. Lubberstedt, M.; Muller-Vieira, U.; Biemel, K.M.; Darnell, M.; Hoffmann, S.A.; Knospel, F.; Wonne, E.C.; Knobeloch, D.; Nussler A.K.; Gerlach, J.C.; et al. Serum-free culture of primary human hepatocytes in a miniaturized hollow-fibre membrane bioreactor for pharmacological in vitro studies. *J. Tissue Eng. Regen. Med.* **2015**, *9*, 1017–1026. [CrossRef] [PubMed]
46. He, Y.T.; Qi, Y.N.; Zhang, B.Q.; Li, J.B.; Bao, J. Bioartificial liver support systems for acute liver failure: A systematic review and meta-analysis of the clinical and preclinical literature. *World J. Gastroenterol.* **2019**, *25*, 3634–3648. [CrossRef]
47. Pellegata, A.F.; Asnaghi, M.A.; Zonta, S.; Zerbini, G.; Mantero, S. A novel device for the automatic decellularization of biological tissues. *Int. J. Artif. Organs* **2012**, *35*, 191–198. [CrossRef] [PubMed]
48. Pellegata, A.F.; Dominioni, T.; Ballo, F.; Maestroni, S.; Asnaghi, M.A.; Zerbini, G.; Zonta, S.; Mantero, S. Arterial Decellularized Scaffolds Produced Using an Innovative Automatic System. *Cells Tissues Organs* **2015**, *200*, 363–373. [CrossRef]

nanomaterials

Review

Biomimetic Scaffolds—A Novel Approach to Three Dimensional Cell Culture Techniques for Potential Implementation in Tissue Engineering

Tomasz Górnicki [1,*], Jakub Lambrinow [1], Afsaneh Golkar-Narenji [2,†], Krzysztof Data [3], Dominika Domagała [3], Julia Niebora [3], Maryam Farzaneh [4], Paul Mozdziak [2], Maciej Zabel [1], Paweł Antosik [5], Dorota Bukowska [6], Kornel Ratajczak [5], Marzenna Podhorska-Okołów [7], Piotr Dzięgiel [1] and Bartosz Kempisty [3,5,8,9,*]

1 Division of Histology and Embryology, Department of Human Morphology and Embryology, Wroclaw Medical University, 50-368 Wroclaw, Poland; jakub.lambrinow@student.umw.edu.pl (J.L.); maciej.zabel@umw.edu.pl (M.Z.); piotr.dziegiel@umw.edu.pl (P.D.)
2 Prestage Department of Poultry Science, North Carolina State University, Raleigh, NC 27607, USA; pemozdzi@ncsu.edu (P.M.)
3 Division of Anatomy, Department of Human Morphology and Embryology, Wroclaw Medical University, 50-368 Wroclaw, Poland; krzysztof.data@umw.edu.pl (K.D.); dominika.domagala@umw.edu.pl (D.D.); julia.niebora@umw.edu.pl (J.N.)
4 Fertility, Infertility and Perinatology Research Center, Ahvaz Jundishapur University of Medical Sciences, Ahvaz P.O. Box 6193673111, Iran; farzaneh-m@ajums.ac.ir
5 Department of Veterinary Surgery, Institute of Veterinary Medicine, Nicolaus Copernicus University in Torun, 87-100 Torun, Poland; pantosik@umk.pl (P.A.); kornel@umk.pl (K.R.)
6 Department of Diagnostics and Clinical Sciences, Institute of Veterinary Medicine, Nicolaus Copernicus University in Torun, 87-100 Torun, Poland; dbukowska@umk.pl
7 Division of Ultrastructure Research, Department of Human Morphology and Embryology, Wroclaw Medical University, 50-368 Wroclaw, Poland; marzenna.podhorska-okolow@umw.edu.pl
8 Physiology Graduate Faculty, North Carolina State University, Raleigh, NC 27613, USA
9 Department of Obstetrics and Gynecology, University Hospital and Masaryk University, 602 00 Brno, Czech Republic
* Correspondence: tomasz.gornicki@student.umw.edu.pl (T.G.); bartosz.kempisty@umw.edu.pl (B.K.)
† Deceased author.

Citation: Górnicki, T.; Lambrinow, J.; Golkar-Narenji, A.; Data, K.; Domagała, D.; Niebora, J.; Farzaneh, M.; Mozdziak, P.; Zabel, M.; Antosik, P.; et al. Biomimetic Scaffolds—A Novel Approach to Three Dimensional Cell Culture Techniques for Potential Implementation in Tissue Engineering. *Nanomaterials* **2024**, *14*, 531. https://doi.org/10.3390/nano14060531

Academic Editors: Daniela Iannazzo, Costantino Del Gaudio, Lucy Di Silvio and Silvia Baiguera

Received: 12 January 2024
Revised: 28 February 2024
Accepted: 14 March 2024
Published: 16 March 2024

Copyright: © 2024 by the authors. Licensee MDPI, Basel, Switzerland. This article is an open access article distributed under the terms and conditions of the Creative Commons Attribution (CC BY) license (https://creativecommons.org/licenses/by/4.0/).

Abstract: Biomimetic scaffolds imitate native tissue and can take a multidimensional form. They are biocompatible and can influence cellular metabolism, making them attractive bioengineering platforms. The use of biomimetic scaffolds adds complexity to traditional cell cultivation methods. The most commonly used technique involves cultivating cells on a flat surface in a two-dimensional format due to its simplicity. A three-dimensional (3D) format can provide a microenvironment for surrounding cells. There are two main techniques for obtaining 3D structures based on the presence of scaffolding. Scaffold-free techniques consist of spheroid technologies. Meanwhile, scaffold techniques contain organoids and all constructs that use various types of scaffolds, ranging from decellularized extracellular matrix (dECM) through hydrogels that are one of the most extensively studied forms of potential scaffolds for 3D culture up to 4D bioprinted biomaterials. 3D bioprinting is one of the most important techniques used to create biomimetic scaffolds. The versatility of this technique allows the use of many different types of inks, mainly hydrogels, as well as cells and inorganic substances. Increasing amounts of data provide evidence of vast potential of biomimetic scaffolds usage in tissue engineering and personalized medicine, with the main area of potential application being the regeneration of skin and musculoskeletal systems. Recent papers also indicate increasing amounts of in vivo tests of products based on biomimetic scaffolds, which further strengthen the importance of this branch of tissue engineering and emphasize the need for extensive research to provide safe for humansbiomimetic tissues and organs. In this review article, we provide a review of the recent advancements in the field of biomimetic scaffolds preceded by an overview of cell culture technologies that led to the development of biomimetic scaffold techniques as the most complex type of cell culture.

Keywords: biomimetic scaffolds; 3D bioprinting; biologically derived materials; 4D materials

1. Introduction

Tissue damage resulting from cancer, congenital defects, and trauma requires new and effective treatments that facilitate tissue regeneration. Tissue engineering presents significant potential in this regard, as it enables the restoration of native tissue architecture and functions through the fusion of cells to specific scaffolds [1]. The goal of tissue engineering (TE) is to restore, preserve, or enhance the structure and function of damaged tissues or organs by integrating biological signals and biological scaffolding strategies [2].

Cell cultures have traditionally been studied on 2D platforms, consisting of cells interacting with a culture dish in a medium environment. However, this method has limitations as the growth of cells on a flat surface does not accurately reproduce their actual functioning in the 3D environment of other cells. The 3D model overcomes these limitations by recreating the in vivo environment. Scientists have provided evidence that scaffolding animal cells can induce the formation of 3D colonies that resemble the natural environment, both molecularly and phenotypically [3]. The extracellular matrix (ECM) plays a crucial role in the transport properties, cell communication, mechanotransduction, and growth factor signaling of the 3D hierarchical microstructure due to its electromechanical nature. This is achieved through the interaction of the ECM with receptors on the cell surface, as well as the binding of growth factors and other signaling molecules [4]. Zhang et al. suggest that the ECM scaffold is a highly promising candidate for tissue engineering applications. Among biomimetic ECM scaffold materials, decellularized ECM scaffolds (dECMs) derived from natural ECM are particularly noteworthy due to their natural components and microenvironment [5].

The goal of regenerative medicine is to repair and replace damaged tissue, and the use of three-dimensional scaffolds is one of the most promising techniques for tissue repair. Hydrogels are one of the most extensively researched types of scaffolds. They have demonstrated positive results in preclinical studies by mimicking the fundamental signals that promote local tissue regeneration [6]. Recent advancements in computer-aided design and 3D printing have helped the production of macroporous hydrogels, enabling the creation of more intricate structures. This approach may lead to the development of fully reconstructed organs. The field of hydrogels is a promising area of research in regenerative medicine, with applications in most tissues of the human body [7].

Biocompatible scaffolds promote cell adhesion, proliferation, and differentiation to facilitate tissue regeneration. These scaffolds provide strength, mechanical stability, flexibility, and an ideal environment for cell growth. Scaffolds can be divided into natural and synthetic categories [5]. Bioscaffolds provide a niche for cell growth, while synthetic scaffolds offer greater control over the size and morphology of regenerated tissues. In this review, we present a comprehensive overview of the current state of knowledge regarding biomimetic scaffolds. We begin by examining recent advancements in scaffold technology and their pivotal role in tissue engineering. Subsequently, we delve into the cell culture techniques that have contributed to the conceptualization and development of biomimetic scaffolds, as well as those evolving in tandem with this technology. In the final section, we highlight recent advancements in biomimetic scaffolds and explore their potential applications in biotechnology and tissue engineering.

2. Scaffolds in Tissue Engineering—Recent Findings and Current Research

In recent years, there has been an observable increase in the diversity of areas studied in the field of tissue engineering. Scientists are seeking more complex opportunities to apply recent technological advancements beyond the classic tissue engineering of the skeletal system [8]. An increasing number of studies provide data indicating usage of scaffolds in tissue engineering of different systems, including endocrine, muscle, genitourinary, digestive and pulmonary systems.

Recent advancements in regenerative endocrinology include the development of novel methods of protecting pancreatic β-cells from destruction by immune cells in hybrid devices consisting of scaffolds made out of polycaprolactone scaffolds and pancreatic β-cells encapsulated in alginate microcapsules. This method prolongs the functionality of the device and potentially eliminates the need for immunosuppressants [9]. Also, the development of prevascularized thyroid organoids opens new possibilities for potential new ways of hypothyroidism treatment [10]. Another advancement in tissue engineering of endocrine glands is development of a 3D-bioprinted functional model of the pancreas using pancreatic and endothelial cell lines with an ability to moderate insulin secretion answering to changes in glucose levels in medium [11]. Also, in the field of human reproductive hormones, there are some significant changes with the introduction of decellularized extracellular matrix providing optimal microenvironment for in vitro spermatogenesis [12].

In the treatment of volumetric muscle loss, scientists have developed a photoreactive hydrogel with the ability to change its stiffness. This allows the determination of the optimal level of hydrogel stiffness, which significantly increases the regeneration of muscle tissue [13].

In the field of muscle tissue regeneration, a recent study has also provided a description of complex artificial muscle tissue consisting of a layer of myofibroblasts connected with motor neurons derived from induced pluripotent stem cells. The motor neurons contracted the layer of myofibroblasts by creating neuromuscular junctions [14].

A study conducted on rabbits demonstrated another potential application of scaffolds in tissue engineering of different systems. It presents the potential usage of adipose tissue engineering as a tool in the treatment of postoperative complications. In order to reduce epidural fibrosis after the procedure of laminectomy, the researchers reconstructed local adipose tissue using a scaffold made from ECM imbued with mesenchymal stem cells [15].

Another area of tissue engineering development is tissue engineering of bile ducts using hydrogels as the scaffold [16].

Scientists have also recently used tissue engineering technologies in the regeneration of the genitourinary system. They created a device consisting of a scaffold and mesenchymal stem cells overexpressing basic fibroblast growth factor. They have provided evidence of the device's potential usage in regenerating full-thickness injuries of the uterus [17].

Tissue engineering is a developing field that has generated interest in creating scaffold-based artificial organs with higher functionality. For instance, a recent study focused on contractile vascular grafts that retain their contractility. The graft is composed of decellularized pulmonary artery and progenitor cardiovascular cells obtained in a bioreactor under physiological flow conditions [18].

3. 2D Cell Cultures—Limitation of Most Common Type of Cell Cultures

Culture vessels with a single surface are used for 2D cell culture. The proliferating cells cover the surface of the culture environment, forming a monolayer cell culture. The crucial factor affecting cell proliferation is confluence, which refers to the degree of surface coverage by cells. As confluence approaches 100%, the metabolism of cultured cells changes and the expression of proliferation markers decreases [19]. The most commonly used technique for culturing cells is on a flat surface due to its simple processing and cost-effectiveness. A monolayer of cells also enables straightforward observation and measurement. There are two types of 2D cultivations: simple cultures and co-cultures. Simple culture involves the interaction of cells with a dish in a culture medium environment. In co-cultures, cells from different tissues are cultured together in a single medium, allowing for direct or indirect interactions, as illustrated in Figure 1. Direct co-culture involves growing different types of cells on the surface of a shared dish, facilitating plenty of interactions. Indirect co-culture, on the other hand, is based on separate cultivation surfaces but with interaction occurring through a common culture medium. One disadvantage of this 2D cultivation model is the lack of representation of real cell surroundings. Although growing cells on a flat, plastic surface covered with dry plasma can enhance cell adhesion, it is not an appropriate method for studying cell metabolism in a natural environment where cells are surrounded by other cells in three dimensions. However, this model is useful for analyzing simple interactions under controlled conditions.

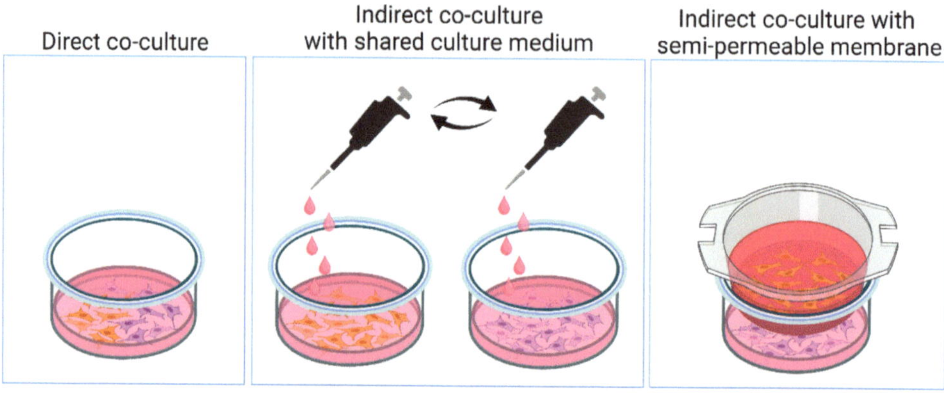

Figure 1. Types of direct and indirect co-culture systems. Created with BioRender.com.

4. 3D Cell Cultures

The next stage in comprehending physiological and pathological processes involving cell cultures is through the use of three-dimensional cell cultures. The interest of researchers in 3D human cell culture has grown rapidly since Hamburger and Salmon published one of the first papers on the subject in 1977 [20,21]. Although 2D cell cultures are the most commonly used method of studying living cells in vitro, there are still some significant drawbacks to this approach. According to research, cells cultured in two-dimensional conditions exhibit less similarity to those found in natural tissues [22]. This is due to changes in cell morphology, reduced cell–cell and cell–ECM interactions, and altered gene and protein expression resulting from the lack of a complex microenvironment [23]. Researchers were compelled to explore 3D culture techniques due to impaired cell polarity, ample access to oxygen and nutrition, and the absence of an external matrix, among other characteristics [24]. In the early 2000s, it became evident that 3D cultures were an emerging technology [25]. While the prediction that 3D culture would replace 2D culture was somewhat misguided, there is a growing trend towards more studies utilizing 3D culture. Three-dimensional cell cultures offer several advantages, the most significant of which is a closer representation of natural tissues [26]. However, 3D cultures also have some disadvantages, including higher costs, lower reproducibility, and greater difficulty in interpretation [23].

In fact, three-dimensional cell culture is a more complex method than two-dimensional culture. 3D cultures often involve co-culturing, where more than one type of cell is cultured simultaneously. In addition, there are various methods for replicating living 3D structures. 3D culture can be categorized as either scaffold-free or scaffold-based [26–29]. In scaffold-free 3D cultures, cells interact with each other and form structures such as spheroids [30]. Scaffold-based 3D cultures are capable of creating more intricate structures by utilizing cell–cell and cell–EMC interactions, which can closely resemble natural tissue [31]. There are two main types of 3D culture methods: those that use scaffolding in the cell culture (scaffold techniques) and those that do not (scaffold-free techniques).

4.1. *Scaffold Free Techniques*

Spheroids

The term 'spheroid' describes the spherical shape of cell colonies, whether they are single-cell or multicellular. In traditional monolayer cultures, cell–flask interactions are the primary type of interaction that contributes to the formation of the culture. In traditional monolayer cultures, cell–flask interactions are the primary type of interaction that contributes to the formation of the culture. However, spheroids emerge due to cell–cell and cell–ECM interactions. The process of forming a spheroid involves three main steps presented on Figure 2. Firstly, the extracellular matrix (ECM) fibers, which are rich in RGD motifs, allow for the binding of integrins expressed on the cell membrane surface, resulting in upregulated cadherin expression. Secondly, cadherins localize on the cell surface. Finally, cadherin-cadherin homophilic bindings between nearby cells cause stronger cell-to-cell adhesion and spheroid formation [24,32,33].

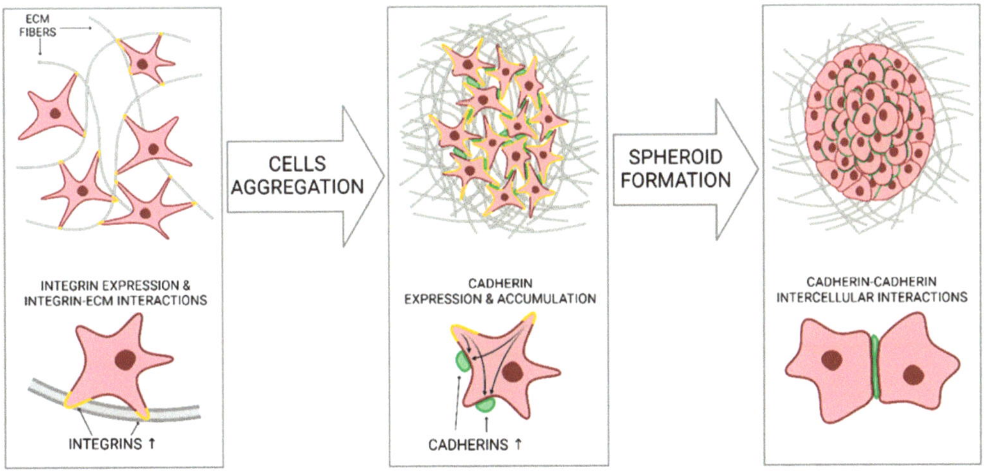

Figure 2. Steps of cell aggregation and spheroid formation (ECM, extracellular matrix). Created with BioRender.com.

There are various methods for manufacturing spheroids. This goal can be achieved using various technologies, such as hanging drop, low-adherence substrate, bioreactor, microwells, magnetic manipulation, and microfluidics [33–36]. The properties of spheroids are more important than the technologies used to manufacture them. Due to their three-dimensional nature, spheroids closely resemble in vivo conditions, including oxygen gradients, nutrition, and metabolic products [35,37].

However, spheroids have some limitations despite their similarities to in vivo processes. This method of cell culture is less well-known than 2D monolayer cultures, and there are also major problems with standardizing spheroid size [33]. Nonetheless, spheroids are an important tool for understanding physiological and pathological processes. For instance, the proliferation rate of spheroid surface cells is significantly higher than that of the inner zone. In fact, due to the oxygen gradient, the inner zone may even consist of a necrotic core instead of living cells [35]. Therefore, spheroids are a superior model to 2D cultures for anti-cancer drug testing [38].

Currently, researchers are exploring the use of spheroids in various fields of bioengineering and medicine. This type of 3D culture is widely studied as an example of cells that mimic cancer tumors. In the case of co-cultures, which are spheroids made up of cancer cells and other cells such as cancer-associated fibroblasts, the similarities to in vivo cancer tumors are even greater [37]. The number of recently published papers on the use of spheroids in cancer research is still increasing. One of the main advantages of using spheroids in cancer research is the ability to measure drug response. Multiple studies have been conducted involving spheroids derived from various types of cancer, including colorectal cancer [39–42], breast cancer [41,43,44], lung cancer [45,46], ovarian cancer [1,47,48] and even Ewing sarcoma cells [49]. It is important to note that 3D cultures, particularly spheroids, are a valuable and effective tool for studying cancer biology. They can be used to assess potential drug resistance, identify new therapeutic targets, and investigate molecular mechanisms [50].

4.2. Scaffold Techniques

4.2.1. Organoids

Organoids are more complex structures than spheroids and are named as such because they represent the complexity of organs, unlike spheroids, which generally represent only one type of tissue [30]. Organoids can be derived from various types of stem cells, including embryonic, induced pluripotent, or tissue-specific cells [24]. Due to their complexity, organoids are valuable models for studying and simulating physiological processes that occur in living organisms' organs. Organoids are self-organising and self-renewing structures, meaning that they have some level of freedom in their organisation. While 2D cultures are easier to control, 3D cultures, especially more complex ones, are less predictable in terms of deterministic processes. Therefore, organoids can closely mimic living organs Similarly, 3D cultures can be scaffold-based or scaffold-free [51]. However, scaffolds are more commonly used in organoid manufacturing to recreate a natural environment [52].

Organoids are often referred to as mini-organs, and current research is focused on creating fully functional organs on a mini-scale. Examples include 'mini-guts' [53], 'mini-hearts' [54], and 'mini-livers' [55]. The name 'organoid' accurately describes a miniaturised model of an entire organ. Organoids are commonly used in cancer research, with patient-derived organoids (PDOs) proving to be a valuable tool in drug testing and the study of cancer biology [56–58]. Additionally, organoids are utilised as models for a range of other diseases, including lung diseases [59], liver cirrhosis [60], inflammatory bowel disease [61], and brain disorders [62].

4.2.2. Hydrogels

Modern 3D cell culture scaffold techniques often rely on the use of various biomaterials, including hydrogels. Hydrogels are hydrophilic, cross-linked polymer chains [63,64] that can be pre-prepared to facilitate 3D cell culture due to their similarity to the extracellular matrix [65]. Although hydrogels can vary greatly in water content, chain composition, and ionic charge [66], their use in 3D cell culture is well-established. The potential applications of hydrogels are vast, as demonstrated by the numerous classifications available. Hydrogels can be classified based on their source (natural or synthetic), size (nano-, macro-, or bulk hydrogels), chain composition, ionic charge, method of crosslinking, response to various stimuli, or biodegradability [67]. Matrigel® is one of the most well-known hydrogels in scientific research [68]. The substance is a natural biomaterial derived from secretions of Engelbreth–Holm–Swarm (EHS) mouse sarcoma cells. Hydrogels can also be obtained naturally by crosslinking collagen [69], gelatin [70], hyaluronic acid [71], or alginate [72]. Synthetic hydrogels, in contrast to natural ones, can be manufactured with greater precision, allowing for better control of desired properties such as water absorption and ionic charge.

4.2.3. Advanced 3D Bioprinting

3D bioprinting is a promising and increasingly popular tool for scaffold production, presenting many advantages. These include high levels of structure customization, which can imitate the structure of desired tissue, as well as the freedom to mix components and impregnate them with additional substances [73]. It allows for the creation of complex tissue structures that scientists around the world are looking to apply in medicine [74]. Current research primarily focuses on using 3D bioprinting to produce advanced scaffolds for skin bioengineering [75], as well as for musculoskeletal, cardiovascular, and neural systems [73].

The recent development of 3D bioprinting technologies has led to the creation of 4D bioprinting, which represents the next generation of biomaterials. These materials are capable of changing their shape, properties, and functions in response to external stimuli such as heat, changes in pH, light, and humidity [76]. A wide range of intelligent polymers and materials are used in this process. For example, iron-sensitive hydrogels, polyethylene glycol that responds to temperature fluctuations, or special ink containing poly(lactic acid) polymer that assembles into tubular shapes under the influence of a magnetic field [77].

This new generation of biomaterials is proposed to have applications in various areas of medicine, such as drug delivery systems (encapsulation devices) and biosensors [78–80]. However, the most significant potential of 4D-printed biomaterials is in tissue engineering [76]. Multi-material 4D-printed implants with heterogeneous morphology can provide an environment that changes over time that promotes cell activity and differentiation, and increases the regenerative capabilities of nearby tissues [81]. Currently, scientists are conducting research on 4D-printed biomaterials in a few applications. This includes the production of functional meniscal implants [82], regeneration of cartilage with chitosan derivatives [83], bone replacement implants for treating losses caused by trauma or genetic diseases [84–87], and tooth implants [88]. Additionally, there is evidence of the potential usage of 4D-printed biomaterials in the engineering of the cardiovascular system. Scientists are currently developing 4D-printed patches to aid heart function after myocardial infarction [89], as well as a model of an artery to study the pathomechanism of cardiovascular diseases [90]. Additionally, there are projects focused on creating 4D-printed heart valve implants [91]. 4D-printed biomaterials may also have applications in otolaryngology, such as craniofacial, tracheal, nasal, and aural implants [77]. Studies have shown that 4D-printed biomaterials can successfully guide stem cell differentiation and fate [92]. The potential applications of 4D-printed biomaterials include regeneration in craniofacial skeletal muscle [93], wound healing [94], implants for bladder disorders [95], and in vitro models of fibroblast remodeling [96].

Bioprinting techniques can be divided into two categories based on the place of synthesis. The most common technique is ex vivo, where the scaffold is combined with cells and other biomolecules outside of the body [76]. The newer approach, in situ, aims to print tissue or organs directly in the human body at the required site of trauma [97]. This technology has a wide variety of potential applications. There are various materials that are considered for use in in situ bioprinting, such as hydrogels based on Gallol-Functionalized Hyaluronic Acid [98], methacrylate-based gelatin [99], and bioinks based on alginate–chitosan and kaolin [100]. Current research focuses on in situ bioprinting applications for direct cartilage repair [101,102] or bioprinting human mesenchymal cells that will transdifferentiate into chondrogenic tissue [103]. In situ bioprinting may also be used for skin wound regeneration, providing accurate coverage of the affected area [104]. Scientists have developed a 3D bioprinter called the 'SkinPen' for skin regeneration, which uses a complex hydrogel controlled by ultrasound and ultraviolet light to enhance adhesive and morphological properties [105]. Currently, studies are being conducted to combine these two goals and create a robot-assisted in situ bioprinter for skin and hair follicle regeneration [106]. The study investigates the use of gelatin methacryloyl (GelMA) with zinc and silicon ions for hair follicle regeneration in a mouse model [107]. Additionally, researchers are exploring the use of robotic technologies for cranial bone regeneration through in situ bioprinting [108]. Robots are also being utilized to develop minimally invasive bioprinting systems for liver tissue bioprinting and regeneration [109]. In endoscopic surgery, in situ bioprinting is seen as a potential new technology for intestine regeneration [110]. Scientists have presented evidence of the potential use of in situ bioprinting for printing neurons, vascular and muscle tissues [111–113]. Intraoperative in situ bioprinting can also be used in orthopaedics for bone tissue regeneration [114]. Additionally, researchers are attempting to bioprint bone tissue with complete vascularisation, adding an additional level of complexity to the constructs created in situ [115]. Scientists see another potential application of in situ bioprinting in functional segmental tracheal reconstruction, where this technique can be used to regenerate destroyed parts of the trachea [116]. In addition, in situ bioprinting may have possible applications in dentistry as a means of regenerating dental pulp [117].

4.2.4. Alternative Materials Employed in the Context of 3D Bioprinting

The three most commonly used techniques in 3D printing are Fused Deposition Modeling (FDM), Stereolithography (SLA), and Selective Laser Sintering (SLS). The crucial difference between these methods is the form of the delivered material and the way the printing material hardens, as shown in Figure 3. FDM is the most commonly used technique where a solid filament is heated, melted, and extruded through a nozzle. The base of the SLA technique is a container with liquid resin that is hardened with a laser beam. Printing with SLS technique also involves a laser beam, but the printing material used is polymeric powder. The materials can be modified and supplemented in various ways due to the printing technique.

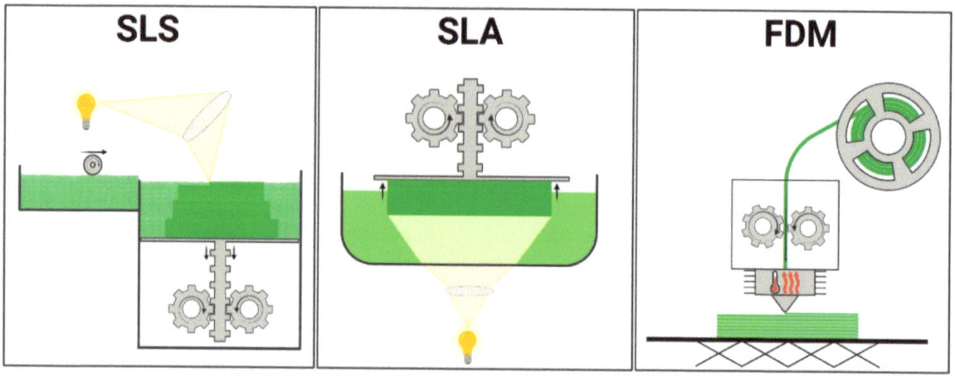

Figure 3. Main 3D printing technologies (abb. SLS, stereolitography; SLA, selective laser sintering; FDM, fused deposition modeling). Created with BioRender.com.

Fused Deposition Modeling (FDM): FDM is a method commonly used for 3D printing with several thermoplastic polymers, including poly (L-lactic acid) (PLA), poly-l-lactide-co-ε-caprolactone (PLCL), and gelatin methacrylate (GelMA), along with various supplements. PLA, which has stiff and cytocompatible mechanical properties, is most often used for bone tissue scaffolds [118]. PLA is obtained through the condensation of lactic acid or the polymerization of lactide. Commercially available polylactic acid (PLA) is typically derived from fermented plant starch, such as corn, cassava, sugarcane, or sugar beet pulp. However, PLA has limitations, including poor mechanical properties, insufficient surface wettability, and a low degradation rate, which restricts its use in biomedical applications [119]. Therefore, there are multiple proposals to enhance the PLA composition or indirectly use PLA for forming bioscaffolds. The use of a PLA printed shape as a frame for low-viscosity bioink can limit the ink from spreading beyond the established shape [120]. This strategy involves using PLA material as a scaffold to form the proper bioscaffold. To mimic tissues, particularly bone tissue, PLA is supplemented to improve biomineralization and physical properties. The addition of hydroxyapatite (HAp) significantly increases the strength and stiffness of PLA. Simultaneously, hydroxyapatite (HAp) contains sites for the deposition of salts and apatite species in the scaffold, which increases the rate of material mineralization in an in vitro study [121]. Bioscaffolds formed from HAp-polylactic acid (PLA) composites can improve bone regeneration in in vivo grafting. As demonstrated in the rabbit model, HAp-PLA scaffolds are biocompatible, degrade over time, and form bone trabeculae and marrow cavities on the surface of the scaffolds. Moreover, the safety and efficacy of HPa-PLA in repairing cranial defects in rabbits are comparable to that of autologous bone transplantation [122]. Bioglass (BG) can also be used to enrich the scaffold and mimic the bone tissue environment. The PLA-BG composite exhibits greater durability than PLA alone [123]. Furthermore, the addition of BG enhances cell viability and the expression of endothelial marker genes in vitro, as demonstrated on human umbilical vein endothelial cells (HUVECs), indicating a positive effect on angiogenesis [124]. Pearl powder [125], graphene oxide [126], and cold argon plasma treatment [119] are all valuable supplements that can influence bone cell differentiation, proliferation, and scaffold strength.

PLCL is a copolymer composed of L-lactic and ε-caprolactone. It is widely used in the production of scaffolds for soft tissue engineering due to its flexibility and rubber-like elasticity. With the adjustment of 3D-printing parameters, it is possible to obtain scaffolds with extensibility comparable to native human tissues, such as vessels, cartilage, and ligaments, while maintaining full cytocompatibility and cell adhesion [127]. A combination of PLA and PLCL has been found to be an effective mimic for cartilage tissue. The inclusion of PLA in the material enhances its mechanical resistance and stiffness, as well as the processability of PLCL for 3D printing. In line with this, the PLCL-PLA scaffold has been shown to improve the proliferation and chondrogenesis of in vitro-seeded chondrocytes more effectively than the PLCL scaffold alone [128]. The hybrid scaffold containing PLCL and decellularized extracellular matrix (dECM), specifically adipose tissue dECM, is a promising technique with potential clinical applications. These adipose-mimicking scaffolds possess mechanical properties comparable to native tissue and have the potential to enhance tissue regeneration. The in vitro test demonstrated that the dECM-PLCL scaffold promotes adipogenesis and angiogenesis, as well as adipose tissue formation, while suppressing apoptosis of human adipose-derived stem cells (hADSC) in vivo [129,130].

Gelatin methacryloyl (GelMA) is a hydrogel made from proteins. It has a porosity of almost 90% and a disordered pore arrangement, which imitates the structure of the extracellular matrix (ECM) due to its high collagen content. This porosity promotes cell viability and is crucial for proper development and metabolism [131]. To cross-link GelMA, it must be exposed to UV light in the presence of photo-initiators. The advantage of this process is its simplicity in modulating mechanical properties by adjusting the time and intensity of UV light and the concentration of photo-initiators. Commonly used photo-initiators include Irgacure 2959 (I2959), lithium phenyl-2,4,6-trimethylbenzoylphosphinate (LAP), and Eosin-Y [132]. However, a disadvantage of photo-crosslinking is the accelerated enzymatic degradation of the scaffold with collagenase, which makes the material less suitable for the regeneration process. The combination of GelMA and tyramine-conjugated 8-arm poly(ethylene glycol) (8PEGTA) reduces degradation while maintaining ECM-mimicking conditions during early tissue regeneration [133]. GelMA can also act as a carrier and support nervous tissue development. A 3D-printed canal filled with 7,8-dihydroxyflavone (7,8-DHF) successfully reconstructed and reconnected a 12-mm nerve defect [134]. The properties of elastic GelMA include the potential to mimic vessel structures. When supplemented with chitin nanocrystals, the material exhibits improved mechanical resistance, cell adhesion, proliferation, and vascularization [135]. GelMA can be used to mimic bone tissue. Supplementing with BG improves mechanical properties, cell adhesion, proliferation, and, most importantly, osteoblast differentiation. This also enhances the release of osteocalcin (OCN), a factor in bone mineralization that orchestrates the osteogenesis process [136].

Stereolitography: Compared to FDM printing, SLA printing is more efficient in printing smaller and more precise models with higher resolution. Stereolithography is a 3D printing process that uses a light source to harden the material. Even when using the same materials, such as GelMA, which is also used in SLA printing, the liquid state of the materials allows for the use of resins such as poly(D,L-lactide) (PDLLA) or poly(propylene fumarate) (PPF). Even when using the same materials, such as GelMA, which is also used in SLA printing, the liquid state of the materials allows for the use of resins such as poly(D,L-lactide) (PDLLA) or poly(propylene fumarate) (PPF).

Factory tests have demonstrated the possibility of utilising GelMA and a low-cost, commercially available printer to conduct three-dimensional in vitro culture. The scaffold produced in this manner is characterised by high printing accuracy and good biocompatible properties [137]. Similar to the FDM method, the enrichment of GelMA with other materials significantly expands and diversifies its potential uses, allowing it to be adapted to mimic various tissues. The combination of PLCL and GelMA has been shown to replicate the properties of small intestine tissue when cultured in vitro [138]. It is important to maintain a clear and logical structure, use precise language, and avoid bias and grammatical errors. Additionally, the supplementation of HAp has successfully aided in the regeneration of bone tissue in vivo [139]. Modifications using the FDM technique may also allow for the replication of other tissue types and material properties.

PDLLA is a flexible polymer with numerous medical applications, including controlled drug delivery [140]. It has a high capacity to mimic bone tissue and has been successfully used for bone regeneration in vivo. The best results in the treatment of bone defects were obtained using a hybrid of PDLLA and PPF. This hybrid also supported the expression of key markers for osteogenesis, such as osteocalcin, collagen, and runt-related transcription factor 2 (RUNX2) [141].

Photo-crosslinkable PPF is a popular choice for bone-mimicking scaffolds due to its superior mechanical resistance, stiffness, and biodegradability. To apply PPF, SLA is necessary using a solvent. Diethyl fumarate (DEF) is a commonly used solvent for PPF, but studies have shown that this combination can have a cytotoxic effect on seeded cells. Ethyl acetate (EA) is a potentially safer alternative. It has been found not to reduce scaffold biocompatibility on preosteoblasts cell-line in vitro [142]. By properly optimizing the PPF material composition, it is possible to create a drug-releasing bioscaffold with the strength and stiffness of native bone [143].

Selective Laser Sintering: Selective laser sintering is advantageous due to its ability to quickly and precisely print complex structures without the need for additional support. The printed elements are supported by powder, which also serves as a printing material. While there is limited research on the use of SLS as a source of scaffolds in medical applications, some materials have been described in this field. Borate-based bioactive glass was used to create bone-mimicking scaffolds that efficiently regenerate bone defects. A powder polymeric binder was used to merge the components, and after the fabrication process, the parts were heated to remove the binder [144]. One type of polymeric binder that can be used is thermoplastic polyurethane (TPU), which is also a suitable scaffold material. TPU is an elastic material with cytocompatible properties that could be used to mimic soft tissues. This enhances biological properties, maintains cell viability, promotes cell proliferation, and differentiates neural stem cells [145].

5. Biomimetic Scaffolds—From Advanced Engineering to Biological Application

Biocomposite structures consist of macromolecules, such as proteins, lipids, polysaccharides, minerals, and polynucleotides, that are naturally present in tissues. The extracellular matrix (ECM) contains various types of proteins, including collagen, elastin, gelatin, and other glycoproteins. Collagen, which is abundant in connective tissues, forms a fibrillary structure that provides strength and structural support. The collagen fibers in the bone tissue are saturated with minerals, which increases their stiffness and mechanical strength. Bioactive domains in collagen are involved in interacting with cell membrane receptors, such as integrins (i.e., $\alpha1\beta1$, $\alpha2\beta1$, $\alpha10\beta1$ and $\alpha11\beta1$). Collagen-binding integrins influence fibroblast activity, regulating differentiation and synthesis of ECM components, which is crucial in tissue wound healing [146]. The presence of elastin in tissues provides elasticity to various organs, such as bladder and artery tissues. The mechanical properties of blood vessels, fibrous connective tissue, and skin are mainly determined by the cooperation of collagen and elastin percentage composition, which provides strength and structural support while also providing elasticity and resilience. Another crucial protein is fibronectin, which has domains that bind to other ECM proteins, such as collagen, heparin, and integrins. Interactions between cells and the extracellular matrix (ECM) play a crucial role in ECM development, homeostasis, and wound healing [147]. As extracting ECM proteins can be expensive, gelatin has been proposed as a cost-effective alternative scaffold base [148]. Polysaccharides, such as chitosan, alginate, dextran, and hyaluronic acid, have advanced scaffold development due to their low cost, ease of commercialization, biocompatibility, and biodegradability. They are similar to the extracellular matrix (ECM), which is rich in glycosaminoglycans, glycoproteins, and glycolipids [149–152]. Tissue engineering scaffolds are generally produced as pre-fabricated or in situ cross-linked hydrogels, with many using 3D printing technology. Pre-fabricated scaffolds primarily focus on the presence of pores and interconnected channels to enhance the viability of seeded cells. Figure 4 present 2 possible concepts of biomimetic scaffolds fabrication that employ various techniques, including traditional methods like decellularization but also electrospinning, rapid prototyping-based microfabrication, and modular hierarchical assembling. Each method provides different characteristics to porous scaffolds. These include degradability to nontoxic materials, production of ultrafine fibers with varying diameters, production of final 2D structures, and production of large and complex 3D structures based on 3D programmed images. In situ gelled hydrogel involves creating biomimetic hydrogel scaffolds using peptide-based biomaterials to repair tissue, as well as controlled drug delivery. Proteins possess molecular properties that enable them to interact with other macromolecules and regulate the hard and soft tissue of an organism.

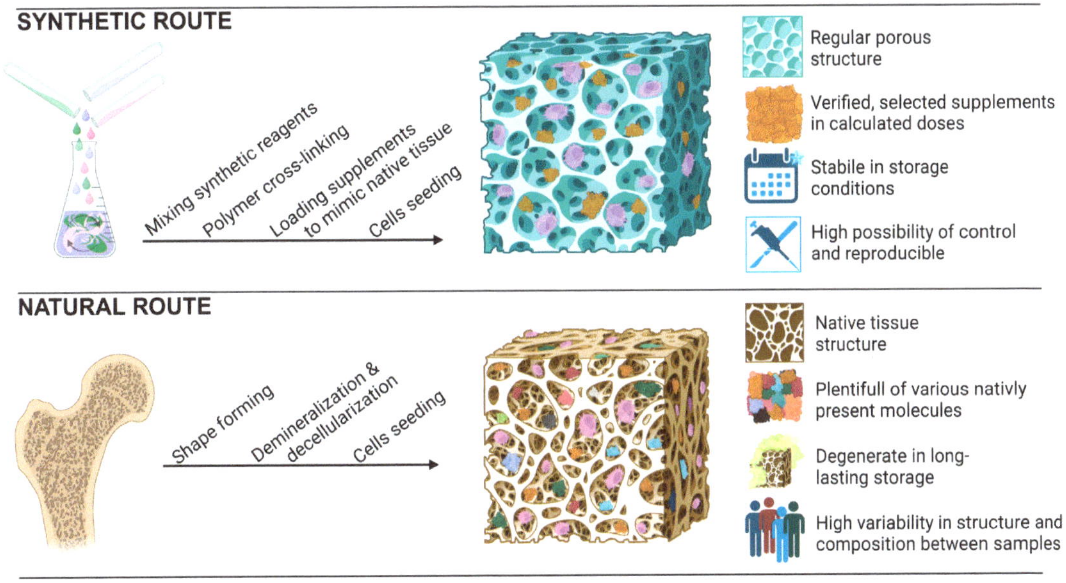

Figure 4. Synthetic and natural route of proceeding biomimetic scaffolds. Created with BioRender.com.

The scaffold is a tool that creates space for tissue formation de novo. It mimics the natural environment to provide optimal conditions for cell growth. The properties of the culturing environments vary depending on the cultured tissue, not only with the material and composition of supplements but also with the mechanical aspects. Tissue engineering employs engineering and life science to generate biological structures to restore, repair, and maintain organ tissue and improve its function. Biological scaffolds can be made from native tissues or synthetically synthesized and can be either degradable or nondegradable [153]. The optimization of scaffold production relies on adjusting the material's biological and chemical properties, such as biocompatibility, immunogenicity, and impact on cell metabolism [154]. The mechanical properties of the bioscaffold are equally important. The adjustment of material cross-linking density enables the regulation of porosity and resistance. A material with a denser cross-linking exhibits greater strength, tear resistance, and stiffness, which may be crucial for supporting connective tissue but can reduce cell viability [155]. Biomechanical resistance testing involves determining ultimate strain and stress and the energy required for failure, particularly under tensile forces [156]. Additionally, scaffolds have been utilized as controlled release tools to maintain therapeutic concentrations of diffusible tissue inductive factors. Loading a bioscaffold with the long-term released proangiogenic factor, leonurine hydrochloride

(LH), can increase the osteogenic differentiation of bone marrow stem cells (BMSC) and promote vessel formation in vivo [157]. The long-term release of the anti-bacterial polyhexamethylenebiguanide (PHMB) efficiently reduces bacterial content in regenerating skin wounds [158]. Cell culture is a primary in vitro biological tool and a crucial component of transgenesis, bioengineering, and regenerative medicine. The extracellular matrix (ECM) plays a critical role in tissue structure and function. For example, tendons contain thick bundles of collagen type 1, which are responsible for their high tensile strength. Collagen and elastin fibrils are responsible for the elasticity of the skin. In terms of tissue function and activity, there are examples such as the Arg-Gly-Asp (RGD) sequence on fibronectin that triggers a binding event or heparin sulfate proteoglycans that facilitate basic fibroblast growth factor (bFGF) activities. ECM provides a degradable environment that is crucial for angiogenic sprouting and remodeling during tissue dynamics, including morphogenesis, homeostasis, and wound healing [159,160]. Embryonic ECM is replaced during morphogenesis to accommodate tissue growth, with an ECM half-life of 7 to 10 h [161]. After tissue development is complete, the physiological degradation of the ECM shifts to maintaining ECM homeostasis. Both morphogenesis and homeostasis strictly involve the p53/laminin pathway [162]. The p53 protein responds specifically to laminin, a key component of the ECM, and regulates the expression of molecules required to establish homeostatic form and dynamics [163]. The production of nitric oxide (NO), an important intercellular signaling molecule [164], also supports another link in the pathway.

Biomimetic scaffolds of the extracellular matrix (ECM) are designed to provide a framework for cell culture to develop tissue with appropriate signal cross-talk. Intercellular communication in 3D co-cultures is more efficient with the involvement of paracrine signaling [165]. An ideal biomimetic scaffold should imitate natural ECM properties and create conditions for positive interaction with cells to increase cell adhesion, growth, migration, and differentiation [166]. Decellularized extracellular matrix (dCEM) scaffolds contain extracellular macromolecules, including collagen, elastin, fibronectin, laminin, and matricellular proteins [158]. These scaffolds are prepared through a decellularization process, which preserves signals and biological performance, providing a 3D biological support structure for subsequent cell seeding on damaged organs. In the model of mice myocardial infarction, a bioengineered cardiac patch was created using decellularized tissue with seeded cells. This approach demonstrated more efficient tissue regeneration compared to using only a scaffold or cell-seeding therapy [167]. Additionally, decellularized grafts were found to provide proper reconstruction of rats' small intestine, with a well-organized structure and intact nervous system in vivo [168]. Studies have shown that dECM scaffolds can create a favourable microenvironment that promotes tissue regeneration. They provide a tissue-specific template for the healing and functional regeneration of various tissues, such as skin [169], cartilage [160], or dense regular connective tissue [170]. During tissue regeneration, fibroblasts and other surrounding cells recreate the tissue environment. It would be advantageous to degrade the used material, making space for newly formed tissue and avoiding the need for surgical removal.

The future of biomimetic scaffolds involves the development of nanobiotechnology or nanotechnology to assemble and control the function of proteins [171]. Molecular biomimetics provides solutions for the control and fabrication of large-scale nanostructures to assemble materials in two and three dimensions. The concept of forming 4D scaffolds relies on adding the dimension of time in 3D biomaterial processing. Under the influence of external factors, such as heat or moisture, 4D biomaterials change their properties and shape in accordance with the controlled treatment process. Bioinspired 4D objects with multiple activities have been introduced as a new approach to producing highly complex smart materials [172]. In drug delivery, small-sized lipid nanoparticles (LNPs) have potential, but their poor stability and intracellular trafficking weaken their effectiveness. The potential of creating hydrophobic scaffolds can be increased by using transformed, self-assembled LNP containing lipids and proteins that have been previously used for gene or drug delivery [173]. The choice of scaffold depends on the specific biological application and intended usage, which can range from biotechnology to regenerative medicine for organ regeneration.

6. Application of Biomimetic Scaffolds in Biotechnology

Stem cells are a fundamental tool for tissue engineering and regenerative medicine due to their pluripotency and differentiation ability. The potential of stem cells is limited by their source, with the fertilized oocyte being the cell with the widest potency, giving rise to cells from all germ layers. Embryonic stem cells (ESCs) and induced pluripotent stem cells (iPSCs) are used in tissue engineering and are influenced by their environment [153]. ESCs are more effective than adult stem cells (ASCs) and can generate any organ through tissue-specific differentiation conditions in vitro. However, ASCs and tissue-specific cells are limited in number, making it difficult to grow a complete organ. The primary sources of stem cells for tissue engineering are bone marrow, umbilical cord blood, and circulating blood [174,175]. Recent stem cell research has shown the potential for nuclear transfer of somatic cells and reprogramming, which can convert fully or partially differentiated cells back to their embryonic state, restoring their stemness [176]. This technology has the advantage of using autologous cells, which helps to overcome issues with immune rejection. The extracellular matrix (ECM) plays a crucial role in determining stem cell fate. The chemical composition, surface chemistry, porosity, degradation behavior, and mechanical strength of a scaffold can all influence the fate of stem cells [177]. Both natural and synthetic biomaterials can create bioactive scaffolds that regulate stem cell differentiation [178]. Another biotool used in biotechnology is gene transfer, which involves managing RNA, DNA, proteins, and other macromolecules, such as lipids, for various purposes. There are different techniques of gene delivery for different biotechnological purposes, including viral and nonviral techniques. Nonviral techniques, such as transfection, can be performed through methods such as injection of naked DNA, electroporation, particle bombardment, and cationic liposomes. Contemporary viral gene transfer occurs in vivo using retroviruses, adenoviruses, or adeno-associated viruses. An alternative strategy involves incorporating DNA directly into a polymeric scaffold or genetically manipulated cells. This technique can protect DNA from degradation, offer better control of transgene and protein levels, and potentially reduce inflammation. In tissue engineering, transfections are utilized to deliver drugs or produce growth factors that can reduce the need for recombinant proteins, thus supporting tissue regeneration [179]. Scaffolds provide a more efficient method for gene delivery to specific cell populations while protecting vectors against extracellular barriers that may reduce their therapeutic efficacy and immune responses. Scaffolds can enhance cell adhesion, maintain cell–cell interaction, and increase protein expression levels [180]. They also offer the potential to alter cell function and fate. Hydrophobic scaffolds have been utilized for drug and gene delivery to treat diseases and develop DNA vaccines due to their increased adsorption capacity for proteins and other supplements [181]. Biopolymers can be used as a carrier for DNA vaccines at the organism scale. Plasmid DNA can be incorporated or surface-adsorbed onto carriers to create a complex that can be encapsulated in layers of other polymers or incorporated into multistructured polymer forms or matrices [182].

7. Application of Biomimetic Scaffolds in Advanced Regenerative and Reconstructive Medicine

For decades, scientists have recognized that the human body provides biological scaffolds and biopolymers that are highly attractive for tissue engineering and regenerative medicine. The human body has the ability to regenerate most organs, so damage to major organs can cause failure and serious disorders. Regenerative medicine combines tissue engineering with cell-based therapy, gene therapy, and immunomodulation. One potential use of dCEM scaffolds is for organ transplantation and tissue repair to regenerate and restore organ function [158]. Primary or genetically engineered cells can be transplanted by seeding them onto tissue-like three-dimensional structures to repair tissue damage. The traditional method of injecting cell solutions into tissues by hand encounters many difficulties, including high mortality rates for implanted cells. Biomimetic scaffolds consist of extracellular matrix (ECM) proteins and growth factors [148], which facilitate the transportation of high-density cells to affected tissues. Tissue engineering, a subfield of regenerative medicine, has the potential to regenerate almost all tissues and organs in the human body. Tissue engineering strategies generally involve implanting a construct into an organism, and delivering growth factors or supplements to enhance cell viability, proliferation, and adhesion to surrounding tissues [160]. Biomimetic materials are often used to replace unhealthy tissue in the body, while sutures are necessary for certain types of wounds. Fabrication methods can improve the production of customizable defect-fillable scaffolds for tissue regeneration in regenerative medicine [130]. Decellularized tissue can effectively replace native tissues as a biomimicking wound dressing [166]. Additionally, sutures for wound sewing can be manufactured with lyophilized decellularized ligament to biomimic the surrounding tissues. The spinning of fibers enabled the creation of thread that is as durable as silk fibers, which has potential use in wound sewing [105]. Biomimetic scaffolds have numerous applications in medicine, including artificial skin, arteries, and joint replacements in the body [183,184].

8. Conclusions and Future Perspectives

Considering the current state of knowledge regarding biomimetic scaffolds, it is difficult to overestimate their potential future applications. They have enhanced traditional 3D cell culture techniques by adding another dimension of modifiable parameters to already complex structures. Further advancement in the field of biomimetic scaffolds may be the branch of tissue engineering that will bring to light concepts like tailor-made organ implants that eliminate the risk of transplant rejection complications, laboratory modelling of patient-specific tumors, and the cultivation of healthy cells in an environment that closely mimics natural conditions. The potential applications of biomimetic scaffolds in regenerative, translational, and personalized medicine are potentially very vast, but due to high complexity of the constructs further extensive studies are still required.

Author Contributions: Conceptualization, T.G., J.L., B.K., M.Z., P.M. and P.A.; investigation, M.F., D.B. and K.R.; resources, A.G.-N. and M.F.; data curation, B.K., P.A. and P.M.; writing—original draft preparation, T.G., J.L., J.N., K.D., D.D. and A.G.-N.; writing—review and editing, B.K., P.M., P.A., M.Z., P.D. and M.P.-O.; visualization, K.D.; supervision: B.K., P.D. and P.A. All authors have read and agreed to the published version of the manuscript.

Funding: This research was funded in part by the USDA Animal Health Project NC 07082.

Data Availability Statement: No new data were created or analyzed in this study.

Conflicts of Interest: The authors declare no conflicts of interest.

References

1. Bertsch, C.; Maréchal, H.; Gribova, V.; Lévy, B.; Debry, C.; Lavalle, P.; Fath, L. Biomimetic Bilayered Scaffolds for Tissue Engineering: From Current Design Strategies to Medical Applications. *Adv. Health Mater.* **2023**, *12*, e2203115. [CrossRef]
2. Naik, R.R.; Singamaneni, S. Introduction: Bioinspired and Biomimetic Materials. *Chem. Rev.* **2017**, *117*, 12581–12583. [CrossRef]
3. Badekila, A.K.; Kini, S.; Jaiswal, A.K. Fabrication techniques of biomimetic scaffolds in three-dimensional cell culture: A review. *J. Cell. Physiol.* **2021**, *236*, 741–762. [CrossRef] [PubMed]
4. Huang, G.; Li, F.; Zhao, X.; Ma, Y.; Li, Y.; Lin, M.; Jin, G.; Lu, T.J.; Genin, G.M.; Xu, F. Functional and Biomimetic Materials for Engineering of the Three-Dimensional Cell Microenvironment. *Chem. Rev.* **2017**, *117*, 12764–12850. [CrossRef] [PubMed]
5. Zhang, Y.; Zhang, C.; Li, Y.; Zhou, L.; Dan, N.; Min, J.; Chen, Y.; Wang, Y. Evolution of biomimetic ECM scaffolds from decellularized tissue matrix for tissue engineering: A comprehensive review. *Int. J. Biol. Macromol.* **2023**, *246*, 125672. [CrossRef]
6. Vinatier, C.; Guicheux, J.; Daculsi, G.; Layrolle, P.; Weiss, P. Cartilage and bone tissue engineering using hydrogels. *Biomed. Mater. Eng.* **2006**, *16* (Suppl. 4), S107–S113.
7. Flégeau, K.; Pace, R.; Gautier, H.; Rethore, G.; Guicheux, J.; Le Visage, C.; Weiss, P. Toward the development of biomimetic injectable and macroporous biohydrogels for regenerative medicine. *Adv. Colloid Interface Sci.* **2017**, *247*, 589–609. [CrossRef] [PubMed]
8. Shin, H.; Jo, S.; Mikos, A.G. Biomimetic materials for tissue engineering. *Biomaterials* **2003**, *24*, 4353–4364. [CrossRef] [PubMed]

9. Mridha, A.R.; Dargaville, T.R.; Dalton, P.D.; Carroll, L.; Morris, M.B.; Vaithilingam, V.; Tuch, B.E. Prevascularized Retrievable Hybrid Implant to Enhance Function of Subcutaneous Encapsulated Islets. *Tissue Eng. Part A* **2022**, *28*, 212–224. [CrossRef] [PubMed]
10. Ogundipe, V.M.; Plukker, J.T.; Links, T.P.; Coppes, R.P. Thyroid Gland Organoids: Current Models and Insights for Application in Tissue Engineering. *Tissue Eng. Part A* **2022**, *28*, 500–510. [CrossRef]
11. A Salg, G.; Poisel, E.; Neulinger-Munoz, M.; Gerhardus, J.; Cebulla, D.; Bludszuweit-Philipp, C.; Vieira, V.; Nickel, F.; Herr, I.; Blaeser, A.; et al. Toward 3D-bioprinting of an endocrine pancreas: A building-block concept for bioartificial insulin-secreting tissue. *J. Tissue Eng.* **2022**, *13*, 20417314221091033. [CrossRef] [PubMed]
12. Gholami, K.; Solhjoo, S.; Aghamir, S.M.K. Application of Tissue-Specific Extracellular Matrix in Tissue Engineering: Focus on Male Fertility Preservation. *Reprod. Sci.* **2022**, *29*, 3091–3099. [CrossRef] [PubMed]
13. Basurto, I.M.; Passipieri, J.A.; Gardner, G.M.; Smith, K.K.; Amacher, A.R.; Hansrisuk, A.I.; Christ, G.J.; Caliari, S.R. Photoreactive Hydrogel Stiffness Influences Volumetric Muscle Loss Repair. *Tissue Eng. Part A* **2022**, *28*, 312–329. [CrossRef]
14. Takahashi, H.; Oikawa, F.; Takeda, N.; Shimizu, T. Contraction Control of Aligned Myofiber Sheet Tissue by Parallel Oriented Induced Pluripotent Stem Cell-Derived Neurons. *Tissue Eng. Part A* **2022**, *28*, 661–671. [CrossRef] [PubMed]
15. Liu, X.; Zhang, F.; Li, L.; He, Y.; Dong, Y. Reconstruction of Epidural Fat to Prevent Epidural Fibrosis After Laminectomy in Rabbits. *Tissue Eng. Part A* **2022**, *28*, 366–372. [CrossRef]
16. Uemoto, Y.; Taura, K.; Nakamura, D.; Xuefeng, L.; Nam, N.H.; Kimura, Y.; Yoshino, K.; Fuji, H.; Yoh, T.; Nishio, T.; et al. Bile Duct Regeneration with an Artificial Bile Duct Made of Gelatin Hydrogel Nonwoven Fabrics. *Tissue Eng. Part A* **2022**, *28*, 737–748. [CrossRef]
17. Xiang, Y.; Wang, W.; Gao, Y.; Zhang, J.; Zhang, J.; Bai, Z.; Zhang, S.; Yang, Y. Production and Characterization of an Integrated Multi-Layer 3D Printed PLGA/GelMA Scaffold Aimed for Bile Duct Restoration and Detection. *Front. Bioeng. Biotechnol.* **2020**, *8*, 971. [CrossRef]
18. Knox, C.; Garcia, K.; Tran, J.; Wilson, S.M.; Blood, A.B.; Kearns-Jonker, M.; Martens, T.P. A Biomimetic Approach Utilizing Pulsatile Perfusion Generates Contractile Vascular Grafts. *Tissue Eng. Part A* **2023**, *29*, 358–371. [CrossRef]
19. Szöőr, Á.; Ujlaky-Nagy, L.; Tóth, G.; Szöllősi, J.; Vereb, G. Cell confluence induces switching from proliferation to migratory signaling by site-selective phosphorylation of PDGF receptors on lipid raft platforms. *Cell. Signal.* **2016**, *28*, 81–93. [CrossRef]
20. Hamburger, A.W.; Salmon, S.E. Primary bioassay of human tumor stem cells. *Science* **1977**, *197*, 461–463. [CrossRef]
21. Jensen, C.; Teng, Y. Is It Time to Start Transitioning From 2D to 3D Cell Culture? *Front. Mol. Biosci.* **2020**, *7*, 33. [CrossRef]
22. Langhans, S.A. Three-Dimensional in Vitro Cell Culture Models in Drug Discovery and Drug Repositioning. *Front. Pharmacol.* **2018**, *9*, 6. [CrossRef] [PubMed]
23. Kapałczyńska, M.; Kolenda, T.; Przybyła, W.; Zajączkowska, M.; Teresiak, A.; Filas, V.; Ibbs, M.; Bliźniak, R.; Łuczewski, Ł.; Lamperska, K. 2D and 3D cell cultures—A comparison of different types of cancer cell cultures. *Arch. Med. Sci.* **2018**, *14*, 910–919. [CrossRef]
24. Biju, T.S.; Priya, V.V.; Francis, A.P. Role of three-dimensional cell culture in therapeutics and diagnostics: An updated review. *Drug Deliv. Transl. Res.* **2023**, *13*, 2239–2253. [CrossRef] [PubMed]
25. Abbott, A. Biology's new dimension. *Nature* **2003**, *424*, 870–872. [CrossRef] [PubMed]
26. Habanjar, O.; Diab-Assaf, M.; Caldefie-Chezet, F.; Delort, L. 3D Cell Culture Systems: Tumor Application, Advantages, and Disadvantages. *Int. J. Mol. Sci.* **2021**, *22*, 12200. [CrossRef]
27. Anthon, S.G.; Valente, K.P. Vascularization Strategies in 3D Cell Culture Models: From Scaffold-Free Models to 3D Bioprinting. *Int. J. Mol. Sci.* **2022**, *23*, 14582. [CrossRef]
28. Sośniak, J.; Opiela, J. 3D Cell Culture Technology—A New Insight Into Research—A Review. *Ann. Anim. Sci.* **2021**, *21*, 1257–1273. [CrossRef]
29. Fontoura, J.C.; Viezzer, C.; dos Santos, F.G.; Ligabue, R.A.; Weinlich, R.; Puga, R.D.; Antonow, D.; Severino, P.; Bonorino, C. Comparison of 2D and 3D cell culture models for cell growth, gene expression and drug resistance. *Mater. Sci. Eng. C Mater. Biol. Appl.* **2020**, *107*, 110264. [CrossRef]
30. Sakalem, M.E.; De Sibio, M.T.; da Costa, F.A.d.S.; de Oliveira, M. Historical evolution of spheroids and organoids, and possibilities of use in life sciences and medicine. *Biotechnol. J.* **2021**, *16*, e2000463. [CrossRef]
31. Arjmand, B.; Rabbani, Z.; Soveyzi, F.; Tayanloo-Beik, A.; Rezaei-Tavirani, M.; Biglar, M.; Adibi, H.; Larijani, B. Advancement of Organoid Technology in Regenerative Medicine. *Regen. Eng. Transl. Med.* **2023**, *9*, 83–96. [CrossRef]
32. Cui, X.; Hartanto, Y.; Zhang, H. Advances in multicellular spheroids formation. *J. R. Soc. Interface* **2017**, *14*, 20160877. [CrossRef] [PubMed]
33. Białkowska, K.; Komorowski, P.; Bryszewska, M.; Miłowska, K. Spheroids as a Type of Three-Dimensional Cell Cultures—Examples of Methods of Preparation and the Most Important Application. *Int. J. Mol. Sci.* **2020**, *21*, 6225. [CrossRef]
34. Liu, D.; Chen, S.; Naing, M.W. A review of manufacturing capabilities of cell spheroid generation technologies and future development. *Biotechnol. Bioeng.* **2021**, *118*, 542–554. [CrossRef]
35. Sant, S.; Johnston, P.A. The production of 3D tumor spheroids for cancer drug discovery. *Drug Discov. Today Technol.* **2017**, *23*, 27–36. [CrossRef]
36. Shyam, R.; Reddy, L.V.K.; Palaniappan, A. Fabrication and Characterization Techniques of In Vitro 3D Tissue Models. *Int. J. Mol. Sci.* **2023**, *24*, 1912. [CrossRef]

37. Madhavan, M.; Jaiswal, D.; Karlberg, S.; Duggan, A.; Almarshad, H.A.; Claffey, K.P.; Hoshino, K. Electron microscopy imaging and mechanical characterization of T47D multicellular tumor spheroids–Older spheroids reduce interstitial space and become stiffer. *PLoS ONE* **2023**, *18*, e0286291. [CrossRef] [PubMed]
38. Tosca, E.M.; Ronchi, D.; Facciolo, D.; Magni, P. Replacement, Reduction, and Refinement of Animal Experiments in Anticancer Drug Development: The Contribution of 3D In Vitro Cancer Models in the Drug Efficacy Assessment. *Biomedicines* **2023**, *11*, 1058. [CrossRef] [PubMed]
39. Yau, J.N.N.; Adriani, G. Three-dimensional heterotypic colorectal cancer spheroid models for evaluation of drug response. *Front. Oncol.* **2023**, *13*, 1148930. [CrossRef]
40. Zhang, Y.; Huo, J.; Yu, S.; Feng, W.; Tuersun, A.; Chen, F.; Lv, Z.; Liu, W.; Zhao, J.; Xu, Z.; et al. Colorectal cancer tissue-originated spheroids reveal tumor intrinsic signaling pathways and mimic patient clinical chemotherapeutic response as a rapid and valid model. *Biomed. Pharmacother.* **2023**, *167*, 115585. [CrossRef]
41. El-Sadek, I.A.; Shen, L.T.-W.; Mori, T.; Makita, S.; Mukherjee, P.; Lichtenegger, A.; Matsusaka, S.; Yasuno, Y. Label-free drug response evaluation of human derived tumor spheroids using three-dimensional dynamic optical coherence tomography. *Sci. Rep.* **2023**, *13*, 15377. [CrossRef] [PubMed]
42. Răileanu, M.; Bacalum, M. Cancer Wars: Revenge of the AMPs (Antimicrobial Peptides), a New Strategy against Colorectal Cancer. *Toxins* **2023**, *15*, 459. [CrossRef] [PubMed]
43. Domingues, M.; Pereira, C.L.; Sarmento, B.; Castro, F. Mimicking 3D breast tumor-stromal interactions to screen novel cancer therapeutics. *Eur. J. Pharm. Sci.* **2023**, *190*, 106560. [CrossRef] [PubMed]
44. Khan, A.H.; Zhou, S.P.; Moe, M.; Quesada, B.A.O.; Bajgiran, K.R.; Lassiter, H.R.; Dorman, J.A.; Martin, E.C.; Pojman, J.A.; Melvin, A.T. Generation of 3D Spheroids Using a Thiol–Acrylate Hydrogel Scaffold to Study Endocrine Response in ER^+ Breast Cancer. *ACS Biomater. Sci. Eng.* **2022**, *8*, 3977–3985. [CrossRef] [PubMed]
45. Klingseisen, V.; Slanovc, J.; Regouc, M.; Hrzenjak, A. Bisdemethoxycurcumin sensitizes the response of cisplatin resistant non-small cell lung carcinoma cell lines by activating apoptosis and autophagy. *J. Nutr. Biochem.* **2022**, *106*, 109003. [CrossRef]
46. Huang, M.; Hou, W.; Zhang, J.; Li, M.; Zhang, Z.; Li, X.; Chen, Z.; Wang, C.; Yang, L. Evaluation of AMG510 Therapy on *KRAS*-Mutant Non–Small Cell Lung Cancer and Colorectal Cancer Cell Using a 3D Invasive Tumor Spheroid System under Normoxia and Hypoxia. *Bioengineering* **2022**, *9*, 792. [CrossRef]
47. Świerczewska, M.; Sterzyńska, K.; Ruciński, M.; Andrzejewska, M.; Nowicki, M.; Januchowski, R. The response and resistance to drugs in ovarian cancer cell lines in 2D monolayers and 3D spheroids. *Biomed. Pharmacother.* **2023**, *165*, 115152. [CrossRef]
48. O'brien, S.; Butticello, M.; Thompson, C.; Wilson, B.; Wyce, A.; Mahajan, V.; Kruger, R.; Mohammad, H.; Fedoriw, A. Inhibiting PRMT5 induces DNA damage and increases anti-proliferative activity of Niraparib, a PARP inhibitor, in models of breast and ovarian cancer. *BMC Cancer* **2023**, *23*, 775. [CrossRef]
49. Fevre, R.; Mary, G.; Vertti-Quintero, N.; Durand, A.; Tomasi, R.F.-X.; Del Nery, E.; Baroud, C.N. Combinatorial drug screening on 3D Ewing sarcoma spheroids using droplet-based microfluidics. *iScience* **2023**, *26*, 106651. [CrossRef] [PubMed]
50. Moskovits, N.; Itzhaki, E.; Tarasenko, N.; Chausky, E.; Bareket-Samish, A.; Kaufman, A.; Meerson, R.; Stemmer, S.M. Establishing 3-Dimensional Spheroids from Patient-Derived Tumor Samples and Evaluating their Sensitivity to Drugs. *J. Vis. Exp.* **2022**, *190*, e64564. [CrossRef]
51. Zhang, W.; Kyritsi, K.; Isidan, A.; Park, Y.; Li, P.; Cross-Najafi, A.A.; Lopez, K.; Kennedy, L.; Sato, K.; Glaser, S.; et al. Development of Scaffold-Free Three-Dimensional Cholangiocyte Organoids to Study the Progression of Primary Sclerosing Cholangitis. *Am. J. Pathol.* **2023**, *193*, 1156–1169. [CrossRef]
52. Yang, Q.; Li, M.; Yang, X.; Xiao, Z.; Tong, X.; Tuerdi, A.; Li, S.; Lei, L. Flourishing tumor organoids: History, emerging technology, and application. *Bioeng. Transl. Med.* **2023**, *8*, e10559. [CrossRef] [PubMed]
53. Xu, Z.-Y.; Huang, J.-J.; Liu, Y.; Zhao, Y.; Wu, X.-W.; Ren, J.-A. Current knowledge on the multiform reconstitution of intestinal stem cell niche. *World J. Stem Cells* **2021**, *13*, 1564–1579. [CrossRef] [PubMed]
54. Kim, H.; Kamm, R.D.; Vunjak-Novakovic, G.; Wu, J.C. Progress in multicellular human cardiac organoids for clinical applications. *Cell Stem Cell* **2022**, *29*, 503–514. [CrossRef] [PubMed]
55. Li, J.; Chu, J.; Lui, V.C.H.; Chen, S.; Chen, Y.; Tam, P.K.H. Bioengineering Liver Organoids for Diseases Modelling and Transplantation. *Bioengineering* **2022**, *9*, 796. [CrossRef] [PubMed]
56. Ma, X.; Wang, Q.; Li, G.; Li, H.; Xu, S.; Pang, D. Cancer organoids: A platform in basic and translational research. *Genes Dis.* **2023**, *11*, 614–632. [CrossRef] [PubMed]
57. Chan, W.S.; Mo, X.; Ip, P.P.C.; Tse, K.Y. Patient-derived organoid culture in epithelial ovarian cancers—Techniques, applications, and future perspectives. *Cancer Med.* **2023**, *12*, 19714–19731. [CrossRef] [PubMed]
58. Li, Y.; Gao, X.; Ni, C.; Zhao, B.; Cheng, X. The application of patient-derived organoid in the research of lung cancer. *Cell. Oncol.* **2023**, *46*, 503–519. [CrossRef]
59. Nizamoglu, M.; Joglekar, M.M.; Almeida, C.R.; Callerfelt, A.-K.L.; Dupin, I.; Guenat, O.T.; Henrot, P.; van Os, L.; Otero, J.; Elowsson, L.; et al. Innovative three-dimensional models for understanding mechanisms underlying lung diseases: Powerful tools for translational research. *Eur. Respir. Rev.* **2023**, *32*, 230042. [CrossRef]
60. Septiana, W.L.; Noviantari, A.; Antarianto, R.D. Induced Pluripotent Stem Cells (Ipscs) Based Liver Organoid: The Benefits and Challenges. *Cell. Physiol. Biochem.* **2023**, *57*, 345–359. [CrossRef]

61. Wright, C.W.; Li, N.; Shaffer, L.; Hill, A.; Boyer, N.; Alves, S.E.; Venkataraman, S.; Biswas, K.; Lieberman, L.A.; Mohammadi, S. Establishment of a 96-well transwell system using primary human gut organoids to capture multiple quantitative pathway readouts. *Sci. Rep.* **2023**, *13*, 16357. [CrossRef]
62. Zhang, X.P.; Wang, X.Y.; Wang, S.N.; Miao, C.Y. The generation and properties of human cortical organoids as a disease model for malformations of cortical development. *Neural Regen. Res.* **2023**, *18*, 2119–2126. [CrossRef]
63. Yan, Z.; Zhang, T.; Wang, Y.; Xiao, S.; Gao, J. Extracellular vesicle biopotentiated hydrogels for diabetic wound healing: The art of living nanomaterials combined with soft scaffolds. *Mater. Today Bio* **2023**, *23*, 100810. [CrossRef]
64. Gan, Z.; Qin, X.; Liu, H.; Liu, J.; Qin, J. Recent advances in defined hydrogels in organoid research. *Bioact. Mater.* **2023**, *28*, 386–401. [CrossRef]
65. Willemse, J.; van der Laan, L.J.W.; de Jonge, J.; Verstegen, M.M.A. Design by Nature: Emerging Applications of Native Liver Extracellular Matrix for Cholangiocyte Organoid-Based Regenerative Medicine. *Bioengineering* **2022**, *9*, 110. [CrossRef]
66. Ho, T.-C.; Chang, C.-C.; Chan, H.-P.; Chung, T.-W.; Shu, C.-W.; Chuang, K.-P.; Duh, T.-H.; Yang, M.-H.; Tyan, Y.-C. Hydrogels: Properties and Applications in Biomedicine. *Molecules* **2022**, *27*, 2902. [CrossRef]
67. Deptuła, M.; Zawrzykraj, M.; Sawicka, J.; Banach-Kopeć, A.; Tylingo, R.; Pikuła, M. Application of 3D- printed hydrogels in wound healing and regenerative medicine. *Biomed. Pharmacother.* **2023**, *167*, 115416. [CrossRef] [PubMed]
68. Kane, K.I.W.; Moreno, E.L.; Lehr, C.M.; Hachi, S.; Dannert, R.; Sanctuary, R.; Wagner, C.; Fleming, R.M.T.; Baller, J. Determination of the rheological properties of Matrigel for optimum seeding conditions in microfluidic cell cultures. *AIP Adv.* **2018**, *8*, 125332. [CrossRef]
69. Tang, C.; Zhou, K.; Zhu, Y.; Zhang, W.; Xie, Y.; Wang, Z.; Zhou, H.; Yang, T.; Zhang, Q.; Xu, B. Collagen and its derivatives: From structure and properties to their applications in food industry. *Food Hydrocoll.* **2022**, *131*, 107748. [CrossRef]
70. Salahuddin, B.; Wang, S.; Sangian, D.; Aziz, S.; Gu, Q. Hybrid Gelatin Hydrogels in Nanomedicine Applications. *ACS Appl. Bio Mater.* **2021**, *4*, 2886–2906. [CrossRef] [PubMed]
71. Saravanakumar, K.; Park, S.; Santosh, S.S.; Ganeshalingam, A.; Thiripuranathar, G.; Sathiyaseelan, A.; Vijayasarathy, S.; Swaminathan, A.; Priya, V.V.; Wang, M.-H. Application of hyaluronic acid in tissue engineering, regenerative medicine, and nanomedicine: A review. *Int. J. Biol. Macromol.* **2022**, *222 Pt B*, 2744–2760. [CrossRef]
72. Roquero, D.M.; Katz, E. "Smart" alginate hydrogels in biosensing, bioactuation and biocomputing: State-of-the-art and perspectives. *Sensors Actuators Rep.* **2022**, *4*, 100095. [CrossRef]
73. Ashammakhi, N.; GhavamiNejad, A.; Tutar, R.; Fricker, A.; Roy, I.; Chatzistavrou, X.; Apu, E.H.; Nguyen, K.-L.; Ahsan, T.; Pountos, I.; et al. Highlights on Advancing Frontiers in Tissue Engineering. *Tissue Eng. Part B Rev.* **2022**, *28*, 633–664. [CrossRef]
74. Banerjee, D.; Singh, Y.P.; Datta, P.; Ozbolat, V.; O'Donnell, A.; Yeo, M.; Ozbolat, I.T. Strategies for 3D bioprinting of spheroids: A comprehensive review. *Biomaterials* **2022**, *291*, 121881. [CrossRef]
75. Jin, R.H.; Zhang, Z.Z.; Xu, P.Q.; Xia, S.Z.; Weng, T.T.; Zhu, Z.K.; Wang, X.G.; You, C.G.; Han, C.M. Effects of three-dimensional bioprinting antibacterial hydrogel on full-thickness skin defect wounds in rats. *Zhonghua Shao Shang Yu Chuang Mian Xiu Fu Za Zhi* **2023**, *39*, 165–174. (In Chinese) [CrossRef]
76. Agarwal, N.; Solanki, V.S.; Ameta, K.L.; Yadav, V.K.; Gupta, P.; Wanale, S.G.; Shrivastava, R.; Soni, A.; Sahoo, D.K.; Patel, A. 4-Dimensional printing: Exploring current and future capabilities in biomedical and healthcare systems—A Concise review. *Front. Bioeng. Biotechnol.* **2023**, *11*, 1251425. [CrossRef]
77. Vyas, J.; Shah, I.; Singh, S.; Prajapati, B.G. Biomaterials-based additive manufacturing for customized bioengineering in management of otolaryngology: A comprehensive review. *Front. Bioeng. Biotechnol.* **2023**, *11*, 1234340. [CrossRef] [PubMed]
78. Trenfield, S.J.; Awad, A.; Madla, C.M.; Hatton, G.B.; Firth, J.; Goyanes, A.; Gaisford, S.; Basit, A.W. Shaping the future: Recent advances of 3D printing in drug delivery and healthcare. *Expert Opin. Drug Deliv.* **2019**, *16*, 1081–1094. [CrossRef] [PubMed]
79. Li, H.; Go, G.; Ko, S.Y.; Park, J.-O.; Park, S. Magnetic actuated pH-responsive hydrogel-based soft micro-robot for targeted drug delivery. *Smart Mater. Struct.* **2016**, *25*, 027001. [CrossRef]
80. Ying, B.; Liu, X. Skin-like hydrogel devices for wearable sensing, soft robotics and beyond. *iScience* **2021**, *24*, 103174. [CrossRef] [PubMed]
81. Chen, A.; Wang, W.; Mao, Z.; He, Y.; Chen, S.; Liu, G.; Su, J.; Feng, P.; Shi, Y.; Yan, C.; et al. Multimaterial 3D and 4D Bioprinting of Heterogenous Constructs for Tissue Engineering. *Adv. Mater.* **2023**, e2307686. [CrossRef]
82. Loverde, J.R.; Piroli, M.; Klarmann, G.J.; Gaston, J.; Wickiser, J.K.; Barnhill, J.; Gilchrist, K.H.; Ho, V.B. Development of a bioreactor for in-vitro compression cycling of tissue engineered meniscal implants. *HardwareX* **2023**, *14*, e00433. [CrossRef] [PubMed]
83. Agarwal, T.; Chiesa, I.; Costantini, M.; Lopamarda, A.; Tirelli, M.C.; Borra, O.P.; Varshapally, S.V.S.; Kumar, Y.A.V.; Reddy, G.K.; De Maria, C.; et al. Chitosan and its derivatives in 3D/4D (bio) printing for tissue engineering and drug delivery applications. *Int. J. Biol. Macromol.* **2023**, *246*, 125669. [CrossRef] [PubMed]
84. Zhang, X.; Yang, Y.; Yang, Z.; Ma, R.; Aimaijiang, M.; Xu, J.; Zhang, Y.; Zhou, Y. Four-Dimensional Printing and Shape Memory Materials in Bone Tissue Engineering. *Int. J. Mol. Sci.* **2023**, *24*, 814. [CrossRef] [PubMed]
85. Kang, X.; Zhang, X.-B.; Gao, X.-D.; Hao, D.-J.; Li, T.; Xu, Z.-W. Bioprinting for bone tissue engineering. *Front. Bioeng. Biotechnol.* **2022**, *10*, 1036375. [CrossRef] [PubMed]
86. Gokdogan, O. Bone grafting in maxillofacial trauma. *Curr. Opin. Otolaryngol. Head Neck Surg.* **2022**, *30*, 260–264. [CrossRef] [PubMed]

87. Yazdanpanah, Z.; Johnston, J.D.; Cooper, D.M.L.; Chen, X. 3D Bioprinted Scaffolds for Bone Tissue Engineering: State-Of-The-Art and Emerging Technologies. *Front. Bioeng. Biotechnol.* **2022**, *10*, 824156. [CrossRef] [PubMed]
88. Chen, X.; Han, S.; Wu, W.; Wu, Z.; Yuan, Y.; Wu, J.; Liu, C. Harnessing 4D Printing Bioscaffolds for Advanced Orthopedics. *Small* **2022**, *18*, 2106824. [CrossRef]
89. Kabirian, F.; Mela, P.; Heying, R. 4D Printing Applications in the Development of Smart Cardiovascular Implants. *Front. Bioeng. Biotechnol.* **2022**, *10*, 873453. [CrossRef]
90. Aizarna-Lopetegui, U.; García-Astrain, C.; Renero-Lecuna, C.; González-Callejo, P.; Villaluenga, I.; del Pozo, M.A.; Sánchez-Álvarez, M.; Henriksen-Lacey, M.; de Aberasturi, D.J. Remodeling arteries: Studying the mechanical properties of 3D-bioprinted hybrid photoresponsive materials. *J. Mater. Chem. B* **2023**, *11*, 9431–9442. [CrossRef]
91. Zhou, Z.; Tang, W.; Yang, J.; Fan, C. Application of 4D printing and bioprinting in cardiovascular tissue engineering. *Biomater. Sci.* **2023**, *11*, 6403–6420. [CrossRef] [PubMed]
92. Zieliński, P.S.; Gudeti, P.K.R.; Rikmanspoel, T.; Włodarczyk-Biegun, M.K. 3D printing of bio-instructive materials: Toward directing the cell. *Bioact. Mater.* **2022**, *19*, 292–327. [CrossRef] [PubMed]
93. Deshpande, M.V.; West, A.J.; Bernacki, S.H.; Luan, K.; King, M.W. Poly(ε-Caprolactone) Resorbable Auxetic Designed Knitted Scaffolds for Craniofacial Skeletal Muscle Regeneration. *Bioengineering* **2020**, *7*, 134. [CrossRef]
94. Wang, Z.; Liang, X.; Wang, G.; Wang, X.; Chen, Y. Emerging Bioprinting for Wound Healing. *Adv. Mater.* **2023**, e2304738. [CrossRef] [PubMed]
95. Borse, K.; Shende, P. 3D-to-4D Structures: An Exploration in Biomedical Applications. *Aaps Pharmscitech* **2023**, *24*, 163. [CrossRef] [PubMed]
96. Douillet, C.; Nicodeme, M.; Hermant, L.; Bergeron, V.; Guillemot, F.; Fricain, J.-C.; Oliveira, H.; Garcia, M. From local to global matrix organization by fibroblasts: A 4D laser-assisted bioprinting approach. *Biofabrication* **2022**, *14*, 025006. [CrossRef] [PubMed]
97. Xie, N.; Shi, G.; Shen, Y.; Xu, Y.; Wang, H.; Feng, H.; Dai, K.; Wang, J.; Cao, Q. Research Progress of Robot Technology in In situ 3D Bioprinting. *Int. J. Bioprinting* **2022**, *8*, 215–222. [CrossRef] [PubMed]
98. Jongprasitkul, H.; Parihar, V.S.; Turunen, S.; Kellomäki, M. pH-Responsive Gallol-Functionalized Hyaluronic Acid-Based Tissue Adhesive Hydrogels for Injection and Three-Dimensional Bioprinting. *ACS Appl. Mater. Interfaces* **2023**, *15*, 33972–33984. [CrossRef]
99. Zhou, B.; Jiang, X.; Zhou, X.; Tan, W.; Luo, H.; Lei, S.; Yang, Y. GelMA-based bioactive hydrogel scaffolds with multiple bone defect repair functions: Therapeutic strategies and recent advances. *Biomater. Res.* **2023**, *27*, 86. [CrossRef]
100. Bhattacharyya, A.; Ham, H.-W.; Sonh, J.; Gunbayar, M.; Jeffy, R.; Nagarajan, R.; Khatun, M.R.; Noh, I. 3D bioprinting of complex tissue scaffolds with in situ homogeneously mixed alginate-chitosan-kaolin bioink using advanced portable biopen. *Carbohydr. Polym.* **2023**, *317*, 121046. [CrossRef]
101. Fani, N.; Peshkova, M.; Bikmulina, P.; Golroo, R.; Timashev, P.; Vosough, M. Fabricating the cartilage: Recent achievements. *Cytotechnology* **2023**, *75*, 269–292. [CrossRef]
102. Kiyotake, E.A.; Thomas, E.E.; Iribagiza, C.; Detamore, M.S. High-stiffness, fast-crosslinking, cartilage matrix bioinks. *J. Biomech.* **2023**, *148*, 111471. [CrossRef]
103. Hafeez, S.; Decarli, M.C.; Aldana, A.; Ebrahimi, M.; Ruiter, F.A.; Duimel, H.; van Blitterswijk, C.; Pitet, L.M.; Moroni, L.; Baker, M.B. In Situ Covalent Reinforcement of a Benzene-1,3,5-Tricarboxamide Supramolecular Polymer Enables Biomimetic, Tough, and Fibrous Hydrogels and Bioinks. *Adv. Mater.* **2023**, *35*, e2301242. [CrossRef]
104. Yang, X.; Li, X.; Wu, Z.; Cao, L. Photocrosslinked methacrylated natural macromolecular hydrogels for tissue engineering: A review. *Int. J. Biol. Macromol.* **2023**, *246*, 125570. [CrossRef]
105. Zhou, F.; Xin, L.; Wang, S.; Chen, K.; Li, D.; Wang, S.; Huang, Y.; Xu, C.; Zhou, M.; Zhong, W.; et al. Portable Handheld "SkinPen" Loaded with Biomaterial Ink for In Situ Wound Healing. *ACS Appl. Mater. Interfaces* **2023**, *15*, 27568–27585. [CrossRef]
106. Chen, H.; Ma, X.; Gao, T.; Zhao, W.; Xu, T.; Liu, Z. Robot-assisted in situ bioprinting of gelatin methacrylate hydrogels with stem cells induces hair follicle-inclusive skin regeneration. *Biomed. Pharmacother.* **2023**, *158*, 114140. [CrossRef]
107. Zhang, F.; Zhang, Z.; Duan, X.; Song, W.; Li, Z.; Yao, B.; Kong, Y.; Huang, X.; Fu, X.; Chang, J.; et al. Integrating zinc/silicon dual ions with 3D-printed GelMA hydrogel promotes in situ hair follicle regeneration. *Int. J. Bioprinting* **2023**, *9*, 200–215. [CrossRef] [PubMed]
108. Fortunato, G.M.; Sigismondi, S.; Nicoletta, M.; Condino, S.; Montemurro, N.; Vozzi, G.; Ferrari, V.; De Maria, C. Analysis of the Robotic-Based In Situ Bioprinting Workflow for the Regeneration of Damaged Tissues through a Case Study. *Bioengineering* **2023**, *10*, 560. [CrossRef] [PubMed]
109. Yang, Y.; Yu, Z.; Lu, X.; Dai, J.; Zhou, C.; Yan, J.; Wang, L.; Wang, Z.; Zang, J. Minimally invasive bioprinting for in situ liver regeneration. *Bioact. Mater.* **2023**, *26*, 465–477. [CrossRef] [PubMed]
110. Thai, M.T.; Phan, P.T.; Tran, H.A.; Nguyen, C.C.; Hoang, T.T.; Davies, J.; Rnjak-Kovacina, J.; Phan, H.; Lovell, N.H.; Do, T.N. Advanced Soft Robotic System for In Situ 3D Bioprinting and Endoscopic Surgery. *Adv. Sci.* **2023**, *10*, e2205656. [CrossRef] [PubMed]
111. Zhang, S.; Qi, C.; Zhang, W.; Zhou, H.; Wu, N.; Yang, M.; Meng, S.; Liu, Z.; Kong, T. In Situ Endothelialization of Free-Form 3D Network of Interconnected Tubular Channels via Interfacial Coacervation by Aqueous-in-Aqueous Embedded Bioprinting. *Adv. Mater.* **2023**, *35*, e2209263. [CrossRef]

112. Dong, H.; Hu, B.; Zhang, W.; Xie, W.; Mo, J.; Sun, H.; Shang, J. Robotic-assisted automated in situ bioprinting. *Int. J. Bioprinting* **2022**, *9*, 98–108. [CrossRef]
113. Li, T.; Hou, J.; Wang, L.; Zeng, G.; Wang, Z.; Yu, L.; Yang, Q.; Yin, J.; Long, M.; Chen, L.; et al. Bioprinted anisotropic scaffolds with fast stress relaxation bioink for engineering 3D skeletal muscle and repairing volumetric muscle loss. *Acta Biomater.* **2023**, *156*, 21–36. [CrossRef]
114. Moncal, K.K.; Yeo, M.; Celik, N.; Acri, T.M.; Rizk, E.; Wee, H.; Lewis, G.S.; Salem, A.K.; Ozbolat, I.T. Comparison of in-situ versus ex-situ delivery of polyethylenimine-BMP-2 polyplexes for rat calvarial defect repair via intraoperative bioprinting. *Biofabrication* **2022**, *15*, 015011. [CrossRef]
115. Shen, M.; Wang, L.; Gao, Y.; Feng, L.; Xu, C.; Li, S.; Wang, X.; Wu, Y.; Guo, Y.; Pei, G. 3D bioprinting of in situ vascularized tissue engineered bone for repairing large segmental bone defects. *Mater. Today Bio* **2022**, *16*, 100382. [CrossRef] [PubMed]
116. Huo, Y.; Xu, Y.; Wu, X.; Gao, E.; Zhan, A.; Chen, Y.; Zhang, Y.; Hua, Y.; Swieszkowski, W.; Zhang, Y.S.; et al. Functional Trachea Reconstruction Using 3D-Bioprinted Native-Like Tissue Architecture Based on Designable Tissue-Specific Bioinks. *Adv. Sci.* **2022**, *9*, e2202181. [CrossRef] [PubMed]
117. Campos, D.F.D.; Zhang, S.; Kreimendahl, F.; Köpf, M.; Fischer, H.; Vogt, M.; Blaeser, A.; Apel, C.; Esteves-Oliveira, M. Hand-held bioprinting for de novo vascular formation applicable to dental pulp regeneration. *Connect. Tissue Res.* **2020**, *61*, 205–215. [CrossRef] [PubMed]
118. Farto-Vaamonde, X.; Diaz-Gomez, L.; Parga, A.; Otero, A.; Concheiro, A.; Alvarez-Lorenzo, C. Perimeter and carvacrol-loading regulate angiogenesis and biofilm growth in 3D printed PLA scaffolds. *J. Control. Release* **2022**, *352*, 776–792. [CrossRef]
119. Zarei, M.; Sayedain, S.S.; Askarinya, A.; Sabbaghi, M.; Alizadeh, R. Improving physio-mechanical and biological properties of 3D-printed PLA scaffolds via in-situ argon cold plasma treatment. *Sci. Rep.* **2023**, *13*, 14120. [CrossRef] [PubMed]
120. Kim, S.J.; Lee, G.; Park, J.-K. Hybrid Biofabrication of Heterogeneous 3D Constructs Using Low-Viscosity Bioinks. *ACS Appl. Mater. Interfaces* **2023**, *15*, 41247–41257. [CrossRef]
121. Custodio, C.L.; Broñola, P.J.M.; Cayabyab, S.R.; Lagura, V.U.; Celorico, J.R.; Basilia, B.A. Powder Loading Effects on the Physicochemical and Mechanical Properties of 3D Printed Poly Lactic Acid/Hydroxyapatite Biocomposites. *Int. J. Bioprint.* **2021**, *7*, 326. [CrossRef] [PubMed]
122. Fan, G.; Yang, L.; Liu, D.; Wang, Y.; Ji, W.; Tukebai; Qin, H.; Wang, Z. Repair of Cranial Defects in Rabbits with 3D-Printed Hydroxyapatite/Polylactic Acid Composites. *BioMed Res. Int.* **2022**, *2022*, 7562291. [CrossRef] [PubMed]
123. Sultan, S.; Thomas, N.; Varghese, M.; Dalvi, Y.; Joy, S.; Hall, S.; Mathew, A.P. The Design of 3D-Printed Polylactic Acid–Bioglass Composite Scaffold: A Potential Implant Material for Bone Tissue Engineering. *Molecules* **2022**, *27*, 7214. [CrossRef]
124. Cichos, S.; Schätzlein, E.; Wiesmann-Imilowski, N.; Blaeser, A.; Henrich, D.; Frank, J.; Drees, P.; Gercek, E.; Ritz, U. A new 3D-printed polylactic acid-bioglass composite for bone tissue engineering induces angiogenesis in vitro and in ovo. *Int. J. Bioprinting* **2023**, *9*, 54–67. [CrossRef] [PubMed]
125. Guo, F.; Wang, E.; Yang, Y.; Mao, Y.; Liu, C.; Bu, W.; Li, P.; Zhao, L.; Jin, Q.; Liu, B.; et al. A natural biomineral for enhancing the biomineralization and cell response of 3D printed polylactic acid bone scaffolds. *Int. J. Biol. Macromol.* **2023**, *242 Pt 1*, 124728. [CrossRef]
126. Jang, H.J.; Kang, M.S.; Kim, W.-H.; Jo, H.J.; Lee, S.-H.; Hahm, E.J.; Oh, J.H.; Hong, S.W.; Kim, B.; Han, D.-W. 3D printed membranes of polylactic acid and graphene oxide for guided bone regeneration. *Nanoscale Adv.* **2023**, *5*, 3619–3628. [CrossRef]
127. Pisani, S.; Mauri, V.; Negrello, E.; Mauramati, S.; Alaimo, G.; Auricchio, F.; Benazzo, M.; Dorati, R.; Genta, I.; Conti, B.; et al. Assessment of different manufacturing techniques for the production of bioartificial scaffolds as soft organ transplant substitutes. *Front. Bioeng. Biotechnol.* **2023**, *11*, 1186351. [CrossRef]
128. Duan, R.; Wang, Y.; Zhang, Y.; Wang, Z.; Du, F.; Du, B.; Su, D.; Liu, L.; Li, X.; Zhang, Q. Blending with Poly(l-lactic acid) Improves the Printability of Poly(l-lactide-*co*-caprolactone) and Enhances the Potential Application in Cartilage Tissue Engineering. *ACS Omega* **2021**, *6*, 18300–18313. [CrossRef]
129. Kim, S.H.; Kim, D.; Cha, M.; Kim, S.H.; Jung, Y. The Regeneration of Large-Sized and Vascularized Adipose Tissue Using a Tailored Elastic Scaffold and dECM Hydrogels. *Int. J. Mol. Sci.* **2021**, *22*, 12560. [CrossRef] [PubMed]
130. Lee, S.; Lee, H.S.; Chung, J.J.; Kim, S.H.; Park, J.W.; Lee, K.; Jung, Y. Enhanced Regeneration of Vascularized Adipose Tissue with Dual 3D-Printed Elastic Polymer/dECM Hydrogel Complex. *Int. J. Mol. Sci.* **2021**, *22*, 2886. [CrossRef] [PubMed]
131. Yin, H.; Zhu, M.; Wang, Y.; Luo, L.; Ye, Q.; Lee, B.H. Physical properties and cellular responses of gelatin methacryloyl bulk hydrogels and highly ordered porous hydrogels. *Front. Soft Matter* **2023**, *2*, 1101680. [CrossRef]
132. Choi, J.R.; Yong, K.W.; Choi, J.Y.; Cowie, A.C. Recent advances in photo-crosslinkable hydrogels for biomedical applications. *BioTechniques* **2019**, *66*, 40–53. [CrossRef]
133. Liang, J.; Wang, Z.; Poot, A.A.; Grijpma, D.W.; Dijkstra, P.J.; Wang, R. Enzymatic post-crosslinking of printed hydrogels of methacrylated gelatin and tyramine-conjugated 8-arm poly(ethylene glycol) to prepare interpenetrating 3D network structures. *Int. J. Bioprint.* **2023**, *9*, 750. [CrossRef]
134. Wu, W.; Dong, Y.; Liu, H.; Jiang, X.; Yang, L.; Luo, J.; Hu, Y.; Gou, M. 3D printed elastic hydrogel conduits with 7,8-dihydroxyflavone release for peripheral nerve repair. *Mater. Today Bio* **2023**, *20*, 100652. [CrossRef]
135. Ling, Z.; Zhao, J.; Song, S.; Xiao, S.; Wang, P.; An, Z.; Fu, Z.; Shao, J.; Zhang, Z.; Fu, W.; et al. Chitin nanocrystal-assisted 3D bioprinting of gelatin methacrylate scaffolds. *Regen. Biomater.* **2023**, *10*, rbad058. [CrossRef]

136. Xu, Z.; Qi, X.; Bao, M.; Zhou, T.; Shi, J.; Xu, Z.; Zhou, M.; Boccaccini, A.R.; Zheng, K.; Jiang, X. Biomineralization inspired 3D printed bioactive glass nanocomposite scaffolds orchestrate diabetic bone regeneration by remodeling micromilieu. *Bioact. Mater.* **2023**, *25*, 239–255. [CrossRef]
137. Pérez-Cortez, J.E.; Sánchez-Rodríguez, V.H.; Gallegos-Martínez, S.; Chuck-Hernández, C.; Rodriguez, C.A.; Álvarez, M.M.; Santiago, G.T.-D.; Vázquez-Lepe, E.; Martínez-López, J.I. Low-Cost Light-Based GelMA 3D Bioprinting via Retrofitting: Manufacturability Test and Cell Culture Assessment. *Micromachines* **2022**, *14*, 55. [CrossRef] [PubMed]
138. Elomaa, L.; Keshi, E.; Sauer, I.M.; Weinhart, M. Development of GelMA/PCL and dECM/PCL resins for 3D printing of acellular in vitro tissue scaffolds by stereolithography. *Mater. Sci. Eng. C Mater. Biol. Appl.* **2020**, *112*, 110958. [CrossRef] [PubMed]
139. Kumar, H.; Sakthivel, K.; Mohamed, M.G.A.; Boras, E.; Shin, S.R.; Kim, K. Designing Gelatin Methacryloyl (GelMA)-Based Bioinks for Visible Light Stereolithographic 3D Biofabrication. *Macromol. Biosci.* **2021**, *21*, e2000317. [CrossRef]
140. Jurczyk, M.; Kasperczyk, J.; Wrześniok, D.; Beberok, A.; Jelonek, K. Nanoparticles Loaded with Docetaxel and Resveratrol as an Advanced Tool for Cancer Therapy. *Biomedicines* **2022**, *10*, 1187. [CrossRef] [PubMed]
141. Sankaralingam, P.; Sakthivel, P.; Thangavel, V.C.; Sankaralingam, P.; Sakthivel, P.; Thangavel, V.C. *Novel Composites for Bone Tissue Engineering. Biomimetics—Bridging the Gap*; IntechOpen: London, UK, 2022. [CrossRef]
142. Guerra, A.J.; Lammel-Lindemann, J.; Katko, A.; Kleinfehn, A.; Rodriguez, C.A.; Catalani, L.H.; Becker, M.L.; Ciurana, J.; Dean, D. Optimization of photocrosslinkable resin components and 3D printing process parameters. *Acta Biomater.* **2019**, *97*, 154–161. [CrossRef]
143. Kondiah, P.J.; Kondiah, P.P.D.; Choonara, Y.E.; Marimuthu, T.; Pillay, V. A 3D Bioprinted Pseudo-Bone Drug Delivery Scaffold for Bone Tissue Engineering. *Pharmaceutics* **2020**, *12*, 166. [CrossRef]
144. Kolan, K.C.R.; Huang, Y.-W.; Semon, J.A.; Leu, M.C. 3D-printed Biomimetic Bioactive Glass Scaffolds for Bone Regeneration in Rat Calvarial Defects. *Int. J. Bioprint.* **2020**, *6*, 274, Erratum in: *Int. J. Bioprint.* **2020**, *6*, 309. [CrossRef] [PubMed]
145. Cheng, K.-C.; Sun, Y.-M.; Hsu, S.-H. Development of double network polyurethane–chitosan composite bioinks for soft neural tissue engineering. *J. Mater. Chem. B* **2023**, *11*, 3592–3606. [CrossRef]
146. Sawant, M.; Wang, F.; Koester, J.; Niehoff, A.; Nava, M.M.; Lundgren-Akerlund, E.; Gullberg, D.; Leitinger, B.; Wickström, S.; Eckes, B.; et al. Ablation of integrin-mediated cell–collagen communication alleviates fibrosis. *Ann. Rheum. Dis.* **2023**, *82*, 1474–1486. [CrossRef]
147. Ahn, S.; Sharma, U.; Kasuba, K.C.; Strohmeyer, N.; Müller, D.J. Engineered Biomimetic Fibrillar Fibronectin Matrices Regulate Cell Adhesion Initiation, Migration, and Proliferation via $\alpha5\beta1$ Integrin and Syndecan-4 Crosstalk. *Adv. Sci.* **2023**, *10*, e2300812. [CrossRef]
148. Tiwari, S.; Patil, R.; Bahadur, P. Polysaccharide Based Scaffolds for Soft Tissue Engineering Applications. *Polymers* **2018**, *11*, 1. [CrossRef] [PubMed]
149. Jung, S.A.; Malyaran, H.; Demco, D.E.; Manukanc, A.; Häser, L.S.; Kučikas, V.; van Zandvoort, M.; Neuss, S.; Pich, A. Fibrin–Dextran Hydrogels with Tunable Porosity and Mechanical Properties. *Biomacromolecules* **2023**, *24*, 3972–3984. [CrossRef]
150. Ma, C.; Wang, T.; Jin, X.; Zhang, W.; Lv, Q. Lineage-specific multifunctional double-layer scaffold accelerates the integrated regeneration of cartilage and subchondral bone. *Mater. Today Bio* **2023**, *23*, 100800. [CrossRef] [PubMed]
151. Ozturk, T.; Erpala, F.; Bozduman, O.; Gedikbas, M.; Eren, M.B.; Zengin, E.C. Arthroscopic Treatment of Femoral Condyle Chondral Lesions: Microfracture Versus Liquid Bioscaffold. *Indian J. Orthop.* **2023**, *57*, 975–982. [CrossRef] [PubMed]
152. Zineh, B.R.; Roshangar, L.; Meshgi, S.; Shabgard, M. 3D printing of alginate/thymoquinone/halloysite nanotube bio-scaffolds for cartilage repairs: Experimental and numerical study. *Med. Biol. Eng. Comput.* **2022**, *60*, 3069–3080. [CrossRef]
153. Patel, D.K.; Dutta, S.D.; Lim, K.-T. Nanocellulose-based polymer hybrids and their emerging applications in biomedical engineering and water purification. *RSC Adv.* **2019**, *9*, 19143–19162. [CrossRef]
154. Konya, P.; Konya, M.N.; Yilmaz, B.K.; Kaga, E.; Kaga, S.; Çetinkol, Y. Comparison of the Therapeutic Efficacy of Antibiotic-Loaded Polymeric Tissue Scaffold and Bone Cement in the Regeneration of Infected Bone Tissue. *Cureus* **2023**, *15*, e46487. [CrossRef]
155. Štiglic, A.D.; Gürer, F.; Lackner, F.; Bračič, D.; Winter, A.; Gradišnik, L.; Makuc, D.; Kargl, R.; Duarte, I.; Plavec, J.; et al. Organic acid cross-linked 3D printed cellulose nanocomposite bioscaffolds with controlled porosity, mechanical strength, and biocompatibility. *iScience* **2022**, *25*, 104263. [CrossRef]
156. Aeberhard, P.; Grognuz, A.; Peneveyre, C.; McCallin, S.; Hirt-Burri, N.; Antons, J.; Pioletti, D.; Raffoul, W.; Applegate, L.A. Efficient decellularization of equine tendon with preserved biomechanical properties and cytocompatibility for human tendon surgery indications. *Artif. Organs* **2020**, *44*, E161–E171. [CrossRef] [PubMed]
157. Tan, Y.; Fan, S.; Wu, X.; Liu, M.; Dai, T.; Liu, C.; Ni, S.; Wang, J.; Yuan, X.; Zhao, H.; et al. Fabrication of a three-dimensional printed gelatin/sodium alginate/nano-attapulgite composite polymer scaffold loaded with leonurine hydrochloride and its effects on osteogenesis and vascularization. *Int. J. Biol. Macromol.* **2023**, *249*, 126028. [CrossRef] [PubMed]
158. Jin, J.; Chen, Z.; Xiang, Y.; Tang, T.; Zhou, H.; Hong, X.; Fan, H.; Zhang, X.; Luo, P.; Ma, B.; et al. Development of a PHMB hydrogel-modified wound scaffold dressing with antibacterial activity. *Wound Repair Regen.* **2020**, *28*, 480–492. [CrossRef] [PubMed]
159. Trapani, G.; Weiß, M.S.; Trappmann, B. Tunable Synthetic Hydrogels to Study Angiogenic Sprouting. *Curr. Protoc.* **2023**, *3*, e859. [CrossRef] [PubMed]

160. Wei, Q.; Liu, D.; Chu, G.; Yu, Q.; Liu, Z.; Li, J.; Meng, Q.; Wang, W.; Han, F.; Li, B. TGF-β1-supplemented decellularized annulus fibrosus matrix hydrogels promote annulus fibrosus repair. *Bioact. Mater.* **2022**, *19*, 581–593. [CrossRef]
161. Matsubayashi, Y.; Sánchez-Sánchez, B.J.; Marcotti, S.; Serna-Morales, E.; Dragu, A.; Díaz-De-La-Loza, M.-D.; Vizcay-Barrena, G.; Fleck, R.A.; Stramer, B.M.; Matsubayashi, Y.; et al. Rapid Homeostatic Turnover of Embryonic ECM during Tissue Morphogenesis. *Dev. Cell* **2020**, *54*, 33–42.e9. [CrossRef] [PubMed]
162. Lee, S.-Y.; Robertson, C.; Diot, A.; Meuray, V.; Bourdon, J.-C.; Bissell, M.J. Δ133p53 coordinates ECM-driven morphogenesis and gene expression in three-dimensional mammary epithelial acini. *J. Cell Sci.* **2022**, *135*, jcs259673. [CrossRef]
163. Ma, M.; Hua, S.; Min, X.; Wang, L.; Li, J.; Wu, P.; Liang, H.; Zhang, B.; Chen, X.; Xiang, S. p53 positively regulates the proliferation of hepatic progenitor cells promoted by laminin-521. *Signal Transduct. Target. Ther.* **2022**, *7*, 290. [CrossRef]
164. Furuta, S.; Ren, G.; Mao, J.-H.; Bissell, M.J. Laminin signals initiate the reciprocal loop that informs breast-specific gene expression and homeostasis by activating NO, p53 and microRNAs. *eLife* **2018**, *7*, e26148. [CrossRef]
165. Rederer, A.; Rose, V.; Krüger, R.; Schmittutz, L.; Swierzy, I.; Fischer, L.; Thievessen, I.; Bauer, J.; Friedrich, O.; Schiffer, M.; et al. Partner, Neighbor, Housekeeper and Dimension: 3D versus 2D Glomerular Co-Cultures Reveal Drawbacks of Currently Used Cell Culture Models. *Int. J. Mol. Sci.* **2023**, *24*, 10384. [CrossRef]
166. Sabaté, M.; Alfonso, F.; Cequier, A.; Romaní, S.; Bordes, P.; Serra, A.; Iñiguez, A.; Salinas, P.; del Blanco, B.G.; Goicolea, J.; et al. Magnesium-Based Resorbable Scaffold Versus Permanent Metallic Sirolimus-Eluting Stent in Patients With ST-Segment Elevation Myocardial Infarction: The MAGSTEMI Randomized Clinical Trial. *Circulation* **2019**, *140*, 1904–1916. [CrossRef] [PubMed]
167. Jiang, Y.; Sun, S.-J.; Zhen, Z.; Wei, R.; Zhang, N.; Liao, S.-Y.; Tse, H.-F. Myocardial repair of bioengineered cardiac patches with decellularized placental scaffold and human-induced pluripotent stem cells in a rat model of myocardial infarction. *Stem Cell Res. Ther.* **2021**, *12*, 13. [CrossRef]
168. Kojima, H.; Ishii, T.; Fukumitsu, K.; Ogiso, S.; Tomofuji, K.; Oshima, Y.; Horie, H.; Ito, T.; Wakama, S.; Makino, K.; et al. In Vivo Regeneration of Tubular Small Intestine With Motility: A Novel Approach by Orthotopic Transplantation of Decellularized Scaffold. *Transplantation* **2023**, *107*, 1955–1964. [CrossRef] [PubMed]
169. Vriend, L.; Sinkunas, V.; Camargo, C.P.; van der Lei, B.; Harmsen, M.C.; van Dongen, J.A. Extracellular Matrix-Derived Hydrogels to Augment Dermal Wound Healing: A Systematic Review. *Tissue Eng. Part B: Rev.* **2022**, *28*, 1093–1108. [CrossRef]
170. Marvin, J.C.; Mochida, A.; Paredes, J.; Vaughn, B.R.; Andarawis-Puri, N. Detergent-Free Decellularization Preserves the Mechanical and Biological Integrity of Murine Tendon. *Tissue Eng. Part C Methods* **2022**, *28*, 646–655. [CrossRef]
171. He, P.; Yang, G.; Zhu, D.; Kong, H.; Corrales-Ureña, Y.R.; Ciacchi, L.C.; Wei, G. Biomolecule-mimetic nanomaterials for photothermal and photodynamic therapy of cancers: Bridging nanobiotechnology and biomedicine. *J. Nanobiotechnol.* **2022**, *20*, 483. [CrossRef]
172. Devillard, C.D.; Mandon, C.A.; Lambert, S.A.; Blum, L.J.; Marquette, C.A. Bioinspired Multi-Activities 4D Printing Objects: A New Approach Toward Complex Tissue Engineering. *Biotechnol. J.* **2018**, *13*, e1800098. [CrossRef]
173. Sato, K.; Kawasaki, A.; Karuo, Y.; Tarui, A.; Kawai, K.; Omote, M. Synthesis of new fluorescent molecules having an aggregation-induced emission property derived from 4-fluoroisoxazoles. *Beilstein J. Org. Chem.* **2020**, *16*, 1411–1417. [CrossRef]
174. Huang, H.-K.; Hsueh, K.-K.; Liao, Y.-T.; Wu, S.-H.; Chou, P.-H.; Yeh, S.-H.; Wang, J.-P. Multilineage differentiation potential in the infant adipose- and umbilical cord-derived mesenchymal stem cells. *J. Chin. Med. Assoc.* **2023**, *86*, 1083–1095. [CrossRef]
175. Marone, R.; Landmann, E.; Devaux, A.; Lepore, R.; Seyres, D.; Zuin, J.; Burgold, T.; Engdahl, C.; Capoferri, G.; Dell'aglio, A.; et al. Epitope-engineered human hematopoietic stem cells are shielded from CD123-targeted immunotherapy. *J. Exp. Med.* **2023**, *220*, e20231235. [CrossRef]
176. Wang, C.-H.; Huang, Y.-F.; Shyu, W.-C.; Jeng, L.-B.; Liu, S.-P. Cbx7 promotes the generation of induced pluripotent stem cells. *Regen. Ther.* **2023**, *24*, 443–450. [CrossRef]
177. Singh, N.K.; Han, W.; Nam, S.A.; Kim, J.W.; Kim, J.Y.; Kim, Y.K.; Cho, D.-W. Three-dimensional cell-printing of advanced renal tubular tissue analogue. *Biomaterials* **2020**, *232*, 119734. [CrossRef]
178. Wilems, T.; Vardhan, S.; Wu, S.; Sakiyama-Elbert, S. The influence of microenvironment and extracellular matrix molecules in driving neural stem cell fate within biomaterials. *Brain Res. Bull.* **2019**, *148*, 25–33. [CrossRef]
179. Zhang, N.; Lin, J.; Lin, V.P.H.; Milbreta, U.; Chin, J.S.; Chew, E.G.Y.; Lian, M.M.; Foo, J.N.; Zhang, K.; Wu, W.; et al. A 3D Fiber-Hydrogel Based Non-Viral Gene Delivery Platform Reveals that microRNAs Promote Axon Regeneration and Enhance Functional Recovery Following Spinal Cord Injury. *Adv. Sci.* **2021**, *8*, 2100805. [CrossRef]
180. Vazirzadeh, M.; Azarpira, N.; Vosough, M.; Ghaedi, K. Galactosylation of rat natural scaffold for MSC differentiation into hepatocyte-like cells: A comparative analysis of 2D vs. 3D cell culture techniques. *Biochem. Biophys. Rep.* **2023**, *35*, 101503. [CrossRef]
181. Yadav, H.O.S.; Kuo, A.-T.; Urata, S.; Funahashi, K.; Imamura, Y.; Shinoda, W. Adsorption characteristics of peptides on ω-functionalized self-assembled monolayers: A molecular dynamics study. *Phys. Chem. Chem. Phys.* **2022**, *24*, 14805–14815. [CrossRef]
182. Franck, C.O.; Fanslau, L.; Popov, A.B.; Tyagi, P.; Fruk, L. Biopolymer-based Carriers for DNA Vaccine Design. *Angew. Chem. Int. Ed.* **2021**, *60*, 13225–13243. [CrossRef] [PubMed]

183. Giovanniello, F.; Asgari, M.; Breslavsky, I.D.; Franchini, G.; Holzapfel, G.A.; Tabrizian, M.; Amabili, M. Development and mechanical characterization of decellularized scaffolds for an active aortic graft. *Acta Biomater.* **2023**, *160*, 59–72. [CrossRef] [PubMed]
184. Kong, D.; Wang, Q.; Huang, J.; Zhang, Z.; Wang, X.; Han, Q.; Shi, Y.; Ji, R.; Li, Y. Design and manufacturing of biomimetic scaffolds for bone repair inspired by bone trabeculae. *Comput. Biol. Med.* **2023**, *165*, 107369. [CrossRef] [PubMed]

Disclaimer/Publisher's Note: The statements, opinions and data contained in all publications are solely those of the individual author(s) and contributor(s) and not of MDPI and/or the editor(s). MDPI and/or the editor(s) disclaim responsibility for any injury to people or property resulting from any ideas, methods, instructions or products referred to in the content.

nanomaterials

Review

Challenges of Periodontal Tissue Engineering: Increasing Biomimicry through 3D Printing and Controlled Dynamic Environment

Ilaria Roato [1,*], Beatrice Masante [1,2,3], Giovanni Putame [2,3], Diana Massai [2,3] and Federico Mussano [1]

1 Bone and Dental Bioengineering Laboratory, CIR-Dental School, Department of Surgical Sciences, University of Turin, 10126 Turin, Italy
2 PolitoBIOMed Lab and Department of Mechanical and Aerospace Engineering, Politecnico di Torino, 10129 Turin, Italy
3 Interuniversity Center for the Promotion of the 3Rs Principles in Teaching and Research, 10129 Turin, Italy
* Correspondence: ilaria.roato@unito.it; Tel.: +39-011-670-3528

Abstract: In recent years, tissue engineering studies have proposed several approaches to regenerate periodontium based on the use of three-dimensional (3D) tissue scaffolds alone or in association with periodontal ligament stem cells (PDLSCs). The rapid evolution of bioprinting has sped up classic regenerative medicine, making the fabrication of multilayered scaffolds—which are essential in targeting the periodontal ligament (PDL)—conceivable. Physiological mechanical loading is fundamental to generate this complex anatomical structure ex vivo. Indeed, loading induces the correct orientation of the fibers forming the PDL and maintains tissue homeostasis, whereas overloading or a failure to adapt to mechanical load can be at least in part responsible for a wrong tissue regeneration using PDLSCs. This review provides a brief overview of the most recent achievements in periodontal tissue engineering, with a particular focus on the use of PDLSCs, which are the best choice for regenerating PDL as well as alveolar bone and cementum. Different scaffolds associated with various manufacturing methods and data derived from the application of different mechanical loading protocols have been analyzed, demonstrating that periodontal tissue engineering represents a proof of concept with high potential for innovative therapies in the near future.

Keywords: periodontitis; mesenchymal stem cell; mechanical loading; periodontal ligament regeneration; tissue engineering

Citation: Roato, I.; Masante, B.; Putame, G.; Massai, D.; Mussano, F. Challenges of Periodontal Tissue Engineering: Increasing Biomimicry through 3D Printing and Controlled Dynamic Environment. *Nanomaterials* **2022**, *12*, 3878. https://doi.org/10.3390/nano12213878

Academic Editor: Jyh-Ping Chen

Received: 29 September 2022
Accepted: 31 October 2022
Published: 2 November 2022

Publisher's Note: MDPI stays neutral with regard to jurisdictional claims in published maps and institutional affiliations.

Copyright: © 2022 by the authors. Licensee MDPI, Basel, Switzerland. This article is an open access article distributed under the terms and conditions of the Creative Commons Attribution (CC BY) license (https://creativecommons.org/licenses/by/4.0/).

1. Introduction

The periodontium is a complex system composed of gingiva, periodontal ligament (PDL), cementum, and alveolar bone, featuring a hierarchically compartmentalized architecture [1]. The homeostasis of this system is maintained by the PDL, a specialized connective tissue, which is located between the cementum and alveolar bone and articulates (gomphosis) the teeth to the jaws [2,3]. Embryologically, PDL derives from the dental follicle cells under the guidance of Hertwig's epithelial root sheath (HERS), which secrete numerous epithelium-derived factors [4] before obliterating almost completely.

From a histological perspective, PDL is an aligned fibrous network with a thickness ranging between 100 and 400 μm and is characterized by an extensive blood supply and a neural network [5]. PDL is constituted by a heterogeneous population of cells (namely PDL cells) that includes periodontal ligament fibroblasts (PDLFs), which represent by far the largest population and are responsible for the deposition and maintenance of the extracellular matrix (ECM) and periodontal ligament stem cells (PDLSCs), showing both osteogenic and tendo/ligamentogenic characteristics. Collagen type I and, in lesser amounts, type III constitute cross-banded fibrils, named Sharpey's fibers, that provide mechanical support and are usually classified as dentinogingival, transseptal, or alveolodental (forming the bulk of proper PDL fibers) [5]. In particular, fibers oblique or perpendicular to the long axis of the tooth are thought to play pivotal roles in eliciting adaptive responses during mastication and occlusion [5]. Among all the fibers, the horizontal ones withstand the greatest loads and exhibit the greatest strain under mastication [6], (Figure 1). The collagen fibers are generally aligned according to a periodic crimped pattern [7] that prevents ligament overextension [8,9]. Sharpey's fibers anchor mostly to acellular cementum, a mineralized layer (50–300 μm thickness) covering the tooth dentin surface. PDL cells are arranged along PDL fibers so that the long cellular axis is parallel to the main fiber bundles of the PDL [10,11]. The presence of a particular type of elastic fibers named oxytytalan, made of fimbrillins, that form a network running parallel to cementum and are thought to interact with vessels and neural fibers is also noteworthy [12]. In a healthy subject, PDL covers the tooth root almost entirely, and a tight epithelial seal within the gingival sulcus prevents microorganisms from reaching the PDL. This delicate system is compromised by the onset of periodontal disease (PD), which affects in its severe form about 10% of adults, ranking sixth among the most prevalent diseases in the world [13]. PD starts as a localized and reversible inflammation of the gingiva (gingivitis) due to dental plaque, and, when untreated, it may become chronic periodontitis, which is characterized by the progressive destruction of the tooth-supporting tissues, i.e., cementum, PDL, and bone [14,15].

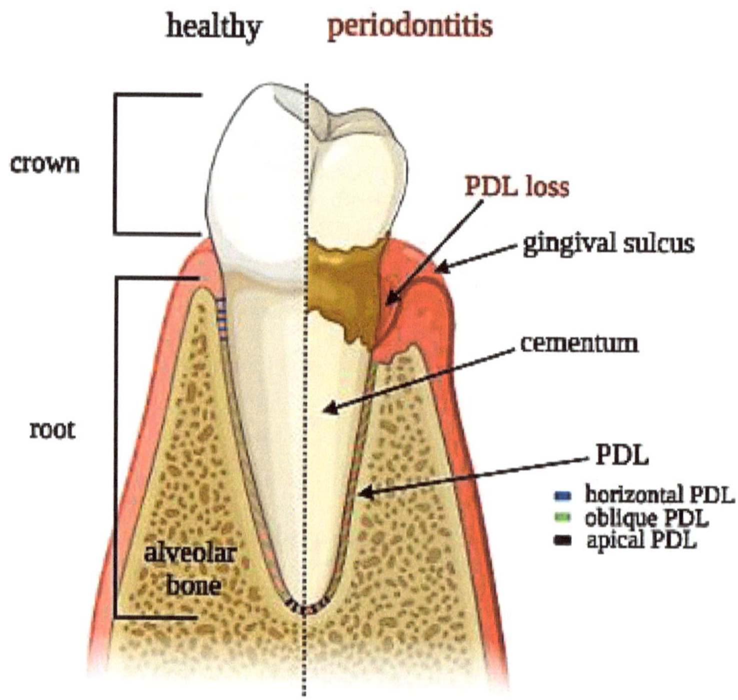

Figure 1. Tooth section showing periodontal ligament anatomy: superficial and deep periodontium along with the typical distribution of PDL fibers in both physiological and pathological conditions.

The complex architecture of the periodontal apparatus (Figure 1), including dual-tissue interfaces (alveolar bone–PDL and PDL–cementum of the tooth root), is difficult to regenerate due to the small dimensions of the PDL and the challenging oral environment [16]. Any strategy aiming at periodontal regeneration should entail studying the specific events that guide the formation and remodeling of the PDL as well as understanding the intimate bond of tissue histology and function. Hence, this review outlines the most recent achievements in the field—with a particular emphasis on the biomimetic approach, foreseeing the use of PDLSCs in combination with biomimetic scaffolds and subjected in vitro to controlled native-like mechanical loadings—showing that periodontal tissue engineering represents a promising strategy for innovative therapies in the near future.

2. The Role of PDL Cells

In 2004, Kawaguchi et al. proposed the autograft of bone-marrow-derived mesenchymal stem cells (BMMSCs) to enhance the healing of periodontal defects, which proved successful in a dog model [17]. This approach highlighted the potential of cellular therapy and paved the way to other pre-clinical studies dealing with BMMSCs [18], adipose derived stem cells (ASCs) [19], and PDLSCs [20]. Among all the possible sources of MSCs, PDLSCs (Figure 2)—which are characterized by the expression of many markers resumed in Table 1—may be selected for PDL regeneration owing to their commitment capacity, as they express scleraxis, i.e., a tendon/ligament-specific transcription factor, more than BMMSCs or dental pulp stem cells (DPSCs) and have the potential to form cementum and PDL-like structures [21,22]. Indeed, the preservation of the PDL is essential to achieving proper regeneration of the periodontium and avoiding the ankylosis of the tooth, i.e., direct contact between the root and the alveolar bone.

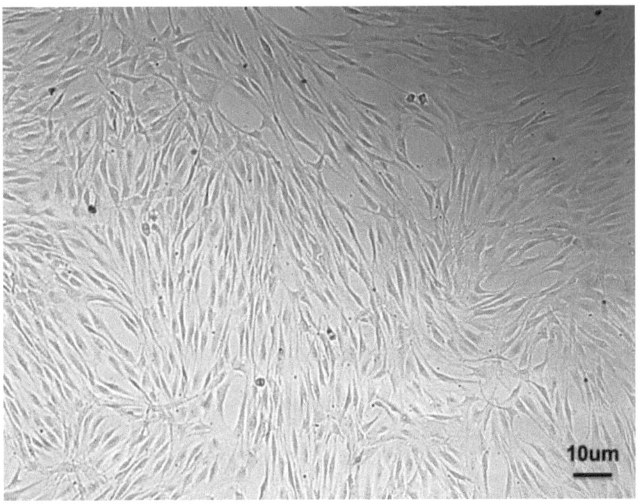

Figure 2. PDLSCs isolated from PDL and expanded in vitro showing the typical spindle shape.

From this perspective, the role of the transforming growth factor–β1 (TGF-β1) signaling becomes interesting since its activation enables the commitment of cementocytes, while its inhibition promotes fibroblastic differentiation of the ligament progenitors [23]. PDLSCs transplanted into a periodontal lesion in a rat model generated typical PDL-like structures in vivo by forming Sharpey's-fiber-like collagen bundles that are connected to cementum-like structures [20]. New insights into the molecular regulation of periodontal attachment have been brought by Bai S et al., who investigated the regulatory mechanism of copine 7 (CPNE7) and cementum attachment protein (CAP) in coordination with cytoskeleton arrangement [24]. Regenerating cementum is regarded as key to promoting new fibrous tissue attachment, preserving the PDL, and avoiding tooth ankylosis [25].

Important issues should be considered in order to implement successful PDLSC-based protocols for PDL regeneration. Owing to the reduced volume of the source tissue, the yield of PDLSCs isolated from a single tooth may be poor, and the in vitro expansion of these cells has been linked to morphological changes and diminished expression of genes associated with pluripotency of embryonic stem cells, such as NANOG and OCT4 [26]. Additionally, reducing the ex vivo manipulation of cells is a goal in every cell transplantation procedure since ex vivo expansion of PDLSCs, as well as of ASCs, causes senescence, decline in multipotency, and is subjected to complex regulatory issues under the compelling requirements of good manufacturing practices (GMP) guidelines [27]. Therefore, it becomes paramount to select the best harvest conditions and methods to maximize the number of cells available. The outgrowth method has been suggested to outperform enzymatic digestion in terms of efficiency and preservation of the commitment capacity regarding both the formation of mineralized nodules and the expression of cementoblast-like genes [28]. Consistently, the paper by Abuarqoub et al. compared the features of PDLSCs expanded using either fetal bovine serum (FBS) or platelet lysate (PL) and concluded that the latter outperformed the former [29].

Regarding the possible sources of heterogeneity-influencing PDLSC behavior pertaining to differentiative potential, donor age plays an important role [30,31]. By analyzing donor age impact on PDLSC function, it was observed that PDLSCs from elderly populations exhibited decreased expression of osteogenesis-related genes, such as osteocalcin (OCN), COLL-1, and runt-related transcription factor 2 (RUNX2) as well as reduced osteogenic activity [30]. This has prompted the possibility of banking cells as soon as teeth are extracted to create a reservoir for any future usage. Little is known, however, about the differences between PDLSCs deriving from deciduous and permanent teeth. The latter seem to be less adipogenic than deciduous teeth according to some authors [32,33], while others disagree [34]. Moreover, PDLSCs from deciduous teeth express higher levels of cytokines that regulate the host immunity and secreting factors involved in tissue degradation and catalytic activities than PDLSCs from permanent teeth [35].

Distinguished researchers [36,37] explored the feasibility of using PDLSCs to form cementum and PDL-like structures in proper animal models, reporting very encouraging results. Unfortunately, the enthusiastic outcome of these in vivo studies does not seem to be fully supported by clinical evidence, which is surprisingly scarce. In fact, only one randomized controlled trial [38] assessed the safety and feasibility of using autologous PDLSCs in combination with bovine-derived bone mineral materials for treating periodontal bone defects, but it could not demonstrate any statistically significant difference "between the Cell group and the Control group ($p > 0.05$)", implying that more studies with incremented numbers will possibly shed light on the matter. It appears that in humans, cells—however important—may alone not be sufficient to regenerate the PDL, which is endowed with a peculiar and rather complex architecture.

Table 1. Main markers expressed by PDLSCs.

	MARKERS	REFERENCE
MSC	CD105, CD73, CD90, CD44	[39,40]
STEMNESS	OCT3/4, NANOG, SOX2	[41–44]
PERICYTE	STRO-1, CD146	[45]
NEUROECTODERMIC	Nestin, SOX10, SLUG CD271/p75NTR	[46,47]

3. Biomimetic Scaffolds to Reproduce the Micro-Environment of PDL

Traditionally, periodontal regeneration achieved through guided tissue regeneration (GTR) has been based on the concept of avoiding epithelial invasion of the bone defect to be treated by means of barrier membranes to allow PDL and bone repopulation of the dental root. Thus, a specific avenue of research has been paved toward the improvement of these membranes from the original non-resorbable expanded polytetrafluoroethylene (e-PTFE) [48] to the high technology level of recent developments, such as that proposed by Nasajpour et al. [49]. In parallel, the promising potential unleashed by tissue engineering, which relies upon combining biomaterials functioning as scaffolds and stem cells, has opened a range of new therapeutic strategies in the periodontal field. In a pioneering proof of concept, Sonoyama et al. proposed PDLSCs, the resident stem cells of PDL, in association with hydroxyapatite (HA) and tricalciumphosphate (TCP) for forming cementum and PDL-like structures [36]. More recently, Shi et al., after culturing PDLSCs in osteogenic conditions and seeding them on a biphasic calcium phosphate scaffold, observed a periodontal regeneration in the recipient animal constituted by new bone formation and PDL-organized fibers correctly inserted into adjacent cementum and bone, along with neo-vascularization, after 12 weeks [37]. Given these premises, it has become increasingly evident that the PDL is itself the key to attaining the complete regeneration of the periodontium, since this thin tissue of less than 500 μm interconnecting dental root and alveolar bone [21,22] through a series of collagen fiber bundles, is obliterated in PD. Many research efforts have been and are currently spent on identifying the best methods for fabricating biomimetic 3D scaffolds able to reproduce the PDL microenvironment. In Table 2, we report some relevant in vivo studies.

Table 2. Representative in vivo studies using different scaffolds and PDLSCs to regenerate the periodontal complex.

Type of Scaffold	Cell Type	Production Method	Outcome	Reference
PGA or PCL+ βTCP	OBs, PDLSCs	Cell sheet technology	Periodontal complex	[50,51]
Human tooth root	PDLSCs, HUVEC	Cell sheet technology	PDL fibers	[52]
PCL	PDLSCs	3D printing of fiber-guided scaffolds	Enhancement of the bone volume fraction and of tissue mineral density	[53,54]
collagen/bioactive glass/chitosan membrane	PDLSCs	electrospinning	Periodontal complex	[55]
PCL/PLGA+BMP-2,-7, CTGF	PDLSCs	3D printing	Cementum-like layer formation	[56]
iTE scaffold (core/shell fibrous super-assembled framework+ BMP-2, bFGF)	PDLSCs	iTE	Periodontal complex	[57]

3.1. Cell Sheet Technology

One of the first tissue engineering approaches the cell sheet technology, based on culturing the cells in hyper-confluency until they produce their own extracellular matrix (ECM) by forming a cell sheet [58]. Proposed by Okano et al., this technique entails the use of poly-N-isopropyl acrylamide (PIPA Am) as a convenient cell substrate capable of both supporting the growth of a cell monolayer at 37 °C and releasing it below 20 °C without any enzymatic degradation [59]. The adhesion of the cell sheet to the root surface was enhanced through the preservation of the integrin–fibronectin complex [60]. In 2009, Iwata et al. isolated canine PDLSCs and seeded them on temperature-responsive culture dishes until sheet formation. Three-layered PDL cell sheets supported with woven poly glycolic acid (PGA) were transplanted to the exposed dental root surfaces, and bone defects were filled with porous β-TCP, inducing both the regeneration of new bone and the connection of cementum with well-oriented collagen fibers [50]. In 2012, Vaquette et al. combined fused deposition modeling with electrospinning, obtaining a biphasic scaffold with compartments through bone and PDL sheets, demonstrating that the presence of cell sheets promoted periodontal fiber attachment and cementum-like cells [51]. Takahashi et al. demonstrated that PIPA Am is useful for fabricating a brush surface with selective patterns capable of supporting cell growth while preserving orientation [61]. To overcome the pitfalls of single cell sheets in large-scale tissue injuries, Raju et al. proposed 3D complex cell sheets composed of multiple types of cells, attaining the functional connection of collagen fibers to the tooth root and alveolar bone [62]. Similarly, the co-culture of PDLSCs and human umbilical vein endothelial cells (HUVECs) allowed the generation of 3D cell sheet constructs that were wrapped around human tooth roots and implanted into the subcutaneous layer of mice. The presence of HUVECs contributed to the regulation of the thickness of PDL, which was thicker than in mice treated with PDLSCs alone [52]. An unsolved issue of PDL cell sheet technology [63], however, is achieving the directional control of the fibrous network within the constructs. To verify this crucial role of the ECM, as a proof of concept, microfibrous scaffolds were prepared by removing the cellular component from tooth slices using sodium dodecyl sulfate and Triton X-100 and supporting the repopulation and differentiation of PDL cells [64].

3.2. The 3D Printing

Among the most promising techniques implemented to physically control the orientations of PDL, researchers have focused recently on additive manufacturing, a technique that allows one to precisely control the macro- and micro-structure of the scaffolds [65,66]. Ideally, this technique can build complex tissues by depositing different materials layer by layer following 3D digital models and can even embed cells directly within the constructs during the fabrication in a process called bio-printing [67]. Hence, the main advantage over traditional tissue engineering protocols relies in the possibility of fine-tuning the creation of tissues to be akin to that of the native cellular micro-environment [68,69]. This approach may be used as a sophisticated mean to reproduce proper fiber orientation, creating specific micro-grooved surfaces for aligning human PDL cells with high predictability [54]. Such is the case of polymeric 3D scaffolds capable of replicating the peculiar micro-patterned histological architecture [70]. In vitro, this arrangement could be maintained for prolonged periods of time in the presence of growing cell populations [25].

3.2.1. Synthetic Polymers and Surface Modifications of Printed Scaffolds

Among the materials suitable for 3D printing, polycaprolactone (PCL) is widely used due to its convenient rheological, mechanical, and biological features [71]. PCL scaffolds endowed with meso/microscale architectural features were also fabricated to form de novo bone–ligament–cementum complexes in vivo [53]. A 3D-printed bone region with grooved pillars seeded with fibroblasts overexpressing bone morphogenetic protein (BMP)-7 was covered with a tooth dentin segment, and was subsequently positioned subcutaneously in a murine model, with a very encouraging outcome [53].

To improve cell adhesion efficiency on PCL, various surface modification treatments aimed at reducing its hydrophobic interface have been proposed, such as graphene oxide (GO), oxygen plasma, and gelatin coatings (Figure 3) [72,73]. Through plasma treatment, it is possible to variate the surface roughness of nanosized PCL scaffolds, conveniently modulating cell adhesion [74]. Moreover, through electrospinning technique, PCL allows the preparation of nanofibrils or nanocellulose membranes, which can be utilized to encapsulate and carry drugs. For instance, membranes made of PCL encapsulated in gelatin nanocellulose were prepared, and magnesium oxide nanoparticles were incorporated inside them. This system showed high biocompatibility and hydrophilicity that promoted PDLSCs proliferation rates [73].

Vera-Sánchez et al. studied the biocompatibility and potential of a composite coating with GO to induce differentiation of human PDLSCs [75], showing that the GO coating technology increases the hydrophilicity of the PCL surface, promoting cell adhesion. Additionally, poly(d,l-lactide-co-glycolide)/hyaluronic acid PLGA/HA biodegradable microcarriers were treated with GO, improving osteogenic differentiation of stem cells [76]. PCL is fundamental in allowing the printability of the scaffold since it can be conveniently integrated with proper hydrogels functioning as cell carriers and possibly other components, such as a mineralized compartment in a multilayered construct. From an anecdotal point of view, a human case of a large periodontal bone defect treated with a 3D-printed PCL-based scaffold and enriched with platelet-derived growth factors has been reported [77]. Unfortunately, the scaffold was removed after 13 months due to exposure and bacterial contamination. This unsuccessful outcome likely depended on the slow resorbability of PCL and the geometry of the construct, which was too bulky and scarcely interconnected.

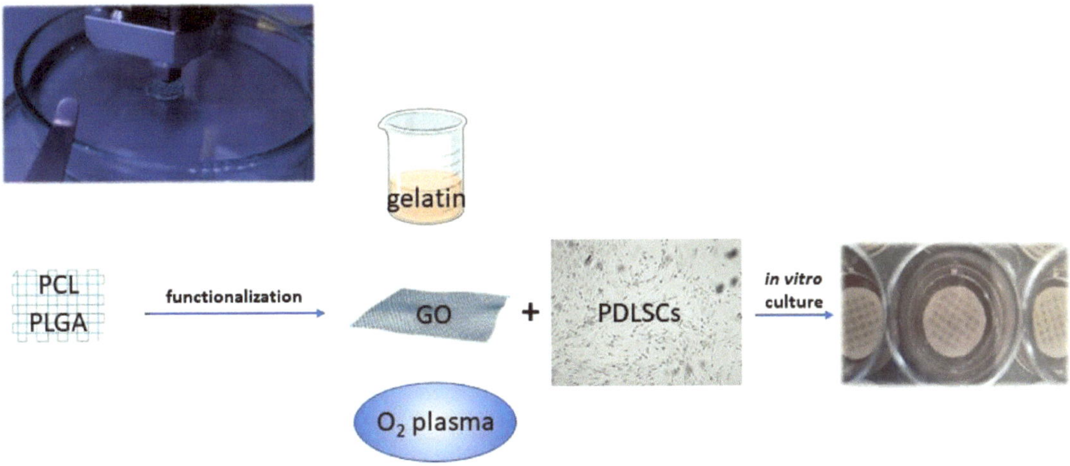

Figure 3. Polymers, such as PCL and PGLA, can be used to print scaffolds functionalized to promote PDLSC adhesion and proliferation. Different strategies of functionalization can be adopted: gelatin nanocellulose can be mixed with PCL or used as an envelope to incorporate other nanoparticles; graphene oxyde (GO) coating increases the hydrophilicity of the PCL surface; oxygen plasma variates the surface roughness.

Melt electrowriting (MEW), a novel technology particularly suitable for PCL [78], is expected to overcome the common limitations of 3D-printed scaffolds, such as porosity not being well matched to tissue, poor resolution, and inflexible shapes [79]. MEW enables the fabrication of micron- to nanodiameter filaments arranged in highly ordered architectures [80] within multicompartmental scaffolds [81] that may mimic the biochemical composition and/or structural organization of the hierarchical structure of the periodontium, incorporating not only PDL but also its interfacial tissues [82]. By presenting selectively regulatory cues within each compartment, multicompartmental scaffolds can guide cells to form the tissue types desired within the anatomical locations designed [83], promoting cell/tissue in-growth [84]. In 2022, the research group led by William V. Giannobile proposed the design of tricompartmental scaffolds obtained via MEW. Thereby, human PDLFs and primary osteoblasts were co-cultured, achieving "a mineral gradient from calcified to uncalcified regions with PDL-like insertions within the transition region" [85]. As the authors claim, their "process effectively recapitulates the key feature of interfacial tissues in periodontium", offering "a fundament for engineering periodontal tissue constructs with characteristic 3D microenvironments similar to native tissues".

3.2.2. Natural Polymers

A viable and natural alternative to PCL is collagen, the primary extracellular PDL protein, which has been employed widely as grafting material owing to its outstanding biocompatibility [86,87]. Unfortunately, collagen alone is not easily printable because of its low viscosity and denaturation temperature [88]. To address this problem, a novel technology called freeform reversible embedding of suspended hydrogels (FRESH) was introduced whereby collagen was deposited in a hydrogel that functioned as a transient mold to be removed non-destructively afterwards [89]. FRESH allowed the printing of collagen parts of human hearts with a satisfactory resolution (20–200 μm) [90].

A good candidate combining antibacterial properties and printability is chitosan, a natural biodegradable polysaccharide already used for guided tissue regeneration [91,92].

A nanohydroxyapatite–chitosan scaffold combined with PDLSCs resulted in effective promotion of bone regeneration in a calvaria bone repair model [93]. In 2018, Varoni et al. [94] prepared a tri-layer scaffold characterized by highly oriented channels aimed at guiding PDL fiber growth through electrochemical deposition. The other compartments were produced with medium- and low-molecular-weight chitosan for regenerating gingiva and bone, respectively. Excellent results supported the feasibility of this resorbable tri-layered structure both in vitro, with high cell survival rates, and in vivo, achieving selective differentiation in terms of mineralized deposits. From this perspective, the development of a new chitosan-based bioink incorporating cellulose nanocrystals, although it has only been tested on murine pre-osteoblasts, is to be regarded as most interesting [95] since it could extend the application of 3D bioprinting for periodontal regeneration to a natural polymer.

3.2.3. Hydrogels

The above-described natural polymers, collagen and chitosan, along with fibrin can also be used as hydrogels, being biodegradable and biocompatible and resembling the original ECM components. The ideal carrier is meant to mimic the ECM, which forms an intricate fibrillar architecture, and can also deeply affect PDLSC colonizing capabilities. Indeed, when seeded on a fibrin sponge, PDLSCs produced abundant ECM, that was positively stained by Alizarin Red S [96]. The effect of a biomimetic electrospun fish–collagen/bioactive–glass/chitosan composite nanofiber membrane (Col/BG/CS) on periodontal regeneration was investigated, showing that the composite membrane promoted cell growth and osteogenic gene expression in vitro, but it was also effective in promoting PDL and bone formation in a canine model [55].

The combination of collagen and methacrylate has garnered growing interest owing to the suitability of the latter for 3D printing; indeed, by adding methacrylate, the collagen may crosslink via UV light in a more controlled way in lieu of using thermal crosslinking. A customized 3D cell-laden hydrogel array with a gradient of gelatin methacrylate (GelMA) and poly(ethylene glycol) (PEG) dimethacrylate compositions showed that the higher the ratio of PEG was, the better the performance of the PDLSCs in cell proliferation and cell spreading on the scaffold [97]. Nonetheless, as inert ECM-based scaffolds alone may be poorly efficient for generating durable tissue repair, they are usually functionalized to release active compounds. PDLSCs sheets combined with platelet-rich plasma were useful for increasing the production of ECM and enhancing cell differentiation [98]. PDLSCs engineered to overexpress platelet-derived growth factor-BB showed increased osteogenic power and were tested in a rat model to induce alveolar bone regeneration [99,100]. The presence of signaling molecules, such as connective tissue growth factor (CTGF), BMP-2, and BMP-7 promote tissue regeneration and cementogenic differentiation [101–103]. These three factors have been incorporated into 3D-printed PLGA microspheres, and the results indicated that BMP-7 triggered thicker cementum-like layers, better integration with the dentin surface, and higher expression of cementum protein-1 [56]. In situ tissue engineering (iTE) allowed the production of a iTE-scaffold made with a PLGA/poly (L-lactic acid (PLLA) shell/core structure and functionalized to allow a sequential delivery of b fibroblast growth factor (bFGF), which promotes regeneration of the periodontium [104] and BMP-2, significantly facilitating stem cell homing, proliferation, and periodontal bone regeneration [57,105]. This iTE-scaffold, implanted in a rat model of periodontal defect, demonstrated an anti-inflammatory response, provided adequate blood supply, and achieved the desired bone repair [106]. Regrettably, over the years, safety concerns have hindered the usage of bioactive molecules such as BMP-2 at the high concentrations required to be effective [107], somehow questioning the classic paradigm of tissue engineering based on cells/scaffolds/signaling cues and favoring the development of smart materials [108].

To avoid the functionalization through bio-active molecules, increasing interest in the synthesis of conductive polymers that are also printable—such as poly(3,4-ethylenedioxythiophene): polystyrene sulfonate (PEDOT:PSS) [109], which can be conveniently enriched by poly(vinyl alcohol) (PVA) forming a hydrogel strain sensor [110]—has been conveyed. Conductive hydrogels may become ideal interfaces with the human body, but rarely simultaneously possess the satisfying electrical, mechanical, and adhesive properties shown by the Ti3C2Tx-polyacrylic acid hydrogel that can be printed into complex geometries with high resolution [111].

4. Mimicking the Physical Micro-Environment of PDL

During normal oral functions, intermittent occlusal contacts accompanied by pressure from the tongue occur, and the PDL periodically undergoes different combinations of mechanical loading (i.e., compression, stretch, fluid-induced shear stress) that contribute to maintain the PDL homeostasis [112]. The mechanical response of PDL to experienced loading is determined by the combination of the oriented collagen fiber bundles and the distribution of the interstitial fluid, which makes the PDL acting as a shock absorber, increasing the tooth's ability to withstand loading via the hydrostatic effect [113,114]. This is accompanied by a mechano-biological response of the PDL that, depending on the location under the tooth, changes structure and functions and induces remodeling in the surrounding tissues [5]. Cells detect and transduce the mechanical signals from their membrane to the nucleus through a molecular process named mechanotransduction [115]. Among the most relevant cell membrane mechanosensors, integrins play a fundamental role, mediating direct contact with the ECM. As transmembrane constituents of the focal adhesions (FA), integrins interact with scaffolding, docking, and signaling proteins linked to the actin cytoskeleton [116]. The variable composition of the FA core depends on ECM and mechanical stimuli. Cells modulate their own cytoskeletal architecture in response to applied forces [117] and remain in a sort of tensional homeostasis, i.e., a basal equilibrium stress state [118]. From a molecular point of view, the role of two actin-binding proteins associating with the cytoplasmic tail of the β1 integrin—talin and filamin A (FLNa) [119]—is noteworthy. By interacting with integrins, talin enhances cell adhesion to the ECM [120]. Without mechanical stimuli, talin is fully structured, but while under increasing force regimes, talin exposes progressively more vinculin binding sites (VBS), thus activating more vinculin proteins [121]. Vinculin may mediate reorganization of cell polarity, helping the cell to adapt to increased tensile forces [122]. FLNa competes with talin for binding to β1 integrin [123], and it is thought to antagonize integrin-mediated cell adhesion [124]. For example, Shifrin et al. showed that through the Rac/Pak/p38 signaling pathway, FLNa may prevent apoptosis in PDL in response to tensile forces [125].

Due to the fundamental role played by mechanical loading in vivo, a compelling strategy for directing cell commitment in periodontal tissue engineering may be represented by the reproduction in vitro of the dynamic environment in which PDL cells operate. Several groups investigated the sensitivity of PDL cells (i.e., cells harvested form the PDL, including not only PDLSCs but also more committed cells and even fibroblasts) to mechanical loading and their involvement in periodontal and bone remodeling in vitro. In the following paragraphs, investigations performed in the last decade on in vitro mechanically stimulated PDL cells and constructs are reported per type of force (i.e., compression, stretch, shear stress; Figure 4 and Table S1) and depending on the adopted in vitro mechanical loading method (Figure 5), highlighting the use of conventional two-dimensional (2D) or more physiological 3D cell culture techniques.

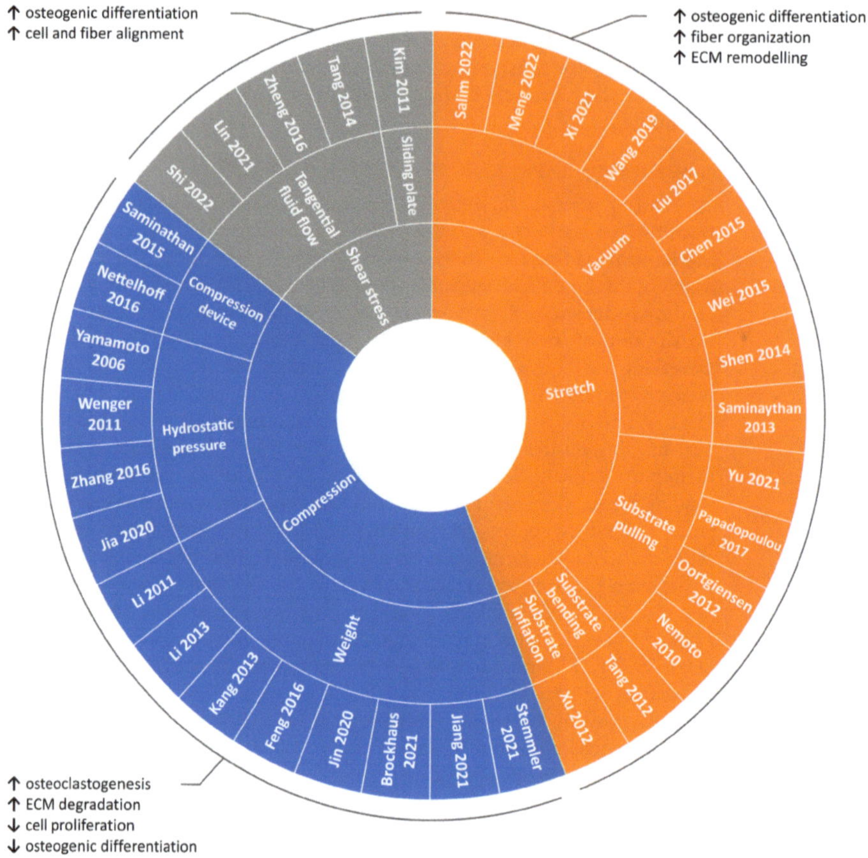

Figure 4. Overview of the reviewed in vitro PDL mechanical loading studies. Schematic summary of the reviewed studies investigating the influence of mechanical loading on PDL cells based on the type of applied stimulation along with main outcomes.

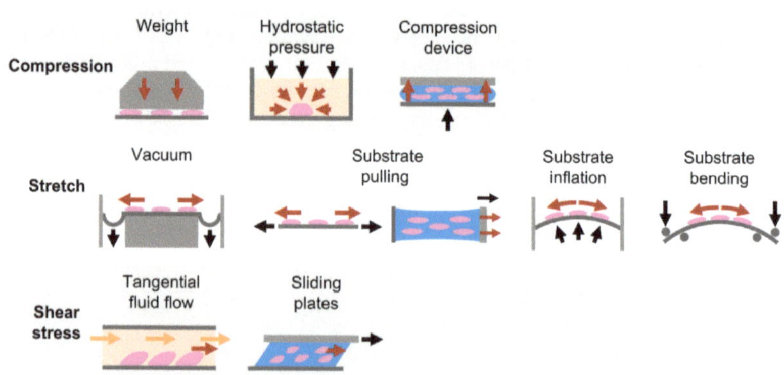

Figure 5. In vitro mechanical loading methods applied for PDL investigations. Schematic representation of the in vitro mechanical loading methods adopted in the reviewed studies for exposing PDL cells cultured in 2D layers or in 3D constructs to compression, stretch, and shear stress stimuli.

4.1. Compression

4.1.1. Weight Method

This method, based on the use of cover glasses or cylinders containing metal granules, which allows the application of tunable static compressive forces to the culture, was widely adopted to investigate in vitro how continuous compression can influence PDL cells. In 2011, Li et al. established a 3D model of PDL tissue based on hPDL cells seeded on a sheet of porous PLGA scaffold and exposed it to static compression (5–35 g/cm^2 for 6–72 h), observing a significant induction of osteoclastogenic genes that did not occur when human gingival fibroblasts were used [126]. Moreover, a predominant upregulation of osteoclastogenesis inducers was observed at the early stage (6 h), while osteoclastogenesis inhibitor genes increased at the late stage (24–72 h), although cell proliferation was reduced [127]. In 2013, a comparison between hPDL cells cultured under compressive forces (2.0 g/cm^2 for 2 or 48 h) in conventional 2D culture dishes or in 3D collagen gel highlighted significant alterations of the expression levels of several genes [128]. In particular, the number of activated integrin–focal adhesion kinase (FAK) was higher in 3D than in 2D culture, supporting that cellular attachment to ECM can strongly influence cellular responses to mechanical forces [128]. In 2016, it was demonstrated that compressive force (1 g/cm^2 for 24 h) applied to hPDLSCs altered cell morphology and repressed collagen expression, which both recovered after force withdrawal [129]. Recently, continuous compression (0–1.5 g/cm^2 for 12 h) on PDLSCs could reduce differentiation ability and increase macrophage migration, osteoclast differentiation, and proinflammatory factor expression. Moreover, a universal upregulation of the subfamily V member 4 of the transient receptor potential calcium channel (TRPV4), which regulated osteoclast differentiation by affecting the system receptor activator of nuclear factor kappa-B ligand (RANKL)/osteoprotegerin (OPG) via extracellular signal-regulated kinase (ERK) signaling (Figure 6a) [130], was shown. Brockhaus et al. reported that hPDLFs cultured under compression (2 g/cm^2 for 24, 48, and 72 h) changed their morphology towards more unstructured, unsorted actin filaments, with a significant reduction of proliferation followed by recovery after 48 h, demonstrating that hPDLFs restore homeostasis and adapt to the compressive force through a lower cell division rate and a slowed cell cycle [131]. Moreover, Stemmler et al. reported that the inflammatory response of hPDLFs caused by periodontal pathogens combined with compressive load (2 g/cm^2 for 6 h) was supported by the growth differentiation factor 15 (GDF15), which modulated the inflammatory response of PDLFs also regulating the levels of the key inflammatory molecule TNFα [132]. Recently, Jiang and colleagues showed a novel cellular mechanism. Indeed, continuous compressive force (0.5–2.5 g/cm^2, 12 h) activated autophagy in hPDLSCs that induced M1 macrophage polarization via the inhibition of the AKT signaling pathway, contributing to the force-induced bone remodeling and tooth movement [133].

4.1.2. Hydrostatic Pressure Method

The hydrostatic pressure method exploits air pressure applied on the culture medium for imposing static or fluctuating compressive forces. Exerting a static compressive stress on PDLSCs (100 kPa, for 1, 6 or 12 h) [134] and on hPDL cells (1 MPa or 6 MPa for 10 or 60 min) [135], the expression of genes regulating osteoclastogenesis and osteoblastogenesis was induced. For mimicking the physiological loading during mastication, cyclic hydrostatic pressure was applied (1 MPa, 0.1 Hz, 3 h/day for 2 days) on hPDLFs, showing that the expression of several integrins, collagens, and metalloproteinases was significantly upregulated [136]. Recently, the inflammatory, osteogenic, and pro-osteoclastic effects of different cyclic compressive loading conditions (50–150 kPa, 0.1 Hz, for 1 h/day for 5 days) were investigated by stimulating hPDL cells in an inflammatory environment using a customized bioreactor [137]. According to the level of cyclic pressure, cells released different levels of inflammatory and pro-osteoclastic factors, modulating the downregulation (with150 kPa) or the upregulation (with 90 kPa) of osteogenic genes (alkaline phosphatase (ALP), collagen type I (COLL-1), RUNX2, OCN, osteopontin (OPN), and osterix (OSX)) [138].

4.1.3. Substrate Deformation Methods

Compression can also be achieved through several commercial and customized devices based on substrate deformation, in which an elastic membrane is deformed by force and the cells/constructs cultured on it are exposed to strain. In 2014, a 3D construct composed of hPDL cells seeded into a matrix of hyaluronan, gelatin, and COLL-1 was exposed to cyclic compression (340.6 g/cm^2 for 1 s every 60 s for 6, 12, and 24 h) by using the commercial Flexercell FX-4000C Strain Unit (Flexcell International Corporation, Hillsborough, NC, USA). Compression increased cell death and the expression of several apoptosis-related genes. ECM genes were mostly upregulated after 6–12 h, but all were downregulated at 24 h, except for the three major ECM-degrading enzymes—MMPs1–3— and the connective tissue growth factor (CTGF), with upregulated matrix metalloproteinase-1 (MMP-1) and tissue inhibitor of metalloproteinases-1 (TIMP-1) protein levels without changes observed in RANKL, OPG, and basic fibroblast growth factor (FGF-2) expression [139]. By using the same device, Nettelhoff et al. exposed hPDLFs to compressive force (5 and 10% for 12 h). The 5% compression induced the highest ALP gene expression and the highest RANKL/OPG ratio, while 10% compression decreased cell viability without promoting apoptosis but resulting in tissue damage [140].

Thus, short-term static compression (almost for 1 h) can promote osteogenic differentiation of PDLSCs, while long-term static compression (for 12 h or longer) can alter the morphology of hPDL cells, may inhibit cell proliferation and the osteogenic differentiation of PDLSCs, and can promote the secretion of osteoclastogenesis-stimulating cytokines and ECM degradation.

4.2. Stretch

4.2.1. Substrate Deformation—Vacuum Approach

The vacuum approach stretches the cell-seeded membrane across a loading post by applying vacuum pressure, delivering a tunable biaxial or uniaxial tensile strain. This method is commonly applied by using commercial devices (e.g., Flexercell tension system (Flexcell International Corporation, Hillsborough, NC, USA). In 2013, Saminathan et al. embedded hPDLFs in an 80–100 μm thick 3D collagen membrane and cultured the construct under equibiaxial cyclic stretching (12%, 0.2 Hz, 5 s every 60 s for 6 h/day up to 21 days) to investigate the influence on ECM homeostasis. Mechanical loading did not affect the cell number, but it significantly upregulated the release of MMP-1 and TIMP-1 in the supernatants, suggesting that fibroblasts were remodeling the surrounding ECM [141]. This was confirmed by Chen et al. who exposed hPDLCs to equibiaxial cyclic stretching (12%, 0.1 Hz for 24 h), showing the upregulation of major periodontal ECM genes, such as COL1A1, COL3A1 and COL5A1 [142]. Some studies reported that hPDLSCs cultured in osteoinductive medium under cyclic stretching enhanced the osteogenic differentiation [143,144], both maintaining the same parameters stimulation and reducing the time of stimulation [145] or keeping the same strain magnitude but halving the frequency [146,147]. In particular, Xi et al. demonstrated that cyclic stretching (10%, 0.5 Hz, for up to 36 h) could increase the generation of reactive oxygen species (ROS), which may lead to the osteogenic differentiation of hPDLSCs (Figure 6b) [146]. In 2017, Liu et al., culturing healthy or pathological donor-derived PDLSCs under static strain (6–14% for 12 h), showed that PDLSCs from patients affected by periodontitis were more sensible to physical load than PDLSCs from healthy patients, likely due to the inflammatory milieu [148]. Recently, Salim et al. cultured human PDL cells under static strain (2.5, 5, and 10% for 24 h) and performed in vivo analyses on teeth with and without orthodontic tooth movement (OTM) [149]. Interestingly, they found that chaperone-assisted selective autophagy (CASA) machinery genes (chaperones HSPA8 and HSPB8, the cochaperones BAG3 and STUB1, and the molecule SYNPO2 interacting with BAG3 for autophagosome membrane formation) were inherently expressed in PDL cells and exhibited transcriptional induction upon in vitro mechanical strain and in vivo after OTM. The role of FLNa was also investigated, pointing out that it acts as a flexible actin crosslinker that is stretched under tension and degraded by CASA when damaged, which is consistent with previous works [150,151], further supporting the importance of the dynamic environment as a key factor of the homeostatic maintenance of PDL both in physiologic and treatment conditions [152].

4.2.2. Substrate Deformation—Pulling Approach

The substrate-pulling approach is based on a system that clamps the cell-seeded membrane and imposes uniaxial stretch by a controlled actuator. Adopting the commercial STB-140 STREX cell stretch system (Strex Co., Osaka, Japan), it was reported that a long-term cyclic stimulation (5%, 60 s/returns, resting time = 29 s for 7 days) could increase collagen mRNA and protein expression, suggesting that cyclic stretch on hPDLFs may contribute to the homeostasis of PDL fibers and to the ECM remodeling [153]. In 2012, a 3D construct based on a collagen film laden with rat PDLFs was exposed to uniaxial cyclic stretch (8%, 1 Hz (15 min stretch + 15 min rest) for 8 h/day for 5 days) using a customized device. After mechanical stimulation, the cells were perpendicularly oriented with respect to the stimulation direction and, analyzing several genes' expression (COLL-1, RUNX2, c-fos, and Cox-2), the authors concluded that PDL cells under loading might tend to have bone-like and, at the same time, tendon-like behavior [154]. Applying the same stimulation for a shorter period (16 h) and in a cyclic or static manner, the cellular orientation could be reached, and three different pathways (ERK, p38 and JNK) were activated [155]. In 2021, Yu et al. demonstrated that exposing a hPDLSCs-laden 3D collagen membrane to uniaxial stretching (20% for 5 days) dramatically enhanced the bioactivity of PDLSC-derived exosomes [156].

4.2.3. Substrate Deformation—Inflation and Bending Approaches

Studies based on the substrate inflation approach were inspired by Howard et al. [157]. A cell-seeded membrane is clamped and deflected by hydrostatic pressure applied to the underside, providing uniform biaxial stretch. In 2012, Xu and colleagues showed that hPDL cells exposed to cyclic stretching (1–20%, 0.1 Hz up to 24 h) appeared aligned perpendicularly to the stretching direction, and the expression of the membrane connexin 43 (Cx43) protein could be modulated in a time- and magnitude-dependent manner via cyclic stretching [158]. By adopting the substrate-bending approach, the osteogenic differentiation of hPDLSCs cultured under tensile stress (3000 μstrain, 0.5 Hz) was reached after 24 h of stimulation, demonstrating upregulation of different osteogenic markers [159].

In summary, the aforementioned studies demonstrated that tensile stress could promote osteogenic differentiation. In particular, cyclic stretch can upregulate the protein and mRNA expression of osteogenic genes and the synthesis of osteoclastogenesis-inhibitory molecules, complemented by an increased expression of the major periodontal ECM genes, leading to the homeostasis and organization of the PDL fibers and to ECM remodeling.

4.3. Shear Stress

The effect of shear stress on PDL cells has been poorly investigated in the literature, even though this stimulus is an important cue in the physiological environment. The most adopted method for investigating the effect of shear stress on PDL behavior is based on tangential fluid provided by parallel plate flow chambers. In 2014, Tang and colleagues cultured hPDL cells in osteogenic medium under steady fluid shear stress (12 dyn/cm^2 for 2 h), showing an early morphologic change and rearrangement of filamentous actin with significant increases in ALP activity and mRNA levels of osteogenic genes and osteoid nodules [160]. Similarly, Zheng et al. exposed hPDL cells to fluid shear stress (6 dyn/cm^2 up to 12 h), reporting a rearranged cell alignment, an inhibited cell proliferation and migration, and osteogenic differentiation [161]. Very recently, Shi et al. observed that fluid shear stress (6 dyn/cm^2 for 4 h) promoted cell proliferation by activating mechanotransduction pathways involving the p38 mitogen-activated protein kinases, angiomotin (AMOT), and Yes-associated protein (YAP) (Figure 6c) [162]. Adopting the sliding plate method where the 3D construct is housed between two parallel plates with one of them connected to an actuator that imposes controlled sliding motion and consequent shear stress on the construct, a static shear stress was applied to a construct composed of hPDL cells embedded in a collagen gel. After 24 h of stimulation, cells and collagen fibers aligned in the direction of the principal strain vector [163]. Recently, a model of PDL regeneration based on a fiber-guiding scaffold seeded with PDL cells and subjected to shear stress in a laminar flow-based bioreactor (6 dynes/cm^2 for 1–4 h) showed increased viability, adhesion, and cytoskeleton arrangement compared to cells in the absence of load [164].

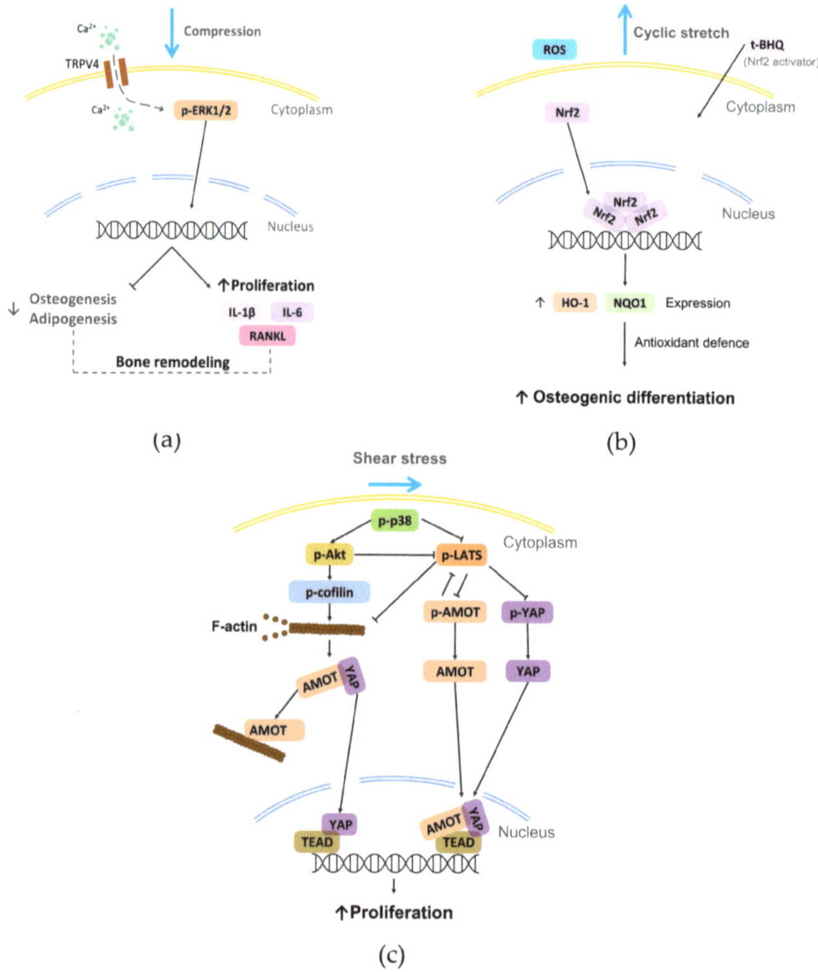

Figure 6. Schematic representation of some mechanotransduction pathways activated on PDL cells by the reported physical stimuli (i.e., compression, stretch, and shear stress). (**a**) External compression applied on rat primary PDLSCs can induce the activation of TRPV4 and consequently modulate bone remodeling [130]; (**b**) external cyclic stretch can promote the osteogenic differentiation of hPDLSCs by activating the Nrf2 [146]; (**c**) shear stress applied to hPDL cells can activate p38, which regulates the nuclear translocation YAP, promoting cell proliferation [162].

Overall, the above-mentioned findings highlight how shear stress can represent an effective approach for guiding cell and fiber alignment and for promoting osteogenic differentiation.

5. Future Perspectives

Exciting advances in engineering and cell biology have begun to address the numerous obstacles hampering the implementation of effective protocols for periodontal regeneration. Despite the flurry of research regarding osteogenic differentiation, however, only a few papers deal with tenogenic differentiation, which is essential to preserve the PDL tissue within the mineralized interfaces of cementum and bone. This lack of knowledge could delay the goal of making available cost-effective, patient-specific treatment options in the future. To this end, it is necessary to assess whether and how mechanical stimulation protocols affect the tenogenic differentiation of PDLSCs, as it is conceivable that various types of physical stimulation, with different application times, are required to achieve the combined modulation of osteogenic and tenogenic activities that allow PDL regeneration. Therefore, in-depth in vitro investigations combined with high-throughput analyses are increasingly necessary to unravel molecular and cellular mechanisms activated by different spatial and physical culture conditions. Finally, from a long-term perspective, developing reliable protocols for periodontal regeneration requires the solution of the following challenges: (1) the definition of the most suitable mechanical stimuli and of the optimal stimulation parameters for promoting tissue regeneration; (2) assessing how the biomechanical load affects the tissues to be regenerated overall; (3) handling the construct at the wound site, avoiding microbial infection/reinfection.

6. Conclusions

Despite the remarkable endeavors done in the last years to identify new regenerative treatments for periodontitis, periodontal regeneration remains a tricky challenge. GTR was proven to be as effective as the open flap debridement [165] and seemingly more efficient in regenerating bone than the PDL, which is the key for a true restitutio ad integrum of the periodontium. Cell sheet engineering is technically demanding owing to difficulties in handling and stabilizing the construct, which has greatly limited its clinical application [81]. Several pre-clinical studies [166] have proven the effectiveness of PDLSC-based therapies in regenerating PDL, but the clinical feasibility of this therapeutic approach is far from being achieved. Indeed, well-designed randomized controlled trials (RCTs) are mandatory for assessing the long-term success of innovative procedures, but they must abide by stringent safety and regulatory issues, delaying the implementation of protocols based on PDLSC transplantation in humans. To the authors' knowledge, only one limited RCT has been retrieved in the scientific literature so far (2022). It reported the clinical safety of the treatment with PDLSCs but with inconclusive evidence regarding PDLSC-based periodontal regeneration [38]. In order to create PDL-cementum complexes, it becomes paramount to tailor the cell environment (mimicking the ECM) by selecting the best scaffold capable of supporting PDLSC growth and differentiation in humans, which implies the adoption of the most up-to-date 3D printing techniques. In parallel, the biomimetic approach demands the use of technological devices for supporting the in vitro maturation of the tissue to be grafted, particularly bioreactors providing controlled physical stimuli to reproduce the native environment [167–169]. Indeed, mechanical stimulation can be essential for promoting specific cellular and tissue processes with the final aim of inducing the correct orientation of the fibers forming the PDL and of maintaining tissue homeostasis. In particular (Figure 4): (i) moderate compressive forces promote active tissue remodeling, while long-term static compression can alter the morphology of hPDL cells, inhibiting the proliferation and the osteogenic differentiation of PDLSCs while inducing osteoclastogenesis and ECM degradation; (ii) tensile stress enhances the osteogenic differentiation of PDLSCs, and cyclic stretching can contribute to the homeostasis and organization of the PDL fibers and to the ECM remodeling; (iii) shear stress promotes osteogenic differentiation and guides cell and PDL fiber alignment. Since PDL is a complex multilayer tissue that is physiologically subjected to all these types of forces, a biomimetic in vitro model should both mimic the PDL microstructure and provide, in a controlled manner, the actual combined mechanical loading experienced in vivo. Such an advanced model would allow the mechano-biological behavior of the PDL to be comprehensively unveiled and the best culture protocols for obtaining reliable results in terms of PDL regeneration to be defined. Finally, synergic collaborations between biologists, biotechnologists, biomaterialists, bioengineers, and dentists become urgent for the development of advanced solutions for more reproducible, standardized, and biomimetic studies that could lead in the near future to a successful production of functionally engineered PDL for clinical applications.

Supplementary Materials: The following supporting information can be downloaded at: https://www.mdpi.com/article/10.3390/nano12213878/s1, Table S1. List of the reviewed in vitro PDL mechanical loading studies.

Author Contributions: Conceptualization, F.M., D.M. and I.R.; methodology, I.R., B.M. and G.P.; formal analysis, F.M.; investigation, I.R., B.M., G.P. and D.M.; resources, I.R. and F.M.; data curation, I.R., B.M. and G.P.; writing—original draft preparation, I.R., F.M., B.M., G.P. and D.M.; writing—review and editing, F.M., I.R., B.M. and D.M.; visualization, I.R., B.M. and G.P.; supervision, F.M. and D.M.; project administration, F.M.; funding acquisition, I.R. and F.M. All authors have read and agreed to the published version of the manuscript.

Funding: This research was funded by Ministero dell'Istruzione, dell'Università e della Ricerca (MIUR), program "Dipartimenti di Eccellenza" ex L.232/2016 of the Dept. of Surgical Sciences, University of Turin. MIUR had no part whatsoever in conducting the research, in the preparation of the article, or in the decision to submit the paper for publication. The APC was funded by a grant of the University of Turin. This work was also supported by CRT Foundations (CRT2020) and Fondazione Ricerca Molinette ONLUS.

Conflicts of Interest: The authors declare no conflict of interest.

References

1. Menicanin, D.; Hynes, K.; Han, J.; Gronthos, S.; Bartold, P.M. Cementum and Periodontal Ligament Regeneration. *Adv. Exp. Med. Biol.* **2015**, *881*, 207–236. [CrossRef] [PubMed]
2. Socransky, S.S.; Haffajee, A.D. Periodontal microbial ecology. *Periodontology 2000* **2005**, *38*, 135–187. [CrossRef]
3. Beertsen, W.; McCulloch, C.A.; Sodek, J. The periodontal ligament: A unique, multifunctional connective tissue. *Periodontology 2000* **1997**, *13*, 20–40. [CrossRef] [PubMed]
4. Itaya, S.; Oka, K.; Ogata, K.; Tamura, S.; Kira-Tatsuoka, M.; Fujiwara, N.; Otsu, K.; Tsuruga, E.; Ozaki, M.; Harada, H. Hertwig's epithelial root sheath cells contribute to formation of periodontal ligament through epithelial-mesenchymal transition by TGF-beta. *Biomed. Res.* **2017**, *38*, 61–69. [CrossRef] [PubMed]
5. de Jong, T.; Bakker, A.D.; Everts, V.; Smit, T.H. The intricate anatomy of the periodontal ligament and its development: Lessons for periodontal regeneration. *J. Periodontal Res.* **2017**, *52*, 965–974. [CrossRef]
6. Ortun-Terrazas, J.; Cegonino, J.; Santana-Penin, U.; Santana-Mora, U.; Perez Del Palomar, A. Approach towards the porous fibrous structure of the periodontal ligament using micro-computerized tomography and finite element analysis. *J. Mech. Behav. Biomed. Mater.* **2018**, *79*, 135–149. [CrossRef]
7. Gathercole, L.J.; Keller, A. Crimp morphology in the fibre-forming collagens. *Matrix* **1991**, *11*, 214–234. [CrossRef]
8. Maceri, F.; Marino, M.; Vairo, G. A unified multiscale mechanical model for soft collagenous tissues with regular fiber arrangement. *J. Biomech.* **2010**, *43*, 355–363. [CrossRef]
9. Szczesny, S.E.; Driscoll, T.P.; Tseng, H.Y.; Liu, P.C.; Heo, S.J.; Mauck, R.L.; Chao, P.G. Crimped Nanofibrous Biomaterials Mimic Microstructure and Mechanics of Native Tissue and Alter Strain Transfer to Cells. *ACS Biomater. Sci. Eng.* **2017**, *3*, 2869–2876. [CrossRef]
10. Cho, M.I.; Garant, P.R. Development and general structure of the periodontium. *Periodontology 2000* **2000**, *24*, 9–27. [CrossRef]
11. Bosshardt, D.D.; Bergomi, M.; Vaglio, G.; Wiskott, A. Regional structural characteristics of bovine periodontal ligament samples and their suitability for biomechanical tests. *J. Anat.* **2008**, *212*, 319–329. [CrossRef]
12. Tsuruga, E.; Irie, K.; Sakakura, Y.; Yajima, T. Expression of fibrillins and tropoelastin by human gingival and periodontal ligament fibroblasts in vitro. *J. Periodontal Res.* **2002**, *37*, 23–28. [CrossRef]
13. Peres, M.A.; Macpherson, L.M.D.; Weyant, R.J.; Daly, B.; Venturelli, R.; Mathur, M.R.; Listl, S.; Celeste, R.K.; Guarnizo-Herreno, C.C.; Kearns, C.; et al. Oral diseases: A global public health challenge. *Lancet* **2019**, *394*, 249–260. [CrossRef]
14. Lourenco, T.G.; Heller, D.; Silva-Boghossian, C.M.; Cotton, S.L.; Paster, B.J.; Colombo, A.P. Microbial signature profiles of periodontally healthy and diseased patients. *J. Clin. Periodontol.* **2014**, *41*, 1027–1036. [CrossRef]

15. Feres, M.; Teles, F.; Teles, R.; Figueiredo, L.C.; Faveri, M. The subgingival periodontal microbiota of the aging mouth. *Periodontology 2000* **2016**, *72*, 30–53. [CrossRef]
16. Sowmya, S.; Mony, U.; Jayachandran, P.; Reshma, S.; Kumar, R.A.; Arzate, H.; Nair, S.V.; Jayakumar, R. Tri-Layered Nanocomposite Hydrogel Scaffold for the Concurrent Regeneration of Cementum, Periodontal Ligament, and Alveolar Bone. *Adv. Healthc. Mater.* **2017**, *6*, 1601251. [CrossRef]
17. Kawaguchi, H.; Hirachi, A.; Hasegawa, N.; Iwata, T.; Hamaguchi, H.; Shiba, H.; Takata, T.; Kato, Y.; Kurihara, H. Enhancement of periodontal tissue regeneration by transplantation of bone marrow mesenchymal stem cells. *J. Periodontol.* **2004**, *75*, 1281–1287. [CrossRef]
18. Du, J.; Shan, Z.; Ma, P.; Wang, S.; Fan, Z. Allogeneic bone marrow mesenchymal stem cell transplantation for periodontal regeneration. *J. Dent. Res.* **2014**, *93*, 183–188. [CrossRef]
19. Mohammed, E.; Khalil, E.; Sabry, D. Effect of Adipose-Derived Stem Cells and Their Exo as Adjunctive Therapy to Nonsurgical Periodontal Treatment: A Histologic and Histomorphometric Study in Rats. *Biomolecules* **2018**, *8*, 167. [CrossRef]
20. Seo, B.M.; Miura, M.; Gronthos, S.; Bartold, P.M.; Batouli, S.; Brahim, J.; Young, M.; Robey, P.G.; Wang, C.Y.; Shi, S. Investigation of multipotent postnatal stem cells from human periodontal ligament. *Lancet* **2004**, *364*, 149–155. [CrossRef]
21. Bartold, P.M.; Gronthos, S.; Ivanovski, S.; Fisher, A.; Hutmacher, D.W. Tissue engineered periodontal products. *J. Periodontal. Res.* **2016**, *51*, 1–15. [CrossRef] [PubMed]
22. Iwata, T.; Yamato, M.; Zhang, Z.; Mukobata, S.; Washio, K.; Ando, T.; Feijen, J.; Okano, T.; Ishikawa, I. Validation of human periodontal ligament-derived cells as a reliable source for cytotherapeutic use. *J. Clin. Periodontol.* **2010**, *37*, 1088–1099. [CrossRef] [PubMed]
23. Lim, J.C.; Bae, S.H.; Lee, G.; Ryu, C.J.; Jang, Y.J. Activation of beta-catenin by TGF-beta1 promotes ligament-fibroblastic differentiation and inhibits cementoblastic differentiation of human periodontal ligament cells. *Stem Cells* **2020**. *Online ahead of print.* [CrossRef] [PubMed]
24. Bai, S.; Lee, J.H.; Son, C.; Lee, D.S.; Park, J.C. CPNE7 regenerates periodontal ligament via TAU-mediated alignment and cementum attachment protein-mediated attachment. *J. Clin. Periodontol.* **2022**, *49*, 609–620. [CrossRef] [PubMed]
25. Park, C.H.; Oh, J.H.; Jung, H.M.; Choi, Y.; Rahman, S.U.; Kim, S.; Kim, T.I.; Shin, H.I.; Lee, Y.S.; Yu, F.H.; et al. Effects of the incorporation of epsilon-aminocaproic acid/chitosan particles to fibrin on cementoblast differentiation and cementum regeneration. *Acta Biomater.* **2017**, *61*, 134–143. [CrossRef]
26. Iwasaki, K.; Komaki, M.; Akazawa, K.; Nagata, M.; Yokoyama, N.; Watabe, T.; Morita, I. Spontaneous differentiation of periodontal ligament stem cells into myofibroblast during ex vivo expansion. *J. Cell. Physiol.* **2019**, *234*, 20377–20391. [CrossRef]
27. Zhu, Y.; Song, X.; Han, F.; Li, Y.; Wei, J.; Liu, X. Alteration of histone acetylation pattern during long-term serum-free culture conditions of human fetal placental mesenchymal stem cells. *PLoS ONE* **2015**, *10*, e0117068. [CrossRef]
28. Shinagawa-Ohama, R.; Mochizuki, M.; Tamaki, Y.; Suda, N.; Nakahara, T. Heterogeneous Human Periodontal Ligament-Committed Progenitor and Stem Cell Populations Exhibit a Unique Cementogenic Property Under In Vitro and In Vivo Conditions. *Stem Cells Dev.* **2017**, *26*, 632–645. [CrossRef]
29. Abuarqoub, D.A.; Aslam, N.; Barham, R.B.; Ababneh, N.A.; Shahin, D.A.; Al-Oweidi, A.A.; Jafar, H.D.; Al-Salihi, M.A.; Awidi, A.S. The effect of platelet lysate in culture of PDLSCs: An in vitro comparative study. *PeerJ* **2019**, *7*, e7465. [CrossRef]
30. Zheng, W.; Wang, S.; Ma, D.; Tang, L.; Duan, Y.; Jin, Y. Loss of proliferation and differentiation capacity of aged human periodontal ligament stem cells and rejuvenation by exposure to the young extrinsic environment. *Tissue Eng. Part A* **2009**, *15*, 2363–2371. [CrossRef]
31. Du, T.; Liu, N.; Gu, B.; Li, L.; Yuan, Y.; Zhang, W.; Zhang, T. Effects of Aging on the Proliferation and Differentiation Capacity of Human Periodontal Ligament Stem Cells. *Chin. Med. Sci. J.* **2017**, *32*, 81–83. [CrossRef]
32. Silverio, K.G.; Rodrigues, T.L.; Coletta, R.D.; Benevides, L.; Da Silva, J.S.; Casati, M.Z.; Sallum, E.A.; Nociti, F.H., Jr. Mesenchymal stem cell properties of periodontal ligament cells from deciduous and permanent teeth. *J. Periodontol.* **2010**, *81*, 1207–1215. [CrossRef]
33. Kim, J.H.; Kim, G.H.; Kim, J.W.; Pyeon, H.J.; Lee, J.C.; Lee, G.; Nam, H. In Vivo Angiogenic Capacity of Stem Cells from Human Exfoliated Deciduous Teeth with Human Umbilical Vein Endothelial Cells. *Mol. Cells* **2016**, *39*, 790–796. [CrossRef]
34. Song, J.S.; Kim, S.O.; Kim, S.H.; Choi, H.J.; Son, H.K.; Jung, H.S.; Kim, C.S.; Lee, J.H. In vitro and in vivo characteristics of stem cells derived from the periodontal ligament of human deciduous and permanent teeth. *Tissue Eng. Part A* **2012**, *18*, 2040–2051. [CrossRef]
35. Kim, K.; Jeon, M.; Lee, H.S.; Park, J.C.; Moon, S.J.; Kim, S.O.; Cho, S.W.; Song, J.S. Comparative analysis of secretory factors from permanent- and deciduous-teeth periodontal ligament cells. *Arch. Oral Biol.* **2016**, *71*, 65–79. [CrossRef]
36. Sonoyama, W.; Liu, Y.; Fang, D.; Yamaza, T.; Seo, B.M.; Zhang, C.; Liu, H.; Gronthos, S.; Wang, C.Y.; Wang, S.; et al. Mesenchymal stem cell-mediated functional tooth regeneration in swine. *PLoS ONE* **2006**, *1*, e79. [CrossRef]
37. Shi, H.; Zong, W.; Xu, X.; Chen, J. Improved biphasic calcium phosphate combined with periodontal ligament stem cells may serve as a promising method for periodontal regeneration. *Am. J. Transl. Res.* **2018**, *10*, 4030–4041.
38. Chen, F.M.; Gao, L.N.; Tian, B.M.; Zhang, X.Y.; Zhang, Y.J.; Dong, G.Y.; Lu, H.; Chu, Q.; Xu, J.; Yu, Y.; et al. Treatment of periodontal intrabony defects using autologous periodontal ligament stem cells: A randomized clinical trial. *Stem Cell Res. Ther.* **2016**, *7*, 33. [CrossRef]

39. Dominici, M.; Le Blanc, K.; Mueller, I.; Slaper-Cortenbach, I.; Marini, F.; Krause, D.; Deans, R.; Keating, A.; Prockop, D.; Horwitz, E. Minimal criteria for defining multipotent mesenchymal stromal cells. The International Society for Cellular Therapy position statement. *Cytotherapy* **2006**, *8*, 315–317. [CrossRef]
40. Bartold, P.M.; Gronthos, S. Standardization of Criteria Defining Periodontal Ligament Stem Cells. *J. Dent. Res.* **2017**, *96*, 487–490. [CrossRef]
41. Menicanin, D.; Mrozik, K.M.; Wada, N.; Marino, V.; Shi, S.; Bartold, P.M.; Gronthos, S. Periodontal-ligament-derived stem cells exhibit the capacity for long-term survival, self-renewal, and regeneration of multiple tissue types in vivo. *Stem Cells Dev.* **2014**, *23*, 1001–1011. [CrossRef] [PubMed]
42. Trubiani, O.; Zalzal, S.F.; Paganelli, R.; Marchisio, M.; Giancola, R.; Pizzicannella, J.; Buhring, H.J.; Piattelli, M.; Caputi, S.; Nanci, A. Expression profile of the embryonic markers nanog, OCT-4, SSEA-1, SSEA-4, and frizzled-9 receptor in human periodontal ligament mesenchymal stem cells. *J. Cell. Physiol.* **2010**, *225*, 123–131. [CrossRef] [PubMed]
43. Pelaez, D.; Huang, C.Y.; Cheung, H.S. Isolation of pluripotent neural crest-derived stem cells from adult human tissues by connexin-43 enrichment. *Stem Cells Dev.* **2013**, *22*, 2906–2914. [CrossRef] [PubMed]
44. Kawanabe, N.; Murata, S.; Murakami, K.; Ishihara, Y.; Hayano, S.; Kurosaka, H.; Kamioka, H.; Takano-Yamamoto, T.; Yamashiro, T. Isolation of multipotent stem cells in human periodontal ligament using stage-specific embryonic antigen-4. *Differentiation* **2010**, *79*, 74–83. [CrossRef] [PubMed]
45. Komaki, M. Pericytes in the Periodontal Ligament. *Adv. Exp. Med. Biol.* **2019**, *1122*, 169–186. [CrossRef]
46. Huang, C.Y.; Pelaez, D.; Dominguez-Bendala, J.; Garcia-Godoy, F.; Cheung, H.S. Plasticity of stem cells derived from adult periodontal ligament. *Regen. Med.* **2009**, *4*, 809–821. [CrossRef]
47. Fortino, V.R.; Chen, R.S.; Pelaez, D.; Cheung, H.S. Neurogenesis of neural crest-derived periodontal ligament stem cells by EGF and bFGF. *J. Cell. Physiol.* **2014**, *229*, 479–488. [CrossRef]
48. Lekovic, V.; Kenney, E.B.; Kovacevic, K.; Carranza, F.A., Jr. Evaluation of guided tissue regeneration in Class II furcation defects. A clinical re-entry study. *J. Periodontol.* **1989**, *60*, 694–698. [CrossRef]
49. Nasajpour, A.; Mandla, S.; Shree, S.; Mostafavi, E.; Sharifi, R.; Khalilpour, A.; Saghazadeh, S.; Hassan, S.; Mitchell, M.J.; Leijten, J.; et al. Nanostructured Fibrous Membranes with Rose Spike-Like Architecture. *Nano Lett.* **2017**, *17*, 6235–6240. [CrossRef]
50. Iwata, T.; Yamato, M.; Tsuchioka, H.; Takagi, R.; Mukobata, S.; Washio, K.; Okano, T.; Ishikawa, I. Periodontal regeneration with multi-layered periodontal ligament-derived cell sheets in a canine model. *Biomaterials* **2009**, *30*, 2716–2723. [CrossRef]
51. Vaquette, C.; Fan, W.; Xiao, Y.; Hamlet, S.; Hutmacher, D.W.; Ivanovski, S. A biphasic scaffold design combined with cell sheet technology for simultaneous regeneration of alveolar bone/periodontal ligament complex. *Biomaterials* **2012**, *33*, 5560–5573. [CrossRef]
52. Panduwawala, C.P.; Zhan, X.; Dissanayaka, W.L.; Samaranayake, L.P.; Jin, L.; Zhang, C. In vivo periodontal tissue regeneration by periodontal ligament stem cells and endothelial cells in three-dimensional cell sheet constructs. *J. Periodontal Res.* **2017**, *52*, 408–418. [CrossRef]
53. Pilipchuk, S.P.; Monje, A.; Jiao, Y.; Hao, J.; Kruger, L.; Flanagan, C.L.; Hollister, S.J.; Giannobile, W.V. Integration of 3D Printed and Micropatterned Polycaprolactone Scaffolds for Guidance of Oriented Collagenous Tissue Formation In Vivo. *Adv. Healthc. Mater.* **2016**, *5*, 676–687. [CrossRef]
54. Park, C.H.; Rios, H.F.; Jin, Q.; Sugai, J.V.; Padial-Molina, M.; Taut, A.D.; Flanagan, C.L.; Hollister, S.J.; Giannobile, W.V. Tissue engineering bone-ligament complexes using fiber-guiding scaffolds. *Biomaterials* **2012**, *33*, 137–145. [CrossRef]
55. Zhou, T.; Liu, X.; Sui, B.; Liu, C.; Mo, X.; Sun, J. Development of fish collagen/bioactive glass/chitosan composite nanofibers as a GTR/GBR membrane for inducing periodontal tissue regeneration. *Biomed. Mater.* **2017**, *12*, 055004. [CrossRef]
56. Cho, H.; Tarafder, S.; Fogge, M.; Kao, K.; Lee, C.H. Periodontal ligament stem/progenitor cells with protein-releasing scaffolds for cementum formation and integration on dentin surface. *Connect. Tissue Res.* **2016**, *57*, 488–495. [CrossRef]
57. Ding, T.; Li, J.; Zhang, X.; Du, L.; Li, Y.; Li, D.; Kong, B.; Ge, S. Super-assembled core/shell fibrous frameworks with dual growth factors for in situ cementum-ligament-bone complex regeneration. *Biomater. Sci.* **2020**, *8*, 2459–2471. [CrossRef]
58. Owaki, T.; Shimizu, T.; Yamato, M.; Okano, T. Cell sheet engineering for regenerative medicine: Current challenges and strategies. *Biotechnol. J.* **2014**, *9*, 904–914. [CrossRef]
59. Okano, T.; Yamada, N.; Sakai, H.; Sakurai, Y. A novel recovery system for cultured cells using plasma-treated polystyrene dishes grafted with poly(N-isopropylacrylamide). *J. Biomed. Mater. Res.* **1993**, *27*, 1243–1251. [CrossRef]
60. Hasegawa, M.; Yamato, M.; Kikuchi, A.; Okano, T.; Ishikawa, I. Human periodontal ligament cell sheets can regenerate periodontal ligament tissue in an athymic rat model. *Tissue Eng.* **2005**, *11*, 469–478. [CrossRef]
61. Takahashi, H.; Nakayama, M.; Itoga, K.; Yamato, M.; Okano, T. Micropatterned thermoresponsive polymer brush surfaces for fabricating cell sheets with well-controlled orientational structures. *Biomacromolecules* **2011**, *12*, 1414–1418. [CrossRef] [PubMed]
62. Raju, R.; Oshima, M.; Inoue, M.; Morita, T.; Huijiao, Y.; Waskitho, A.; Baba, O.; Inoue, M.; Matsuka, Y. Three-dimensional periodontal tissue regeneration using a bone-ligament complex cell sheet. *Sci. Rep.* **2020**, *10*, 1656. [CrossRef] [PubMed]
63. Tsumanuma, Y.; Iwata, T.; Washio, K.; Yoshida, T.; Yamada, A.; Takagi, R.; Ohno, T.; Lin, K.; Yamato, M.; Ishikawa, I.; et al. Comparison of different tissue-derived stem cell sheets for periodontal regeneration in a canine 1-wall defect model. *Biomaterials* **2011**, *32*, 5819–5825. [CrossRef] [PubMed]
64. Son, H.; Jeon, M.; Choi, H.J.; Lee, H.S.; Kim, I.H.; Kang, C.M.; Song, J.S. Decellularized human periodontal ligament for periodontium regeneration. *PLoS ONE* **2019**, *14*, e0221236. [CrossRef] [PubMed]

65. Chen, H.; Han, Q.; Wang, C.; Liu, Y.; Chen, B.; Wang, J. Porous Scaffold Design for Additive Manufacturing in Orthopedics: A Review. *Front. Bioeng. Biotechnol.* **2020**, *8*, 609. [CrossRef] [PubMed]
66. Park, C.H.; Kim, K.H.; Rios, H.F.; Lee, Y.M.; Giannobile, W.V.; Seol, Y.J. Spatiotemporally controlled microchannels of periodontal mimic scaffolds. *J. Dent. Res.* **2014**, *93*, 1304–1312. [CrossRef]
67. Oliveira, N.K.; Salles, T.H.C.; Pedroni, A.C.; Miguita, L.; D'Avila, M.A.; Marques, M.M.; Deboni, M.C.Z. Osteogenic potential of human dental pulp stem cells cultured onto poly-epsilon-caprolactone/poly (rotaxane) scaffolds. *Dent. Mater.* **2019**, *35*, 1740–1749. [CrossRef]
68. Thattaruparambil Raveendran, N.; Vaquette, C.; Meinert, C.; Samuel Ipe, D.; Ivanovski, S. Optimization of 3D bioprinting of periodontal ligament cells. *Dent. Mater.* **2019**, *35*, 1683–1694. [CrossRef]
69. Zhang, L.; Yang, G.; Johnson, B.N.; Jia, X. Three-dimensional (3D) printed scaffold and material selection for bone repair. *Acta Biomater.* **2019**, *84*, 16–33. [CrossRef]
70. Park, C.H. Biomaterial-Based Approaches for Regeneration of Periodontal Ligament and Cementum Using 3D Platforms. *Int. J. Mol. Sci.* **2019**, *20*, 4364. [CrossRef]
71. Niinomi, M. Metallic biomaterials. *J. Artif. Organs* **2008**, *11*, 105–110. [CrossRef]
72. Park, J.; Park, S.; Kim, J.E.; Jang, K.J.; Seonwoo, H.; Chung, J.H. Enhanced Osteogenic Differentiation of Periodontal Ligament Stem Cells Using a Graphene Oxide-Coated Poly(epsilon-caprolactone) Scaffold. *Polymers* **2021**, *13*, 797. [CrossRef]
73. Peng, W.; Ren, S.; Zhang, Y.; Fan, R.; Zhou, Y.; Li, L.; Xu, X.; Xu, Y. MgO Nanoparticles-Incorporated PCL/Gelatin-Derived Coaxial Electrospinning Nanocellulose Membranes for Periodontal Tissue Regeneration. *Front. Bioeng. Biotechnol.* **2021**, *9*, 668428. [CrossRef]
74. Jeon, H.; Lee, H.; Kim, G. A surface-modified poly(varepsilon-caprolactone) scaffold comprising variable nanosized surface-roughness using a plasma treatment. *Tissue Eng. Part C Methods* **2014**, *20*, 951–963. [CrossRef]
75. Vera-Sanchez, M.; Aznar-Cervantes, S.; Jover, E.; Garcia-Bernal, D.; Onate-Sanchez, R.E.; Hernandez-Romero, D.; Moraleda, J.M.; Collado-Gonzalez, M.; Rodriguez-Lozano, F.J.; Cenis, J.L. Silk-Fibroin and Graphene Oxide Composites Promote Human Periodontal Ligament Stem Cell Spontaneous Differentiation into Osteo/Cementoblast-Like Cells. *Stem Cells Dev.* **2016**, *25*, 1742–1754. [CrossRef]
76. Fu, C.; Yang, X.; Tan, S.; Song, L. Enhancing Cell Proliferation and Osteogenic Differentiation of MC3T3-E1 Pre-osteoblasts by BMP-2 Delivery in Graphene Oxide-Incorporated PLGA/HA Biodegradable Microcarriers. *Sci. Rep.* **2017**, *7*, 12549. [CrossRef]
77. Rasperini, G.; Pilipchuk, S.P.; Flanagan, C.L.; Park, C.H.; Pagni, G.; Hollister, S.J.; Giannobile, W.V. 3D-printed Bioresorbable Scaffold for Periodontal Repair. *J. Dent. Res.* **2015**, *94*, 153S–157S. [CrossRef]
78. Kade, J.C.; Dalton, P.D. Polymers for Melt Electrowriting. *Adv. Healthc. Mater.* **2021**, *10*, e2001232. [CrossRef]
79. Latimer, J.M.; Maekawa, S.; Yao, Y.; Wu, D.T.; Chen, M.; Giannobile, W.V. Regenerative Medicine Technologies to Treat Dental, Oral, and Craniofacial Defects. *Front. Bioeng. Biotechnol.* **2021**, *9*, 704048. [CrossRef]
80. Eichholz, K.F.; Hoey, D.A. Mediating human stem cell behaviour via defined fibrous architectures by melt electrospinning writing. *Acta Biomater.* **2018**, *75*, 140–151. [CrossRef]
81. Vaquette, C.; Pilipchuk, S.P.; Bartold, P.M.; Hutmacher, D.W.; Giannobile, W.V.; Ivanovski, S. Tissue Engineered Constructs for Periodontal Regeneration: Current Status and Future Perspectives. *Adv. Healthc. Mater.* **2018**, *7*, e1800457. [CrossRef] [PubMed]
82. Ivanovski, S.; Vaquette, C.; Gronthos, S.; Hutmacher, D.W.; Bartold, P.M. Multiphasic scaffolds for periodontal tissue engineering. *J. Dent. Res.* **2014**, *93*, 1212–1221. [CrossRef] [PubMed]
83. Gonzalez-Fernandez, T.; Rathan, S.; Hobbs, C.; Pitacco, P.; Freeman, F.E.; Cunniffe, G.M.; Dunne, N.J.; McCarthy, H.O.; Nicolosi, V.; O'Brien, F.J.; et al. Pore-forming bioinks to enable spatio-temporally defined gene delivery in bioprinted tissues. *J. Control. Release* **2019**, *301*, 13–27. [CrossRef] [PubMed]
84. Abbasi, N.; Abdal-Hay, A.; Hamlet, S.; Graham, E.; Ivanovski, S. Effects of Gradient and Offset Architectures on the Mechanical and Biological Properties of 3-D Melt Electrowritten (MEW) Scaffolds. *ACS. Biomater. Sci. Eng.* **2019**, *5*, 3448–3461. [CrossRef] [PubMed]
85. Yao, Y.; Raymond, J.E.; Kauffmann, F.; Maekawa, S.; Sugai, J.V.; Lahann, J.; Giannobile, W.V. Multicompartmental Scaffolds for Coordinated Periodontal Tissue Engineering. *J. Dent. Res.* **2022**, *101*, 1457–1466. [CrossRef]
86. Lee, C.H.; Singla, A.; Lee, Y. Biomedical applications of collagen. *Int. J. Pharm.* **2001**, *221*, 1–22. [CrossRef]
87. van den Bos, T.; Tonino, G.J. Composition and metabolism of the extracellular matrix in the periodontal ligament of impeded and unimpeded rat incisors. *Arch. Oral. Biol.* **1984**, *29*, 893–897. [CrossRef]
88. Bozec, L.; Odlyha, M. Thermal denaturation studies of collagen by microthermal analysis and atomic force microscopy. *Biophys. J.* **2011**, *101*, 228–236. [CrossRef]
89. Hinton, T.J.; Jallerat, Q.; Palchesko, R.N.; Park, J.H.; Grodzicki, M.S.; Shue, H.J.; Ramadan, M.H.; Hudson, A.R.; Feinberg, A.W. Three-dimensional printing of complex biological structures by freeform reversible embedding of suspended hydrogels. *Sci. Adv.* **2015**, *1*, e1500758. [CrossRef] [PubMed]
90. Lee, A.; Hudson, A.R.; Shiwarski, D.J.; Tashman, J.W.; Hinton, T.J.; Yerneni, S.; Bliley, J.M.; Campbell, P.G.; Feinberg, A.W. 3D bioprinting of collagen to rebuild components of the human heart. *Science* **2019**, *365*, 482–487. [CrossRef] [PubMed]
91. Mota, J.; Yu, N.; Caridade, S.G.; Luz, G.M.; Gomes, M.E.; Reis, R.L.; Jansen, J.A.; Walboomers, X.F.; Mano, J.F. Chitosan/bioactive glass nanoparticle composite membranes for periodontal regeneration. *Acta Biomater.* **2012**, *8*, 4173–4180. [CrossRef]

92. Qasim, S.B.; Delaine-Smith, R.M.; Fey, T.; Rawlinson, A.; Rehman, I.U. Freeze gelated porous membranes for periodontal tissue regeneration. *Acta Biomater.* **2015**, *23*, 317–328. [CrossRef]
93. Ge, S.; Zhao, N.; Wang, L.; Yu, M.; Liu, H.; Song, A.; Huang, J.; Wang, G.; Yang, P. Bone repair by periodontal ligament stem cellseeded nanohydroxyapatite-chitosan scaffold. *Int. J. Nanomed.* **2012**, *7*, 5405–5414. [CrossRef]
94. Varoni, E.M.; Vijayakumar, S.; Canciani, E.; Cochis, A.; De Nardo, L.; Lodi, G.; Rimondini, L.; Cerruti, M. Chitosan-Based Trilayer Scaffold for Multitissue Periodontal Regeneration. *J. Dent. Res.* **2018**, *97*, 303–311. [CrossRef]
95. Maturavongsadit, P.; Narayanan, L.K.; Chansoria, P.; Shirwaiker, R.; Benhabbour, S.R. Cell-Laden Nanocellulose/Chitosan-Based Bioinks for 3D Bioprinting and Enhanced Osteogenic Cell Differentiation. *ACS Appl. Bio Mater.* **2021**, *4*, 2342–2353. [CrossRef]
96. Trubiani, O.; Orsini, G.; Zini, N.; Di Iorio, D.; Piccirilli, M.; Piattelli, A.; Caputi, S. Regenerative potential of human periodontal ligament derived stem cells on three-dimensional biomaterials: A morphological report. *J. Biomed. Mater. Res. A* **2008**, *87*, 986–993. [CrossRef]
97. Ma, Y.; Ji, Y.; Huang, G.; Ling, K.; Zhang, X.; Xu, F. Bioprinting 3D cell-laden hydrogel microarray for screening human periodontal ligament stem cell response to extracellular matrix. *Biofabrication* **2015**, *7*, 044105. [CrossRef]
98. Xu, Q.; Li, B.; Yuan, L.; Dong, Z.; Zhang, H.; Wang, H.; Sun, J.; Ge, S.; Jin, Y. Combination of platelet-rich plasma within periodontal ligament stem cell sheets enhances cell differentiation and matrix production. *J. Tissue Eng. Regen. Med.* **2017**, *11*, 627–636. [CrossRef]
99. Pan, J.; Deng, J.; Luo, Y.; Yu, L.; Zhang, W.; Han, X.; You, Z.; Liu, Y. Thermosensitive Hydrogel Delivery of Human Periodontal Stem Cells Overexpressing Platelet-Derived Growth Factor-BB Enhances Alveolar Bone Defect Repair. *Stem Cells Dev.* **2019**, *28*, 1620–1631. [CrossRef]
100. Safi, I.N.; Al-Shammari, A.M.; Ul-Jabbar, M.A.; Hussein, B.M.A. Preparing polycaprolactone scaffolds using electrospinning technique for construction of artificial periodontal ligament tissue. *J. Taibah. Univ. Med. Sci.* **2020**, *15*, 363–373. [CrossRef]
101. Duan, X.; Ji, M.; Deng, F.; Sun, Z.; Lin, Z. Effects of connective tissue growth factor on human periodontal ligament fibroblasts. *Arch. Oral Biol.* **2017**, *84*, 37–44. [CrossRef] [PubMed]
102. Thomadakis, G.; Ramoshebi, L.N.; Crooks, J.; Rueger, D.C.; Ripamonti, U. Immunolocalization of Bone Morphogenetic Protein-2 and -3 and Osteogenic Protein-1 during murine tooth root morphogenesis and in other craniofacial structures. *Eur. J. Oral Sci.* **1999**, *107*, 368–377. [CrossRef] [PubMed]
103. Pitaru, S.; Pritzki, A.; Bar-Kana, I.; Grosskopf, A.; Savion, N.; Narayanan, A.S. Bone morphogenetic protein 2 induces the expression of cementum attachment protein in human periodontal ligament clones. *Connect. Tissue Res.* **2002**, *43*, 257–264. [CrossRef] [PubMed]
104. Kono, K.; Maeda, H.; Fujii, S.; Tomokiyo, A.; Yamamoto, N.; Wada, N.; Monnouchi, S.; Teramatsu, Y.; Hamano, S.; Koori, K.; et al. Exposure to transforming growth factor-beta1 after basic fibroblast growth factor promotes the fibroblastic differentiation of human periodontal ligament stem/progenitor cell lines. *Cell Tissue Res.* **2013**, *352*, 249–263. [CrossRef]
105. Kang, W.; Liang, Q.; Du, L.; Shang, L.; Wang, T.; Ge, S. Sequential application of bFGF and BMP-2 facilitates osteogenic differentiation of human periodontal ligament stem cells. *J. Periodontal Res.* **2019**, *54*, 424–434. [CrossRef]
106. Ding, T.; Kang, W.; Li, J.; Yu, L.; Ge, S. An in situ tissue engineering scaffold with growth factors combining angiogenesis and osteoimmunomodulatory functions for advanced periodontal bone regeneration. *J. Nanobiotechnology* **2021**, *19*, 247. [CrossRef]
107. James, A.W.; LaChaud, G.; Shen, J.; Asatrian, G.; Nguyen, V.; Zhang, X.; Ting, K.; Soo, C. A Review of the Clinical Side Effects of Bone Morphogenetic Protein-2. *Tissue Eng. Part B Rev.* **2016**, *22*, 284–297. [CrossRef]
108. Qu, M.; Jiang, X.; Zhou, X.; Wang, C.; Wu, Q.; Ren, L.; Zhu, J.; Zhu, S.; Tebon, P.; Sun, W.; et al. Stimuli-Responsive Delivery of Growth Factors for Tissue Engineering. *Adv. Healthc. Mater.* **2020**, *9*, e1901714. [CrossRef]
109. Yuk, H.; Lu, B.; Lin, S.; Qu, K.; Xu, J.; Luo, J.; Zhao, X. 3D printing of conducting polymers. *Nat. Commun.* **2020**, *11*, 1604. [CrossRef]
110. Shen, Z.; Zhang, Z.; Zhang, N.; Li, J.; Zhou, P.; Hu, F.; Rong, Y.; Lu, B.; Gu, G. High-Stretchability, Ultralow-Hysteresis ConductingPolymer Hydrogel Strain Sensors for Soft Machines. *Adv. Mater.* **2022**, *34*, e2203650. [CrossRef]
111. Zhao, W.; Cao, J.; Wang, F.; Tian, F.; Zheng, W.; Bao, Y.; Zhang, K.; Zhang, Z.; Yu, J.; Xu, J.; et al. 3D Printing of Stretchable, Adhesive and Conductive Ti3C2Tx-Polyacrylic Acid Hydrogels. *Polymers* **2022**, *14*, 1992. [CrossRef]
112. Gauthier, R.; Jeannin, C.; Attik, N.; Trunfio-Sfarghiu, A.M.; Gritsch, K.; Grosgogeat, B. Tissue Engineering for Periodontal Ligament Regeneration: Biomechanical Specifications. *J. Biomech. Eng.* **2021**, *143*, 030801. [CrossRef]
113. Yang, L.; Yang, Y.; Wang, S.; Li, Y.; Zhao, Z. In vitro mechanical loading models for periodontal ligament cells: From two-dimensional to three-dimensional models. *Arch. Oral Biol.* **2015**, *60*, 416–424. [CrossRef]
114. Natali, A.N.; Pavan, P.G.; Scarpa, C. Numerical analysis of tooth mobility: Formulation of a non-linear constitutive law for the periodontal ligament. *Dent. Mater.* **2004**, *20*, 623–629. [CrossRef]
115. Martino, F.; Perestrelo, A.R.; Vinarsky, V.; Pagliari, S.; Forte, G. Cellular Mechanotransduction: From Tension to Function. *Front. Physiol.* **2018**, *9*, 824. [CrossRef]
116. Schiller, H.B.; Fassler, R. Mechanosensitivity and compositional dynamics of cell-matrix adhesions. *EMBO Rep.* **2013**, *14*, 509–519. [CrossRef]
117. Webster, K.D.; Ng, W.P.; Fletcher, D.A. Tensional homeostasis in single fibroblasts. *Biophys. J.* **2014**, *107*, 146–155. [CrossRef]
118. Brown, R.A.; Prajapati, R.; McGrouther, D.A.; Yannas, I.V.; Eastwood, M. Tensional homeostasis in dermal fibroblasts: Mechanical responses to mechanical loading in three-dimensional substrates. *J. Cell. Physiol.* **1998**, *175*, 323–332. [CrossRef]

119. Kim, C.; Ye, F.; Ginsberg, M.H. Regulation of integrin activation. *Annu. Rev. Cell. Dev. Biol.* **2011**, *27*, 321–345. [CrossRef]
120. Calderwood, D.A.; Zent, R.; Grant, R.; Rees, D.J.; Hynes, R.O.; Ginsberg, M.H. The Talin head domain binds to integrin beta subunit cytoplasmic tails and regulates integrin activation. *J. Biol. Chem.* **1999**, *274*, 28071–28074. [CrossRef]
121. Haining, A.W.; von Essen, M.; Attwood, S.J.; Hytonen, V.P.; Del Rio Hernandez, A. All Subdomains of the Talin Rod Are Mechanically Vulnerable and May Contribute to Cellular Mechanosensing. *ACS Nano.* **2016**, *10*, 6648–6658. [CrossRef] [PubMed]
122. Carisey, A.; Tsang, R.; Greiner, A.M.; Nijenhuis, N.; Heath, N.; Nazgiewicz, A.; Kemkemer, R.; Derby, B.; Spatz, J.; Ballestrem, C. Vinculin regulates the recruitment and release of core focal adhesion proteins in a force-dependent manner. *Curr. Biol.* **2013**, *23*, 271–281. [CrossRef] [PubMed]
123. Kiema, T.; Lad, Y.; Jiang, P.; Oxley, C.L.; Baldassarre, M.; Wegener, K.L.; Campbell, I.D.; Ylanne, J.; Calderwood, D.A. The molecular basis of filamin binding to integrins and competition with talin. *Mol. Cell* **2006**, *21*, 337–347. [CrossRef] [PubMed]
124. Das, M.; Ithychanda, S.S.; Qin, J.; Plow, E.F. Migfilin and filamin as regulators of integrin activation in endothelial cells and neutrophils. *PLoS ONE* **2011**, *6*, e26355. [CrossRef] [PubMed]
125. Shifrin, Y.; Pinto, V.I.; Hassanali, A.; Arora, P.D.; McCulloch, C.A. Force-induced apoptosis mediated by the Rac/Pak/p38 signalling pathway is regulated by filamin A. *Biochem. J.* **2012**, *445*, 57–67. [CrossRef]
126. Li, Y.; Zheng, W.; Liu, J.S.; Wang, J.; Yang, P.; Li, M.L.; Zhao, Z.H. Expression of osteoclastogenesis inducers in a tissue model of periodontal ligament under compression. *J. Dent. Res.* **2011**, *90*, 115–120. [CrossRef]
127. Li, Y.; Li, M.; Tan, L.; Huang, S.; Zhao, L.; Tang, T.; Liu, J.; Zhao, Z. Analysis of time-course gene expression profiles of a periodontal ligament tissue model under compression. *Arch. Oral Biol.* **2013**, *58*, 511–522. [CrossRef]
128. Kang, K.L.; Lee, S.W.; Ahn, Y.S.; Kim, S.H.; Kang, Y.G. Bioinformatic analysis of responsive genes in two-dimension and three-dimension cultured human periodontal ligament cells subjected to compressive stress. *J. Periodontal Res.* **2013**, *48*, 87–97. [CrossRef]
129. Feng, L.; Yang, R.; Liu, D.; Wang, X.; Song, Y.; Cao, H.; He, D.; Gan, Y.; Kou, X.; Zhou, Y. PDL Progenitor-Mediated PDL Recovery Contributes to Orthodontic Relapse. *J. Dent. Res.* **2016**, *95*, 1049–1056. [CrossRef]
130. Jin, S.S.; He, D.Q.; Wang, Y.; Zhang, T.; Yu, H.J.; Li, Z.X.; Zhu, L.S.; Zhou, Y.H.; Liu, Y. Mechanical force modulates periodontal ligament stem cell characteristics during bone remodelling via TRPV4. *Cell Prolif.* **2020**, *53*, e12912. [CrossRef]
131. Brockhaus, J.; Craveiro, R.B.; Azraq, I.; Niederau, C.; Schroder, S.K.; Weiskirchen, R.; Jankowski, J.; Wolf, M. In Vitro Compression Model for Orthodontic Tooth Movement Modulates Human Periodontal Ligament Fibroblast Proliferation, Apoptosis and Cell Cycle. *Biomolecules* **2021**, *11*, 932. [CrossRef]
132. Stemmler, A.; Symmank, J.; Steinmetz, J.; von Brandenstein, K.; Hennig, C.L.; Jacobs, C. GDF15 Supports the Inflammatory Response of PdL Fibroblasts Stimulated by P. gingivalis LPS and Concurrent Compression. *Int. J. Mol. Sci.* **2021**, *22*, 13608. [CrossRef]
133. Jiang, N.; He, D.; Ma, Y.; Su, J.; Wu, X.; Cui, S.; Li, Z.; Zhou, Y.; Yu, H.; Liu, Y. Force-Induced Autophagy in Periodontal Ligament Stem Cells Modulates M1 Macrophage Polarization via AKT Signaling. *Front. Cell Dev. Biol.* **2021**, *9*, 666631. [CrossRef]
134. Zhang, L.; Liu, W.; Zhao, J.; Ma, X.; Shen, L.; Zhang, Y.; Jin, F.; Jin, Y. Mechanical stress regulates osteogenic differentiation and RANKL/OPG ratio in periodontal ligament stem cells by the Wnt/beta-catenin pathway. *Biochim. Biophys. Acta* **2016**, *1860*, 2211–2219. [CrossRef]
135. Yamamoto, T.; Kita, M.; Kimura, I.; Oseko, F.; Terauchi, R.; Takahashi, K.; Kubo, T.; Kanamura, N. Mechanical stress induces expression of cytokines in human periodontal ligament cells. *Oral Dis.* **2006**, *12*, 171–175. [CrossRef]
136. Wenger, K.H.; El-Awady, A.R.; Messer, R.L.; Sharawy, M.M.; White, G.; Lapp, C.A. Pneumatic pressure bioreactor for cyclic hydrostatic stress application: Mechanobiology effects on periodontal ligament cells. *J. Appl. Physiol. (1985)* **2011**, *111*, 1072–1079. [CrossRef]
137. Yousefian, J.; Firouzian, F.; Shanfeld, J.; Ngan, P.; Lanese, R.; Davidovitch, Z. A new experimental model for studying the response of periodontal ligament cells to hydrostatic pressure. *Am. J. Orthod. Dentofac. Orthop.* **1995**, *108*, 402–409. [CrossRef]
138. Jia, R.; Yi, Y.; Liu, J.; Pei, D.; Hu, B.; Hao, H.; Wu, L.; Wang, Z.; Luo, X.; Lu, Y. Cyclic compression emerged dual effects on the osteogenic and osteoclastic status of LPS-induced inflammatory human periodontal ligament cells according to loading force. *BMC Oral Health.* **2020**, *20*, 7. [CrossRef]
139. Saminathan, A.; Sriram, G.; Vinoth, J.K.; Cao, T.; Meikle, M.C. Engineering the periodontal ligament in hyaluronan-gelatin-type I collagen constructs: Upregulation of apoptosis and alterations in gene expression by cyclic compressive strain. *Tissue Eng. Part A* **2015**, *21*, 518–529. [CrossRef]
140. Nettelhoff, L.; Grimm, S.; Jacobs, C.; Walter, C.; Pabst, A.M.; Goldschmitt, J.; Wehrbein, H. Influence of mechanical compression on human periodontal ligament fibroblasts and osteoblasts. *Clin. Oral Investig.* **2016**, *20*, 621–629. [CrossRef]
141. Saminathan, A.; Vinoth, K.J.; Low, H.H.; Cao, T.; Meikle, M.C. Engineering three-dimensional constructs of the periodontal ligament in hyaluronan-gelatin hydrogel films and a mechanically active environment. *J. Periodontal Res.* **2013**, *48*, 790–801. [CrossRef] [PubMed]
142. Chen, Y.; Mohammed, A.; Oubaidin, M.; Evans, C.A.; Zhou, X.; Luan, X.; Diekwisch, T.G.; Atsawasuwan, P. Cyclic stretch and compression forces alter microRNA-29 expression of human periodontal ligament cells. *Gene* **2015**, *566*, 13–17. [CrossRef] [PubMed]
143. Shen, T.; Qiu, L.; Chang, H.; Yang, Y.; Jian, C.; Xiong, J.; Zhou, J.; Dong, S. Cyclic tension promotes osteogenic differentiation in human periodontal ligament stem cells. *Int. J. Clin. Exp. Pathol.* **2014**, *7*, 7872–7880. [PubMed]
144. Wei, F.; Liu, D.; Feng, C.; Zhang, F.; Yang, S.; Hu, Y.; Ding, G.; Wang, S. microRNA-21 mediates stretch-induced osteogenic differentiation in human periodontal ligament stem cells. *Stem Cells Dev.* **2015**, *24*, 312–319. [CrossRef] [PubMed]

145. Wang, H.; Feng, C.; Jin, Y.; Tan, W.; Wei, F. Identification and characterization of circular RNAs involved in mechanical force-induced periodontal ligament stem cells. *J. Cell. Physiol.* **2019**, *234*, 10166–10177. [CrossRef]
146. Xi, X.; Zhao, Y.; Liu, H.; Li, Z.; Chen, S.; Liu, D. Nrf2 activation is involved in osteogenic differentiation of periodontal ligament stem cells under cyclic mechanical stretch. *Exp. Cell Res.* **2021**, *403*, 112598. [CrossRef]
147. Meng, X.; Wang, W.; Wang, X. MicroRNA-34a and microRNA-146a target CELF3 and suppress the osteogenic differentiation of periodontal ligament stem cells under cyclic mechanical stretch. *J. Dent. Sci.* **2022**, *17*, 1281–1291. [CrossRef]
148. Liu, J.; Li, Q.; Liu, S.; Gao, J.; Qin, W.; Song, Y.; Jin, Z. Periodontal Ligament Stem Cells in the Periodontitis Microenvironment Are Sensitive to Static Mechanical Strain. *Stem Cells Int.* **2017**, *2017*, 1380851. [CrossRef]
149. Salim, C.; Muders, H.; Jager, A.; Konermann, A. Role of chaperone-assisted selective autophagy (CASA) in mechanical stress protection of periodontal ligament cells. *J. Orofac. Orthop.* **2022**, *83*, 1–12. [CrossRef]
150. Ulbricht, A.; Eppler, F.J.; Tapia, V.E.; van der Ven, P.F.; Hampe, N.; Hersch, N.; Vakeel, P.; Stadel, D.; Haas, A.; Saftig, P.; et al. Cellular mechanotransduction relies on tension-induced and chaperone-assisted autophagy. *Curr. Biol.* **2013**, *23*, 430–435. [CrossRef]
151. Ehrlicher, A.J.; Nakamura, F.; Hartwig, J.H.; Weitz, D.A.; Stossel, T.P. Mechanical strain in actin networks regulates FilGAP and integrin binding to filamin A. *Nature* **2011**, *478*, 260–263. [CrossRef]
152. Huang, H.; Yang, R.; Zhou, Y.H. Mechanobiology of Periodontal Ligament Stem Cells in Orthodontic Tooth Movement. *Stem Cells Int.* **2018**, *2018*, 6531216. [CrossRef]
153. Nemoto, T.; Kajiya, H.; Tsuzuki, T.; Takahashi, Y.; Okabe, K. Differential induction of collagens by mechanical stress in human periodontal ligament cells. *Arch. Oral Biol.* **2010**, *55*, 981–987. [CrossRef]
154. Oortgiesen, D.A.; Yu, N.; Bronckers, A.L.; Yang, F.; Walboomers, X.F.; Jansen, J.A. A three-dimensional cell culture model to study the mechano-biological behavior in periodontal ligament regeneration. *Tissue Eng. Part C Methods* **2012**, *18*, 81–89. [CrossRef]
155. Papadopoulou, A.; Iliadi, A.; Eliades, T.; Kletsas, D. Early responses of human periodontal ligament fibroblasts to cyclic and static mechanical stretching. *Eur. J. Orthod.* **2017**, *39*, 258–263. [CrossRef]
156. Yu, W.; Su, X.; LI, M.; Wan, W.; Li, A.; Zhou, H.; Xu, F. Three-dimensional mechanical microenvironment enhanced osteogenic activity of mesenchymal stem cells-derived exosomes. *Chem. Eng. J.* **2021**, *417*, 128040. [CrossRef]
157. Howard, P.S.; Kucich, U.; Taliwal, R.; Korostoff, J.M. Mechanical forces alter extracellular matrix synthesis by human periodontal ligament fibroblasts. *J. Periodontal Res.* **1998**, *33*, 500–508. [CrossRef]
158. Xu, C.; Fan, Z.; Shan, W.; Hao, Y.; Ma, J.; Huang, Q.; Zhang, F. Cyclic stretch influenced expression of membrane connexin 43 in human periodontal ligament cell. *Arch. Oral Biol.* **2012**, *57*, 1602–1608. [CrossRef]
159. Tang, N.; Zhao, Z.; Zhang, L.; Yu, Q.; Li, J.; Xu, Z.; Li, X. Up-regulated osteogenic transcription factors during early response of human periodontal ligament stem cells to cyclic tensile strain. *Arch. Med. Sci.* **2012**, *8*, 422–430. [CrossRef]
160. Tang, M.; Peng, Z.; Mai, Z.; Chen, L.; Mao, Q.; Chen, Z.; Chen, Q.; Liu, L.; Wang, Y.; Ai, H. Fluid shear stress stimulates osteogenic differentiation of human periodontal ligament cells via the extracellular signal-regulated kinase 1/2 and p38 mitogen-activated protein kinase signaling pathways. *J. Periodontol.* **2014**, *85*, 1806–1813. [CrossRef]
161. Zheng, L.; Chen, L.; Chen, Y.; Gui, J.; Li, Q.; Huang, Y.; Liu, M.; Jia, X.; Song, W.; Ji, J.; et al. The effects of fluid shear stress on proliferation and osteogenesis of human periodontal ligament cells. *J. Biomech.* **2016**, *49*, 572–579. [CrossRef] [PubMed]
162. Shi, Q.; Zheng, L.; Na, J.; Li, X.; Yang, Z.; Chen, X.; Song, Y.; Li, C.; Zhou, L.; Fan, Y. Fluid shear stress promotes periodontal ligament cells proliferation via p38-AMOT-YAP. *Cell. Mol. Life Sci.* **2022**, *79*, 551. [CrossRef] [PubMed]
163. Kim, S.G.; Kim, S.G.; Viechnicki, B.; Kim, S.; Nah, H.D. Engineering of a periodontal ligament construct: Cell and fibre alignment induced by shear stress. *J. Clin. Periodontol.* **2011**, *38*, 1130–1136. [CrossRef] [PubMed]
164. Lin, H.H.; Chao, P.G.; Tai, W.C.; Chang, P.C. 3D-Printed Collagen-Based Waveform Microfibrous Scaffold for Periodontal Ligament Reconstruction. *Int. J. Mol. Sci.* **2021**, *22*, 7725. [CrossRef]
165. Tsai, S.J.; Ding, Y.W.; Shih, M.C.; Tu, Y.K. Systematic review and sequential network meta-analysis on the efficacy of periodontal regenerative therapies. *J. Clin. Periodontol.* **2020**, *47*, 1108–1120. [CrossRef]
166. Tassi, S.A.; Sergio, N.Z.; Misawa, M.Y.O.; Villar, C.C. Efficacy of stem cells on periodontal regeneration: Systematic review of pre-clinical studies. *J. Periodontal Res.* **2017**, *52*, 793–812. [CrossRef]
167. Putame, G.; Gabetti, S.; Carbonaro, D.; Meglio, F.D.; Romano, V.; Sacco, A.M.; Belviso, I.; Serino, G.; Bignardi, C.; Morbiducci, U.; et al. Compact and tunable stretch bioreactor advancing tissue engineering implementation. Application to engineered cardiac constructs. *Med. Eng. Phys.* **2020**, *84*, 1–9. [CrossRef]
168. Lim, D.; Renteria, E.S.; Sime, D.S.; Ju, Y.M.; Kim, J.H.; Criswell, T.; Shupe, T.D.; Atala, A.; Marini, F.C.; Gurcan, M.N.; et al. Bioreactor design and validation for manufacturing strategies in tissue engineering. *Bio-Des. Manuf.* **2022**, *5*, 43–63. [CrossRef]
169. Gabetti, S.; Masante, B.; Cochis, A.; Putame, G.; Sanginario, A.; Armando, I.; Fiume, E.; Scalia, A.C.; Daou, F.; Baino, F.; et al. An automated 3D-printed perfusion bioreactor combinable with pulsed electromagnetic field stimulators for bone tissue investigations. *Sci. Rep.* **2022**, *12*, 13859. [CrossRef]

nanomaterials

MDPI

Review

Topography: A Biophysical Approach to Direct the Fate of Mesenchymal Stem Cells in Tissue Engineering Applications

Xingli Cun and Leticia Hosta-Rigau *

DTU Health Tech, Centre for Nanomedicine and Theranostics, Technical University of Denmark, Nils Koppels Allé, Building 423, 2800 Kgs. Lyngby, Denmark; xincu@dtu.dk

* Correspondence: leri@dtu.dk; Tel.: +45-4525-8155

Received: 21 September 2020; Accepted: 16 October 2020; Published: 20 October 2020

Abstract: Tissue engineering is a promising strategy to treat tissue and organ loss or damage caused by injury or disease. During the past two decades, mesenchymal stem cells (MSCs) have attracted a tremendous amount of interest in tissue engineering due to their multipotency and self-renewal ability. MSCs are also the most multipotent stem cells in the human adult body. However, the application of MSCs in tissue engineering is relatively limited because it is difficult to guide their differentiation toward a specific cell lineage by using traditional biochemical factors. Besides biochemical factors, the differentiation of MSCs also influenced by biophysical cues. To this end, much effort has been devoted to directing the cell lineage decisions of MSCs through adjusting the biophysical properties of biomaterials. The surface topography of the biomaterial-based scaffold can modulate the proliferation and differentiation of MSCs. Presently, the development of micro- and nano-fabrication techniques has made it possible to control the surface topography of the scaffold precisely. In this review, we highlight and discuss how the main topographical features (i.e., roughness, patterns, and porosity) are an efficient approach to control the fate of MSCs and the application of topography in tissue engineering.

Keywords: biomaterials; differentiation; mesenchymal stem cells; scaffolds; tissue engineering; topography

---

1. Introduction

Stem cells (SCs) can differentiate into several types of cells and, as such, they have the capacity to repair injured parts of organs and tissues, thus holding a lot of potential in regenerative medicine.

In vivo, self-renewal (the maintenance of potency), proliferation, and differentiation toward a particular cell lineage, are regulated by the SCs but also by their niche (their specific microenvironment). During tissue development, SCs fate is influenced not only by biochemical factors (in the form of growth factors, low molecular weight chemicals, hormones, etc.) but also by the biophysical cues of their extracellular matrix (ECM) microenvironment [1–3]. The ECM is a key component since, to undergo fundamental biological processes, cells must adhere to the underlying ECM where they receive and respond to complex molecular signals [4]. Such biophysical cues located within the ECM can be classified into surface topography, matrix stiffness and mechanical forces [5–7]. In particular, the effect of topography on the SCs fate was reported as early as 1964 by Curtis et al. First, the influence of topographical features in the micron range was identified and, since then, the effect of microtopography onto the cells fate has been thoroughly studied [2,3,8]. However, the ECM that cells are exposed to, also displays nanometer range features. Thanks to the recent advances in nanofabrication techniques [3,9,10], it has been also possible to demonstrate how engineered substrates with analogous nanoscale features to those of the natural ECM, are able to direct cell behavior. For example, by manipulating nanoscale topography alone, it has been possible to elicit different responses from several cell types including, osteoblasts (OBs), or mesenchymal stem cells (MSCs) [3,11–15].

However, despite the enormous potential that SCs hold, their use in tissue engineering (TE) is still very limited as a result of their low differentiation efficiency and the difficulties associated with guiding their differentiation towards a specific cell lineage. Growth factors and other smaller chemical compounds can effectively promote SCs differentiation; however, making use of biochemical cues has its limitations and risks. For example, supra-physiological dosages need to be administered to obtain a desired phenotype. This results in high costs, as well as potential off-target effects and challenges in attaining optimal release kinetics. Additionally, upon removal of such biochemical cues as often occurs following in vivo transplantation [16], the long-term maintenance of the SCs differentiation is rapidly lost [17]. What is more, providing the required induction factors in vivo is rather challenging. It involves the design of implants that can incorporate a combination of molecules, as well as the ability to release such chemicals with controlled manner in terms of dose and kinetics over extended periods (e.g., days or even weeks) [18].

Importantly, SCs also respond to biophysical stimuli within their surrounding ECM. Therefore, to address the limitations of the soluble factors, biomaterials-based approaches have emerged as a powerful tool to modulate the proliferation, self-renewal and differentiation of SCs into multiple lineages. Thanks to the advent of micro- and nanofabrication techniques, scaffolds mimicking the structural complexity of the ECM, including its specialized textured topography both at the micro- and nano-range, can now be fabricated [17,19]. In general, such biophysical stimuli are introduced by modifying the substrate/surface with cell-matrix interactions which will ultimately influence the mechanics of the cell cytoskeleton and, in turn, the expression of genes and proteins [20–22]. What is more, biophysical factors, such as topography, can be applied continuously with their effects propagating faster than the diffusion of chemical entities [18,23].

While a variety of SCs have been investigated for biomedical applications, MSCs are of particular interest [22]. MSCs are multipotent stromal SCs able to differentiate into several cell types, including adipocytes, OBs, chondrocytes; therefore, they can form fat [24], bone [25], and cartilage tissues [22,26]. Additionally, MSCs have excellent accessibility, versatility and low risk of teratoma formation [22]. Since they can be extracted from adult tissue, they are free from ethical concerns as compared to, for example, embryonic SCs. Furthermore, MSCs also display proangiogenic capabilities, immune-regulatory and anti-inflammatory potential.

In this review, we discuss several prior studies regarding the application of topography to regulate the differentiation of MSCs. Different biomaterials-based scaffolds with control over the substrate roughness, patterns and porosity are evaluated in terms of their reported ability to either maintain the MSCs self-renewal abilities or to direct them to the cell lineage of choice. The overview of the topics covered in this review is presented in Scheme 1.

We start by providing an overview of the isolation from different tissues and characterization in terms of the differentiation potential of MSCs. The different therapeutic applications of MSCs will be also discussed. Next, we review the possible mechanisms behind the cell-surface topography interactions. The main part of this Review will be devoted to discuss how the main topographical features (i.e., roughness, surface patterns and porosity) of the biomaterials-based scaffolds control the cell-fate decisions of MSCs. Emphasis will be put in the examples examining the potential of MSCs and topography in the context of TE and the associated challenges.

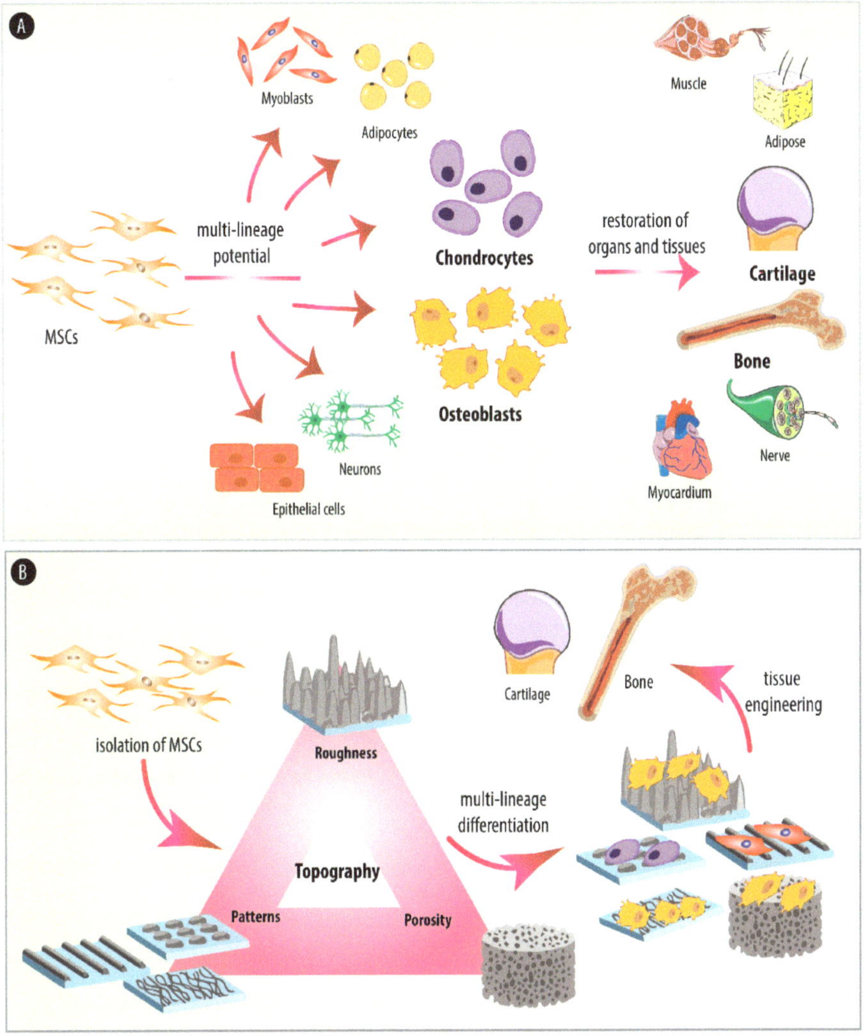

Scheme 1. (**A**) Schematic diagram of the multilineage differentiation of mesenchymal stem cells (MSCs) and their application in tissue engineering. (**B**) The topographical properties (roughness, patterns, porosity) of the scaffolds influence the cell lineage decisions of MSCs and their application in tissue engineering. In this review, we focus on how the roughness, patterns, and porosity of biomaterials-based scaffolds direct the fate of MSCs and their application in bone and cartilage tissue engineering.

2. Origin and Relevance of MSCs

Being the most multipotent adult SCs in the human body, MSCs have attracted a lot of interest [27]. These spindle-shaped type of cells reside in the stromal compartment of bone marrow (BM) and were identified in 1966 in a study conducted by Friedenstein and Petrakova [27,28]. In 1991, this kind of BM-isolated cells were officially named "mesenchymal stem cells (MSCs)" [29]. Interestingly, in Caplan's latest work, he proposed to change the name of "mesenchymal stem cells" to "medicinal signaling cells" to emphasize their therapeutic effects after transplantation, which is based on the homing of MSCs to the injured or diseased site where they secrete cocktails of proteins to modulate the immune response [30]. To date, several names have been proposed for this type of cells which include: medicinal signaling cells, mesenchymal stromal cells, multipotent stromal cells, mesodermal stem cells or marrow stromal cells [30,31]. However, although the name of "mesenchymal stem cells" has raised some controversy regarding the stemness of MSCs, it is still the most commonly used name and we decide to use this name throughout this manuscript [31,32].

MSCs are an excellent cell source for tissue regeneration since they can be isolated from several adult tissues, such as BM, but also adipose tissue, where they can be easily extracted in with a minimally invasive procedure [33]. Importantly, their multipotency allows them to differentiate towards multiple cell types of the mesodermal lineage, including adipocytes, osteocytes, and chondrocytes. MSCs also circumvent complications associated with immune rejection since they can be directly extracted from a patient, induced to differentiate into a specific cell lineage to be transplanted back into the same patient [27,34].

As a result of the abovementioned properties, MSCs have attracted a lot of attention and pre-clinical and clinical studies have already demonstrated their therapeutic value [27].

2.1. Isolation and Characterization

MSCs can be isolated from many and rich sources. Tissues where MSCs can be extracted from include BM, adipose, synovium tissues and also blood from the umbilical cord (UC) [27,33]. The protocols to isolate the extracted MSCs involve a digestion step. Next, the cells are cultured for three to five days, the non-adherent cells discarded and the remaining adherent cells (the MSCs) are cultured to the desired passage [33]. However, since MSCs are obtained from donors, their quality differs from donor to donor and by factors, such as aging or other disorders [27,35,36]. Therefore, it is important to thoroughly characterize them to assure their quality [27]. Since MSCs do not have one specific identification marker, three criteria were stablished in 2006 by the International Society for Cellular Therapy [37]. First, the cells must be capable to adhere to plastic when cultured in standard conditions. Secondly, the cells' surface must be positive (>95%) for particular markers, such as cluster of differentiation (CD)105, CD90, and CD73, and negative (<2%) for other markers, such as CD34, CD45, CD14, or CD79α, CD19, CD11b, and surface molecules of the human leukocyte antigen–DR isotype (HLA-DR), which is class cell surface receptor present in professional antigen-presenting cells. Thirdly, the cells must have the ability of in vitro differentiation into chondrocytes, adipocytes and OBs.

Since MSCs were first isolated from BM, BM-derived MSCs (BM-MSCs) are, so far, the most studied and thoroughly characterized MSCs. Additionally, they are regarded as the standard MSCs and used for comparison with MSCs derived from other sources. The different tissues and organs where MSCs can be extracted from show different advantages and disadvantages [33,38,39]. For example, while the extraction of MSCs from adipose tissue involves a minimally invasive procedure, UC-derived MSCs (UC-MSCs) can be harvested in a painless method [40]. Additionally, UC-MSCs renew faster than their BM-derived counterparts [41]. Lastly, an alternative approach to obtain MSCs is to derive them from induced pluripotent stem cells (iPSCs). iPSCs-derived MSCs (iPSC-MSCs) have shown similar morphology and antigen profile to BM-MSCs but have also demonstrated enhanced survival following transplantation [27,38,42].

2.2. Differentiation Potential

The ability to differentiate is the key feature of MSCs. However, the different tissue sources where MSCs can be obtained from, have an effect on their proliferation and differentiation potential [33]. For example, BM-MSCs have shown enhanced ability to differentiate into the osteogenic and chondrogenic lineages [43]. In contrast, MSCs derived from synovium tissues (S-MSCs) display enhanced proliferation and chondrogenic differentiation than their adipose-derived counterparts (AD-MSCs) [33,44]. UC-MSCs show several advantages over MSCs derived from other tissues, including the ability to be cultured for longer periods, large-scale expansion, and higher anti-inflammatory effects [33,45]. Therefore, depending on the application in mind, MSCs will be derived from one tissue or another.

2.3. Therapeutic Applications

MSCs have been broadly studied and applied in TE and regenerative medicine for the reconstruction of several tissues, including bone, cartilage, or musculoskeletal tissues.

Due to their ability to differentiate into OBs, MSCs have been widely evaluated in the context of bone regeneration. Specifically, BM-MSCs, UC-MSCs, and AD-MSCs have been extensively studied for bone TE applications. While BM-MSCs have been so far the most used type for OBs differentiation [33], UC-MSCs are also a very interesting cell type due to their angiogenic potential favoring the formation of neovasculature and promoting blood supply during the process of bone regeneration [46]. AD-MSCs are also a relevant cell type for bone TE applications since they can be isolated in good quantity and high purity with a minimally invasive procedure (lipoaspirate) [43,47]. The ability of MSCs to repair bone defects has been widely investigated both in in vitro and in vivo studies, with some of the approaches being already evaluated in clinical trials [48]. For example, biomaterials-based scaffolds pre-seeded with MSCs have been employed by dentists for the regeneration of alveolar cleft and jaw defects and for the augmentation of maxillary sinus [33,49–51]. Local implantation of MSCs with or without scaffold into non-human tubular bone defects has also been evaluated in clinical trials with the outcome demonstrating the safety and efficiency of such a procedure [52]. For example, patients with distal tibial fractures were injected with BM-MSCs in the fracture site, and the results showed a median time to union of 1.5 months as compared to 3 months for the control group [33,52].

Owing to their ability to differentiate into chondrocytes, MSCs have also been evaluated in the context of cartilage repair [33]. BM-MSCs, AD-MSCs but also synovium-derived MSCs (S-MSCs) [53] have been broadly studied in this context [53,54]. While BM-MSCs have shown the best results in terms of chondrogenic differentiation both in vitro and in vivo; S-MSCs have demonstrated enhanced proliferation [33,53]. Although the first clinical trial employing MSCs to treat a cartilage defect was already conducted in 1994, currently, there are no commercially available products for cartilage reconstruction using MSCs-based therapies [33,55]. Implants of expanded autologous MSCs have been used in patients with defects in knee cartilages demonstrating both the safety and the ability to alleviate some symptoms of such approaches [56,57]. However, so far, reparation of cartilage defects through MSCs-based therapy still remains to be demonstrated [33]. Due to their ability to differentiate into chondrocytes, MSCs have been explored for the regeneration of intervertebral disks. Specifically, BM-MSCs have been implanted into patients with lumbar disc degeneration demonstrating better prognosis as compared to the current treatments, such as spinal fusion or total disc replacement [58]. Such a procedure has also resulted in decreased pain, while also highlighting the safety and feasibility of intradiscal BM-MSCs therapy [59].

More recently, MSCs-based therapies have been evaluated to regenerate other musculoskeletal tissues, such as meniscus, tendons, or ligaments [33]. Clinical trials where MSCs have been administered for meniscal regeneration in osteoarthritis patients have been already conducted, and positive outcomes in terms of the safety of the procedure and pain relief have been obtained [60]. However, despite these encouraging results, there is still a lack of evidence regarding the ability of MSCs to repair and form tissues similar to the meniscus [61]. MSCs have also been explored in the context of tendon injury since they can differentiate into the tendogenic lineage following mechanical stimulation. Tendogenic differentiation under the influence of uniaxial cyclic stretching has been reported both in vitro and in vivo, with some reports showing the ability of MSCs in delaying lesion progression [62].

MSCs have also neuronal differentiation potential, making them a good cell source to repair damaged neurons of the central nervous system [63]. Clinical studies have already been conducted in patients with spinal cord and/or brain injury [64,65]. The outcome of these studies has demonstrated the feasibility and safety of such a procedure, however, with limited efficacy. MSCs have also been employed for the treatment of neurological disorders, including multiple sclerosis, ischemic stroke, and Parkinson's disease [66–68]. The different clinical trials have shown diverse degrees of remission which have been attributed to the immunomodulatory and neuroprotective properties of MSCs [33].

Administration of MSCs has been used to improve cardiac function in the context of several cardiovascular diseases (e.g., myocardial infarction or cardiomyopathy) [69]. Clinical trials have demonstrated the ability of MSCs to improve myocardial perfusion and function at the ischaemic region, which is attributed to their ability to differentiate into cardiomyocytes [33,69].

Liver regeneration with MSCs has also been considered due to their ability to exert potent immunosuppressive and anti-inflammatory effects [70]. MSCs have also shown to be able to differentiate into hepatocytes, further highlighting their applicability in liver-specific therapeutics [71]. However, so far, liver reconstruction using differentiated MSCs is in the early stages, and its potential has only been assessed in animal models [33]. Other tissues where MSCs have shown potential applicability include cornea, trachea, or skin regeneration, which include successful outcomes in clinical trials [33].

3. Cell-Implant Surface Interactions

Within our body, every tissue is characterized by a distinct ECM composed of a main backbone of fibrous proteins (i.e., collagens, fibronectin, elastin, and fibrillin) which provides tensile strength and elasticity to the tissue. Numerous glycoproteins and highly hydrated proteoglycans are also embedded within this fibrous ECM [72]. Proteoglycans accumulate and release growth factors and are also responsible for the stiffness of the connective tissue. Other tissue-specific constituents of the ECM include lipoproteins, phosphoproteins, and other post-translationally modified proteins [72].

Within the tissues, there are three different aspects of the ECM that are sensed by the cells, thus having an influence in modulating their behavior: (a) Biochemical aspects: the chemical nature of the constituting biomacromolecules, such as charge, hydrophobicity, and molecular composition; (b) mechanical aspects: the elasticity and resilience of the ECM; and (c) topographical aspects: the three dimensional (3D) architecture, including its micro- and nano-scale structure [72]. Therefore, the surface topography of the implanted scaffold will have an influence in regulating the cells behavior, including the SCs commitment towards defined cell lineages. At the microscale, the extent of cell spreading directly correlates to the total substratum coverage with ECM-proteins, regardless of the geometrical pattern [73]. In contrast, at the nanoscale, several reports are suggesting that the different cell types react specifically and in a differential manner to the topographical features in that range [72]. The different surface topographic elements at the nanoscale (i.e., roughness, patterns, and porosity) have the ability to promote changes in cytoskeletal organization, cell shape, focal contact formation, motility, differentiation, and gene expression [72]. Such a reaction to the substrate topography is partly attributed to adhesion-related mechanisms [74]. The cells can sense the different topographic features of the scaffold by means of integrin molecules, which are cell-surface transmembrane receptors having a central role in the transduction of "outside-in signaling" [75].

Once implanted into the host body, plasma proteins rapidly adsorb onto the surface of the scaffold before cell arrival. The chemical composition, topography or wettability of the scaffold, will have an impact on the type, amount and also conformation of the adsorbed proteins [74,76,77]. Next, the extracellular domains of the cell integrins will bind to the different peptide motifs of the adsorbed proteins (i.e., tripeptide Arg-Gly-Asp (RGD)). This integrin-ligand binding will result in the interaction and clustering of cytoplasmatic proteins, such as focal adhesion kinase (FAK), vinculin, or paxillin. These clusters of over 150 structural and signaling proteins, which are known as focal adhesion (FA) complexes, regulate intracellular signaling cascades (e.g., mitogen-activated protein kinases (MAPK) signals) [75,78]. Important processes, including proliferation, differentiation, or motility, are also associated with integrins [78–81].

Apart from integrin-ligand binding, it has also been suggested by several reports that, after the initial contact, the cells produce discrete nanolength projections (i.e., "nanopodia") that transiently sense the surface to gain optimum anchorage and spreading [82,83]. Thus, the cells experience contact guidance from the different nanotopographical features gathering geometric information via nanoscale adhesion-localized structures [72,84]. Subsequent integrin-related FAs results in linkage to actin fibers (stress fibers) that are associated with cell spreading and enhanced cell movement [72,85]. On one hand, these acting stress fibers deform the nuclear membrane, which, in turn, results in the opening of nucleopores [78]. As a result of this nucleopore opening, there is enhanced mRNA transport to the cytosol, thus increasing protein translation. On the other hand, the cytoskeletal tension can promote the stretching of the cell membrane. This enhanced spreading of the cell membrane opens membrane-bound calcium ion channels allowing the transport of calcium ions to the cytosol [78]. This results in positive feedback with myosin light-chain kinase which translates into further cellular contractility [78]. Additionally, recent evidence is showing that cytoskeletal remodeling in response to surface topography can also affect nuclear morphology (nucleoskeleton arrangement), which will, in turn, result in chromosomal translocation and epigenetic DNA to regulate key genes that are involved in cell growth and functions [74,86].

4. The Effect of Substrate Topography

The fact that cell behaviors can be regulated by topographical cues from the underlying substrates was demonstrated as early as 1911 [87]. Now, it is widely acknowledged that, by contact guidance, cells can adjust their orientation and align along with the topographical features that they are cultured on. Since cells can respond to topographical cues down to 5 nm, the development of micro- and nanofabrication methods able to replicate the topographical cues of the cell niche in a controllable and reproducible fashion, has been crucial for the progress of the field [4]. Thus, we review and discuss the effect of the three main features of topography (i.e., roughness, surface patterns, and porosity) onto the cell-fate decisions of MSCs.

4.1. Substrate Roughness

Despite the enormous regenerative capacity of bone tissue, large bone defects that frequently accompany trauma, infection or tumor resection surgeries require therapeutic intervention. While autologous bone grafting is currently the gold standard therapeutic strategy, such a procedure has many drawbacks. For example, there is a limited supply of autologous bone, additional surgery is needed which leads to donor site morbidity, and the associated risk of infection [33,88]. Since alternative approaches are highly sought after, MSC-based bone regeneration has been explored [33,89].

4.1.1. Ceramic-Based Scaffolds

Currently, the most common approaches to induce osteogenic differentiation of MSCs make use of growth factors (e.g., bone morphogenetic protein 2 (BMP-2)) or small molecular compounds (e.g., dexamethasone). However, the use of soluble factors entails some limitations and risks, such as an inherently short-term availability or the fact that they can have complex interactions with the medium. Thus, the use of biomaterials-based substrates with control over their different biophysical features (i.e., topography, matrix stiffness, and mechanical forces) has emerged as an alternative strategy to promote the formation of bone tissue in a more efficient and safe manner [90].

Roughness topography can be modulated to control the osteogenic differentiation of MSCs, thus being of particular interest for bone TE (BTE) [90–92]. Such an effect has been attributed to the fact that roughness topography may replicate the physical features left by osteoclast (OC) activity on the morphology of the bone surface during the process of bone resorption [92,93]. Thus, scaffolds with tailored surface roughness able to direct and continuously support the differentiation of MSCs, resulting on a robust and inexpensive approach have been extensively investigated [94]. For example, in a recent study, Yang and collaborators fabricated hydroxyapatite (HA)-based scaffolds with varying surface roughness [90]. HA is a very interesting material for BTE since it is the main inorganic component present in human bone. The authors prepared HA disks with varying average roughness (Ra) (ranging from 0.2 to 1.65 μm) and mean distance between peaks (RSm) (ranging from 89.7 to 18.6 μm) and evaluated the osteogenic differentiation of pre-seeded human BM-MSCs (hBM-MSCs). The expression levels of osteogenic genes (i.e., runt-related transcription factor 2 (RUNX2), osteopontin (OPN), alkaline phosphatase (ALP), and collagen type I (COLI)), together with the mineralization by conducting Alizarin Red staining (ARS), were assessed. ALP is a membrane-bound enzyme that hydrolyzes phosphate esters during mineralization, and is considered a marker of early osteogenic differentiation. In contrast, ARS is used to locate calcium deposits and it is an end point measure of matrix mineralization. Optimal osteogenic differentiation in the presence of osteogenic medium was obtained for scaffolds displaying a Ra from 0.7 to 1.0 μm and RSm ranging from 53.9 to 39.3 μm. Cell attachment, spread, and F-actin arrangement were also optimal for scaffolds displaying these particular Ra and RSm parameters, suggesting that surface topography might increase osteogenesis by affecting cell attachment and cytoskeletal tension [90]. Mechanistic studies were also conducted indicating that the expression of the nuclear transcriptional factor Yes-associated protein (YAP)/TAZ was involved in the osteogenic differentiation of the cells. This is not surprising since YAP/TAZ is known to be regulated by the actin cytoskeleton and to participate in controlling the osteogenic differentiation of MSCs [90,95,96].

Gradient surfaces are of particular interest in this context, since they allow for systematic studies by continuously varying a surface parameter per experiment. Faia-Torres and coworkers fabricated polycaprolactone (PCL)-based substrates with gradient Ra from the sub-micron to the micrometer range to the osteogenic differentiation of MSCs [92]. PCL was chosen as the material since it is a resorbable polymer that has already been approved for craniofacial bone implants by the United States Food and Drug Administration (FDA) [97]. The substrates, which were created by applying a hot-embossing replication technique, displayed Ra gradients from ~0.5 to 4.7 μm and RSm values gradually varying from ~214 μm to 33 μm. The results indicated that surface roughness displaying Ra and RSm values in the range of 2.1–3.1 μm and 71.1–48.1 μm, respectively, had the biggest impact on the osteogenic differentiation of human BM-MSCs. These Ra and RSm values resulted in a faster osteogenic commitment and stronger osteogenic expression as shown by the ALP values which were more than 10 times higher than the control material (tissue culture polystyrene substrate). Robust COLI deposition was also observed for such surface roughness. Interestingly, the optimal RSm values for these PCL-based scaffolds (i.e., in the range from 71.1–48.1 μm) are the closest ones to the reported RSm values found in native bones (i.e., ~40 μm) [92,93]. While this first work was conducted in the presence of growth factors, on a follow-up study, the impact of surface roughness on the osteogenic differentiation of MSCs was assessed in the absence of osteogenic inducers [94]. In this study, Ra and RSm values of ~1.53 μm and ~79 μm, respectively, supported the highest ALP and COLI expression, as well as the highest mineralization for human hBM-MSCs cultured in the presence of osteogenic inducers. In contrast, Ra and RSm values of ~0.93 μm and ~135 μm, respectively, supported the highest ALP and COLI expression and mineralization when using basal growth medium only. Importantly, a Ra of 0.93 μm was able to direct the differentiation of BM-MSCs towards the osteogenic lineage even in the total absence of cell-differentiating factors [94]. These results demonstrated, for the first time, that specific surface roughness on its own was able to induce a strong osteogenic differentiation of hBM-MSCs. It was hypothesized by the authors that such specific Ra values promoting the commitment of BM-MSCs towards the osteogenic lineage, might mimic the pits left by osteoclastic bone resorption. Surface roughness gradients have also been explored onto catecholic polyglycerol coatings, which were the highest formation of FAs and filopodia, as well as the highest cellular tension, was observed for a Ra of 278.64 nm [98]. Such a Ra value also enhanced the osteogenic differentiation of MSCs. Interestingly, substrates with a Ra of 1050 nm were able to maintain the phenotype of MSCs by minimizing cell adhesion.

4.1.2. Titanium (Ti)-Based Scaffolds

Due to the load-bearing function of bone implants, metal materials have been widely explored in BTE. Ti is a particularly important material in this context. Ti is biocompatible, light-weighted but stronger than other metals, such as steel or cobalt alloys, which are also used as medical implants [99,100]. The excellent biocompatibility of Ti is attributed to the natural Ti dioxide deposited on its surface [99]. Although Ti-based implants have been widely employed as hip replacements, bone plates or in dental surgeries, pure Ti materials lack a good chemical bonding to the bone. As a result, about 5–20% of Ti-based implants fail in the early post-implantation period [101,102]. An efficient strategy to improve the biological performance of Ti implants is to implement topographical structures, including roughness. However, the reported studies show substantial discrepancies, a fact that can be attributed to the lack of a high-throughput strategy which could have been achieved by using surface roughness gradients [98]. For example, a first study conducted in 2009 showed how Ti surfaces with small Ra of 15 nm in height enhanced the adhesion and osteogenesis of MSCs as compared to Ras of 55 and 100 nm [103]. However, on a different report, Ras of 150 and 450 nm enhanced the osteogenic differentiation of MSCs as compared to Ra of 20 nm [104]. A different study showed no difference in terms of OBs adhesion was observed for Ti substrates with different roughness [105].

The impact of surface roughness employing Ti substrates has been extensively studied by the Cai group. The authors developed a sub-micrometer topography composed of nanosheet-pore structures onto pure Ti surfaces [101,106,107]. The surfaces, which were prepared by a simple vapor alkaline-treatment method, showed a Ra value of 0.05 μm with pore sizes ranging from 100 to 400 nm and depths ranging from 450 to 750 nm. In a first study, it was shown how these were able to promote the adhesion and osteogenic differentiation of MSCs as shown by protein adsorption, cell adhesion, differentiation and the expression of bone formation-related proteins (i.e., RUNX2 and Osterix (OSX)) [107]. Importantly, enhanced osteogenic differentiation of MSCs was also observed when cultured in medium without osteogenic inducers [107]. In a follow-up study, strontium ions (Sr^{2+}) were incorporated into the substrates [106]. The aim was to combine both biochemical and biophysical cues to enhance the osteogenic differentiation of MSCs. Sr^{2+}, which was incorporated into the Ti substrates via ion-change, is an inorganic bioactive element widely employed in BTE applications due to its ability in promoting proliferation, differentiation and mineralization of OBs [106,108]. The results revealed enhanced ALP expression and mineralization level, as well as up-regulation of RUNX2 and OSX genes for MSCs cultured onto the Sr^{2+}-incorporating substrates. Additionally, the mRNA expression of ALP, COLI, OPN, and osteocalcin (OCN) was also dramatically enhanced for these substrates. Finally, the osteoinductive potential was also investigated in vivo using a rabbit femur defect model where a larger amount of new dense bone tissue was formed for the Sr^{2+}-incorporating Ti substrates following 4 and 12 weeks post-implantation. Following on, the authors reported a new type of surface-modified Ti substrates by applying a multilayered coating of chitosan-catechol (Chi-C), gelatin (Gel) and HA nanofibers. The aim was to create a coating able to promote both osteogenesis and angiogenesis (Figure 1A) [101]. By scanning electron microscopy (SEM) and atomic force microscopy (AFM) measurements, the root mean square (RMS) surface roughness was evaluated, showing values of ~69.6 nm and ~23.5 nm for the multicoated and the uncoated Ti substrates, respectively (Figure 1B). Additionally, needle-like structures were observed for the multilayered substrates which was attributed to the incorporation of the HA nanofibers. The multilayered substrates were employed to evaluate the adhesion, morphology and recruitment of both AD-MSCs but also human umbilical vein endothelial cells (HUVECs). Due to the interrelation between osteogenesis and angiogenesis, studying the communication between OBs and endothelial cells is becoming a crucial aspect to understand the process of bone healing. Following co-culture with AD-MSCs, the multilayered substrates were able to significantly increase the invasion and vasculogenic network formation of HUVECs (Figure 1C). Significant enhancement in the stromal cell-derived factor 1a (SDF-1a) and vascular endothelial growth factor (VEGF) expression of AD-MSCs was also observed. These results were further corroborated in vivo following implantation for 7 days on a rat model. The histological examination of the rat femoral tissue showed a markedly increased angiogenic response together with the engraftment of autologous MSCs into the peri-implant tissue at an early stage of bone healing (Figure 1D). The only limitation of this study is that such a design, where both surface chemistry and roughness are coupled together, does not allow to discern which the individual contributions to the enhanced angiogenesis and osteogenesis.

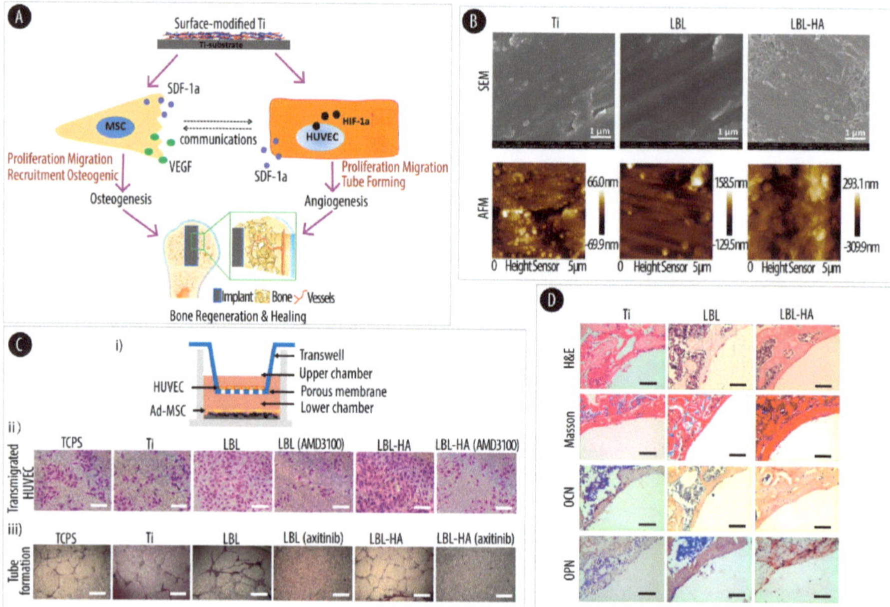

Figure 1. Multilayered coating of titanium (Ti) implants promoting coupled osteogenesis and angiogenesis in vitro and in vivo. (**A**) Schematic illustration of coupled osteogenesis and angiogenesis by surface-modified Ti substrates for enhanced bone healing. (**B**) Scanning electron microscopy (SEM) and atomic force microscopy (AFM) images of the Ti substrate, gelatin (Gel)/chitosancatechol (Chi-C) multilayer Ti-coated substrate (LBL) and the hydroxyapatite (HA) nanofiber-embedded LBL substrate (LBL-HA). (scale bar: 1 μm). (**C**) Effect of the substrates on the angiogenic factors secreted by adipose-derived mesenchymal stem cells (AD-MSCs) and enhanced regulation of human umbilical vein endothelial cells (HUVECs) functions. (i) Scheme of AD-MSCs and HUVECs co-culture system. (ii) Images of transmigrated HUVECs that were stained by crystal violet. LBL (AMD3100) and LBL-HA (AMD3100) groups were treated with C-X-C chemokine receptor type 4 (CXCR4) inhibitor AMD3100. (iii) Images of vasculogenic networks formed by HUVECs. For the LBL (axitinib) and LBL-HA (axitinib) groups, the vascular endothelial growth factor receptor (VEGFR) inhibitor axitinib was added to the co-culture system. (scale bar: 200 μm). (**D**) Rat histopathological detection of bone formation after implantation for 28 days. Hematoxylin and eosin (H&E) staining (scale bars: 200 μm), Masson staining (scale bar: 200 μm), osteocalcin (OCN) staining (scale bar: 100 μm), and osteopontin (OPN) staining (scale bar: 100 μm) of femoral tissue. Adapted from [101], with permission from Elsevier, 2018.

Worth mentioning in the context of Ti implants is the fabrication of coating surfaces displaying osteogenic inducing activity but also antibacterial properties [109]. Inhibiting bacteria adhesion is an important feature since infection is a major problem often times leading to the failure of the implant [110]. Once a bacterial bio-film has been formed, it becomes very difficult to treat and it can lead to implant removal. Therefore, a promising strategy to reduce implant-associated infections would be to fabricate implants displaying topographical features able to minimize initial bacterial adhesion. Such a strategy was accomplished by Huang et al. [109]. By sandblasting and consecutive acid etching procedures, the authors were able to fabricate a hierarchical micro/submicron/nano-scale structure on a Ti substrate. The Ra of the substrates, which could be fine-tuned by the exposure time of the second acid etching step, ranged from 2.25 μm to 1.75 μm for the longest incubation time. By cell counting and SEM it was confirmed that the substrate with the lowest Ra (i.e., Ra of about 2.25 μm) displayed the lowest bacterial adhesion. Furthermore, such a substrate demonstrated enhanced proliferation and osteogenic differentiation of pre-OBs [109].

While the regulatory effect of substrate topography on the process of osteogenesis has been widely studied, the impact of implant surface topography on osteoclastogenesis or OCs function remains largely unexplored. In a pioneering study, the Boyan group explored the effect of microrough Ti surfaces on the communication among MSCs, OBs and OCs [111]. For that, the authors collected the conditioned media from MSCs and OBs that had been cultured on three Ti surfaces with different topographies: smooth, hydrophobic-microrough, and hydrophilic-microrough for 7 days. The collected conditioned media from was then used to culture OCs for 2 days, and the OCs activity was subsequently evaluated by quantifying the amount of released collagen followed by mRNA quantification. The results indicated that both MSCs and OBs were able to suppress OCs activity. The OCs activity was also influenced by the surface roughness of the Ti substrate. This study demonstrates, for the first time, that the implant surface properties have an effect in the communication among the different cells involved in the remodeling of primary bone.

All in all, these discussed examples demonstrate how, by fine-tuning the surface roughness of an implant, it is possible to influence the cell lineage decisions of MSCs. Thus, this strategy could potentially be employed in orthopedic applications as a robust and inexpensive alternative to the less safe soluble osteogenic inducers.

4.2. Substrate Patterns

Surface modifications with patterns at both micro- and nanorange have also demonstrated the ability to influence the commitment of MSCs towards a selected cell lineage [112]. This is of particular relevance for BTE applications since, by engineering implants with surface modifications (i.e., surface roughness or micro/nanopatterns), it will be possible to control the interaction between the scaffold and the host tissue [113].

4.2.1. Ceramic-Based Scaffolds

Bioceramics, in particular, HA-based scaffolds, have been widely explored in this context. However, although 70% of the inorganic material of native bones is constituted by HA, HA-based scaffolds lack osteoinductive ability. This fact negatively impacts their repair capacity for large bone defects, non-unions and follow-up function restoration [114,115]. Thus, micro- and nanopatterned topographies have been implemented to enhance their osteoinductive properties. For example, HA-based scaffolds displaying nanopatterns fabricated with nanorods, have been evaluated. Specifically, the effects of orientation, spacing and diameter of the nanorods have been assessed and optimized [116–119]. The different studies have shown how quasi-upright HA nanorods are much more efficient in increasing the viability and proliferation of OBs as compared to the bent or parallel oriented counterparts [116]. The effect of the spacing between the quasi-upright HA nanorods was also evaluated for a range from 30 to 300 nm [117,118]. The results showed how a 70 nm interrod spacing promoted the highest osteogenic effect on pre-seeded MSCs, while also boosting the in vivo osteointegration of the scaffold. The impact of the nanorod diameter on osteogenesis was also evaluated keeping a constant interrod spacing of 70 nm [119]. From the three different studied nanorods (i.e., with 30, 70, and 150 nm in diameter), a 70 nm diameter was the best in terms of promoting the osteogenic differentiation of MSCs. Specifically, the 70 nm-diameter nanorods were able to direct the differentiation of MSCs towards the osteogenic lineage even in the absence of osteogenic supplements. Optimal osteointegration, evaluated by histological inspection employing Van Gieson's staining, was also observed for the 70 nm-diameter nanorods. The biomechanical strength following integration of the bone-implant, which was assessed by a pull-out test, was also optimal for the 70 nm-diameter nanorods. The authors suggested that the 70 nm-diameter nanorods favored FA formation on the pre-seeded MSCs and the subsequent induced mechanotransduction [119].

Surface patterns employing nanorods constituted by bulk metallic glass (BMGs) have also been evaluated [120]. BMG is an interesting material since it can be easily molded both at the micro and nano length scale with good precision and high aspect ratios [120–123]. In contrast to conventional metals and alloys, BMGs can also be processed like plastics above their glass transition temperatures [124]. In a recent study, flat and nanopatterned BMG surfaces were assessed in the context of MSCs differentiation [120]. The nanopatterns, consisting in 200 nm-diameter nanorods fabricated by thermoplastic nanomolding, were able to direct the differentiation of MSCs towards the adipogenic lineage. MSCs seeded in the nanopatterned surfaces, displayed a rounded morphology while also forming few FA points. In contrast, MSCs seeded onto flat BMGs showed enhanced formation of FAs and a commitment towards the osteogenic lineage. The cellular localization of the YAP nuclear transducer was also evaluated. Assessing YAP localization in MSCs is important in this context since it is know that the presence of YAP in the nuclear region indicates osteogenesis while a loss of nuclear YAP takes place in MSCs undergoing differentiation towards the adipogenic lineage [120,125]. The results confirmed that, when cultured onto the nanopatterned substrates, mainly cytoplasmatic YAP could be observed while flat BMGs surfaces induced the nuclear localization of YAP. The authors also compared the osteogenic ability of flat BMGs and the grade 5 Ti substrate, showing a more pronounced differentiation for the former surface. Since BMGs have an elastic modulus in the GPa range, the authors speculate that BMG could be considered as an alternative to Ti in load-bearing surgical interventions.

HA-based scaffolds with ordered micropatterns have also been explored. Using a micropatterned nylon sieve as a template, quadrate convex micropatterns of different sizes were fabricated [126]. With this method, the micropattern height and width could be easily regulated by tailoring the meshes of the nylon template. The differentiation of BM-MSCs was evaluated and the highest osteogenic response was observed for BM-MSCs cultured onto micropatterns with sizes close to that of a cell size. In a different study, the same group also evaluated the effect of three different surface topographies on the osteogenic differentiation of BM-MSCs [112,114]. Figure 2A shows SEM images of the three different patterns namely, nanosheets, nanorods and the hybrid of both nanorod and microrod (micro-nano-hybrid). The authors speculated that, since native bones display hierarchical features both in the micro- and nanorange, creating scaffolds combining both micro- and nano-structural features could result in additive or even synergistic effects [112]. The results, as shown by cell adhesion studies (Figure 2B), proliferation, and gene expression of osteogenic genes (Figure 2C, demonstrated a synergistic effect for the micro/nano hybrid structures on the osteogenesis of human BM-MSCs. Such a synergistic effect was attributed to the fact that the nanorod and micropattern structures resulted in the activation of different integrin subunits, BMP-2 downstream receptors and gap-junction related connexin 43 (Cx43) proteins. These facts indicate that micro- and nanostructures have different roles in regulating the fate of SCs. This is an important aspect that needs to be taken into account when designing biomaterials-based scaffolds for BTE. Importantly, in vivo studies were also conducted in a rat critical-sized calvarial defect model [114]. While the three different topographies resulted in enhanced new bone formation and mineralization as compared to scaffolds with flat surfaces, the micro-nano-hybrid combination was the structure promoting the best results both in vitro and in vivo.

Several other studies have also evaluated the ability of hierarchical micro/nano structured surfaces in stimulating the osteogenic commitment of MSCs [112,127–129]. For example, α-tricalcium phosphate (TCP) has been employed as a precursor to fabricate HA scaffolds with nanosheet, nanorod and a hierarchical micro/nano structure consisting of a hybrid of nanorods and microrods [129]. With such an approach, the authors were able to demonstrate that the hierarchical micro-/nano-topography surfaces not only significantly enhance cell attachment and viability of pre-seeded MSCs but also the ALP activity and the mRNA expression levels of both osteogenic and angiogenic markers. Follow up studies by the same group demonstrated that these hierarchical structures could significantly enhance the regeneration of new bone in a rat critical-sized calvarial defect model [128].

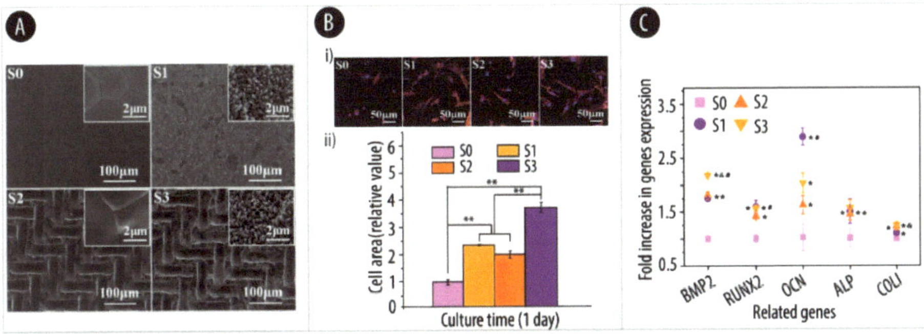

Figure 2. Micro-pattern and nano-topography of hydroxyapatite bioceramics stimulating the osteogenic differentiation of MSCs. (**A**) Field emission scanning electron microscope (FESEM) images of the control sample S0 with a smooth ceramic surface and the HA bioceramics with nanorod surface structure (S1), micropatterned surface structure (S2) and micropattern/nanorod hybrid surface structure (S3). The insets in the top-right corner of S0, S1, S2 and S3 correspond to high magnification images. (**B**) (i) Confocal images showing cytoskeleton of human bone marrow MSCs (hBM-MSCs) cultured on the samples S0–S3 for 24 h; (ii) Quantification of cell spreading area. * $p < 0.05$, ** $p < 0.01$. (**C**) Osteogenic genes expression of hBM-MSCs cultured on S0–S3 for 7 days. * indicates significant differences between S0 and the other samples, & indicates significant differences between S1 and the other samples, # indicates significant differences between S2 and the other samples, $p < 0.05$. Adapted from [112], with permission from Elsevier, 2018.

In a very recent and original study, the effect of hierarchical micro/nano structured surfaces was evaluated, not only in the context of the osteogenic differentiation of MSCs but also in terms of macrophage response [127]. This is an important aspect since, the reaction of the host immune system, has a central role in mediating the performance of the implant [127,130]. The inflammation elicited by the implanted scaffolds, will inevitably affect the processes of wound healing and tissue remodeling. As such, the ability to promote a positive immune microenvironment would ensure the long-term success of the implant [131]. In this context, assessing the macrophages' response is of particular interest since macrophages are one of the predominant immune cells in our body having a central role in the inflammatory response [127,132,133]. Depending on the microenvironment, macrophages polarize towards two main phenotypes, including pro-inflammatory (M1) and pro-healing (M2) polarization, which will inevitably affect the healing of the tissue [134–136]. The macrophages response was evaluated on HA-based substrates with three different and well-defined patterned hierarchical micro/nano structures (Figure 3A) [127]. By combining photolithography and hydrothermal techniques substrates with three circular micropatterns of 4 μm, 12 μm, and 36 μm in diameter willed with similar nanoneedle structures were fabricated (Figure 3B). Such a size range was chosen based on the sizes of the mouse macrophage cell line RAW 264.7 (of about 10 μm in diameter when fully spread). The proportion of M1- and M2-polarized macrophages depending on the studied hierarchical micro/nano structured surface was evaluated by measuring the fluorescence intensity signal of the M1-polarized (CD80) and M2-polarized (CD206) markers. The results, as shown by flow cytometry measurements, demonstrated that the structures of 12 and 36 μm-diameter induced much less M1 polarization and much more M2 polarization than the 4 μm-diameter or the flat structures (Figure 3C). While RAW 246.7 cells cultured onto 36 μm-diameter structures showed the highest fluorescence for CD206, the smallest 4 μm-diameter yielded the strongest fluorescent signal for CD80. These results were corroborated by qRT-PCR by measuring the expression of the pro-inflammatory gene CCR7, iNOS and TNF-α. Next, the effect of the immune response elicited by the macrophages on osteogenic and angiogenic differentiation of MSCs and HUVEC, respectively, was also evaluated. To do so, the expression of osteogenic and angiogenic genes in MSCs and HUVECs, respectively, was evaluated after being cultured in conditioned media collected from the structure-activated RAW 264.7 cells

(Figure 3D). Interestingly, the results show significantly enhanced expression for all the osteogenic genes for the 12 and 36 μm-diameter surface structure-stimulated RAW 264.7 cells. In contrast, the 4 μm-diameter group showed significant inhibition as compared to the flat group. Similarly, the expression of the endothelial nitric oxide synthase (eNOS), VEGF and basic fibroblast growth factor (BFGF) angiogenic genes was also up-regulated for HUVEC cells cultured in conditioned media from the 12 and 36 μm-diameter surface structure-stimulated RAW 264.7 cells. Thus, all in all, these results indicate that specific micro/nano hierarchical structures are also able to promote both osteogenesis and angiogenesis due to their ability in regulating the immune microenvironment. As such, this study provides a new perspective of how micro/nano hierarchical structures may promote bone regeneration by the manipulation of the macrophage's polarization.

Bioceramics have also been used in combination with different biomacromolecules, such as collagen or chitosan, to prepare composite scaffolds to be used as bone implants [137–140]. Amongst them, silk fibroin (SF) appears as an interesting material for BTE due to its excellent mechanical and functional properties, low immunogenicity and good biocompatibility [137,141]. Thus, SF has been employed in combination with HA to render SF/HA composite scaffolds with excellent biocompatibility, osteoconductivity, and osteoinductivity [142,143]. MSCs were seeded onto such composite scaffolds which were fabricated by using air-plasma-treated SF films. Following in vivo implantation into a rat subcutaneous model, the air-plasma-treated SF/HA films promoted the most efficient formation of bone.

4.2.2. Ti-Based Scaffolds

Surface patterns on Ti implants have also demonstrated the ability to significantly enhance the adhesion and osteoblastic differentiation of MSCs.

Research efforts have focused on constructing TiO_2 nanostructures onto the implant's surface to control SCs fate and improve implant performance. TiO_2 is a native oxide formed on the Ti metal surface which bonds directly to the adjacent bone. In particular, many studies have been focused on TiO_2 nanotube arrays fabricated by anodization onto Ti substrates [144,145]. Examples include the work by Lv et al., demonstrating that TiO_2 nanotubes of 70 nm in diameter are the optimal size to promote the osteogenic differentiation of MSCs [146]. Oh et al., also showed how nanotubes ranging from 70 to 100 nm-diameter promoted a dramatic elongation of pre-seeded MSCs, which translated into cytoskeletal stress and osteogenic differentiation [14]. In contrast, smaller TiO_2 nanotubes of about 30 nm-diameter did not promote the differentiation of MSCs [14]. Additionally, nanostructured surfaces consisting of a regular array of columnar TiO_2 nanotubes have also shown the ability to promote osteoblastic cell differentiation [14,147,148], which then translated into enhanced bone apposition and fixation following implantation into rat tibias [147].

Research efforts have also focused on understanding the mechanistic aspects of the osteogenic differentiation of MSCs promoted by TiO_2 nanopatterns. Hou et al. recently postulated that the such differentiation was mediated by the transient receptor potential vanilloid 4 (TRPV4) via the regulation of the nuclear factor of activated T-cells, cytoplasmic 1 (NFATc1) and wingless-related integration site (Wnt)/β-catenin signaling [149]. They conducted a study to elucidate whether calcium ions (Ca^{2+}), which are the central element of calcified tissues, participate in the osteoblastic differentiation promoted by the nanotubes. To do so, MSCs were cultured onto Ti disks containing Ti nanotubes (TNT) with diameters of ~30 and ~100 nm (designated as TNT-30 and TNT-100, respectively) making use of polished Ti disks (PT) as the control. The authors chose to study TRPV4 since this ion channel is responsible for the intracellular transfer of Ca^{2+} [150]. To assess whether the TNTs promoted the osteogenic differentiation of MSCs via TRPV4, MSCs were cultured onto TNT-30, TNT-100, and PT for 7 days, and the expression of TRPV4, NFATc1, and Wnt/β-catenin, as well as their cellular distribution, was evaluated. The results showed that both TNT-30 and TNT-100 substrates promoted the expression and activation of TRPV4. Additionally, when inhibiting TRPV4 in MSCs cultured on TNT-100, reduced expression of osteoblastic genes and protein levels of NFATc1 and Wnt3a/β-catenin was also observed. As such, these results revealed that the osteoblastic differentiation of)BM-MSCs is mediated by TRPV4 [149].

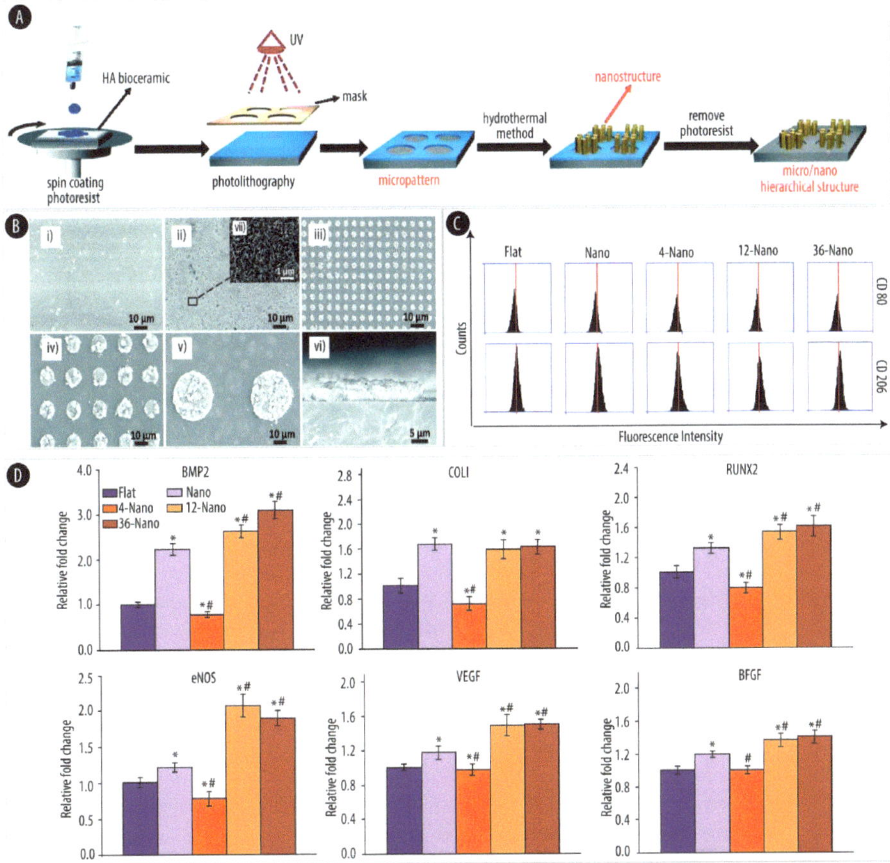

Figure 3. Micro/nano hierarchical hydroxyapatite (HA) stimulation of osteogenetic and angiogenetic differentiation via macrophage immunomodulation. (**A**) Schematic of the preparation of HA bioceramics with micro/nano hierarchical surfaces using photolithography and hydrothermal techniques. (**B**) SEM images of HA bioceramics with different sized micro/nano hierarchical structures. (i) Control sample with a flat surface (Flat group). (ii) Only nanoneedle surface (Nano-group). (iii) 4 μm micro-dot/nanoneedle hierarchical surface (4-Nano group). (iv) 12 μm micro-dot/nanoneedle hierarchical surface (12-Nano group). (v) 36 μm micro-dot/nanoneedle hierarchical surface (36-Nano group). (vi) Cross-section of the 36-Nano group. (vii) The corresponding high magnification image of the 36-Nano-group. (scale bar: 10 μm (i, ii, iii, iv, v), 5 μm (vi), 1 μm (vii)). (**C**) Fluorescence intensity of cluster of differentiation (CD)80 (top) and CD206 (bottom) of RAW 264.7 (mouse mononuclear macrophages) cells on HA bioceramics with different topographical surfaces. (**D**) RT-qPCR results of the osteogenic gene expression (bone morphogenetic protein 2 (BMP-2), collagen type I (COLI), and runt-related transcription factor 2 (RUNX2)) of hBM-MSCs and angiogenic gene expression (endothelial nitric oxide synthase (eNOS), vascular endothelial growth factor (VEGF), and vascular endothelial growth factor (BFGF)) of human umbilical vein endothelial cells (HUVECs) cultured in RAW 264.7 cell-conditioned medium. * indicates significant difference ($p < 0.05$) compared to the Flat group. # indicates a significant difference ($p < 0.05$) compared to the Nano group. Adapted from [127] with permission from the Royal Society of Chemistry, 2019.

Apart from nanotubes, TiO_2 nanorods have also demonstrated the ability to influence the lineage commitment of MSCs [151]. In a first study, clustered TiO_2 nanorods were fabricated on an acid-etched microstructured Ti surface employing hydrogen peroxide. By PCR, ALP expression and ARS, together with OPN and OCN immunofluorescence analyses, it was shown that such a topography was able to promote the osteogenic differentiation of pre-seeded MSCs. In a more recent study, a TiO_2 nanorod array was fabricated employing a hydrothermal method using tetrabutyl titanate as a precursor [145]. The behavior of MSCs was evaluated and compared to a well-polished surface of TiO_2 ceramic. Studies on the proliferation, adhesion, morphology, and differentiation of the pre-seeded MSCs demonstrated commitment towards the osteogenic lineage for the TiO_2 nanorod array surface. In contrast, the control TiO_2 ceramic with a smooth surface promoted the self-renewal of MSCs. Thus, overall, this study reveals an approach where, by using only different surface topographies, it is possible to control self-renewal or differentiation behaviors of MSCs.

4.2.3. Polymeric Substrates

The use of nanopatterns onto polymeric substrates to direct the differentiation of MSCs was pioneered by Dalby and collaborators [152]. In such a pioneering study, the authors employed electron beam lithography (EBL) to fabricate ultra-precise nanotopographies onto a poly(methyl methacrylate) (PMMA) and demonstrated that the substrates displaying deliberately disordered pits in a square arrangement promoted the differentiation of MSCs into OBs [152]. In contrast, when employing nanopatterns displaying pits arranged in a square lattice symmetry, the results were different, and the pre-seeded MSCs were able to maintain their stem-cell phenotype and self-renewal for up to 8 weeks [153].

Differentiation of MSCs into adipogenic lineages by nanopatterns was also evaluated by Abagnale et al. [154]. In this study, microgrooved patterns were fabricated using polyimide (PI), which is a biocompatible, resistant, and long-lasting polymer [155–157]. The authors conducted a study employing a combination of 25 different structures with a systematic variation of the width of grooves and ridges ranging from 15 μm-sized structures down to the sub-micron range. By varying one specific parameter at the time, the authors aimed at understanding the impact of individual topographic features on cellular functions. The results showed that MSCs were aligned parallel to the main axis of the grooves for all the studied 25 patterns. In the presence of differentiation media and while keeping constant the groove-width (to 5 μm), 15 μm ridges were able to enhance adipogenic differentiation. In contrast, 2 μm ridges resulted in increased commitment towards the osteogenic lineage. This fact was attributed to the direct impact of the physical size of the ridges on cellular morphology. As such, while elongated morphology might enhance osteogenesis, a round morphology might drive the cells towards adipogenic lineage. Interestingly, when making use of nano-patterns, the results were different and PI substrates with a periodicity of 650 nm increased differentiation towards both osteogenic and adipogenic lineages. This fact was demonstrated by the up-regulation of both adipogenic (adiponectin (ADIPOQ), peroxisome proliferator-activated receptor gamma (PPARG), and fatty acid binding protein 4 (FABP4)) and osteogenic (secreted protein acidic and rich in cysteine (SPARC), secreted phosphoprotein 1 (SPP1), and bone gamma-carboxyglutamic acid-containing protein (BGLAP)) marker genes. Thus, these results suggest that the support of nanotopography on in vitro differentiation is not necessarily lineage-specific. Mechanistic studies were also conducted. Western blot fragmentation analysis did not show increased translocation of YAP to the cytoplasm for MSCs cultured on PI 650 nm as compared to flat PI substrates. This fact reveals that, in this case, YAP, which has been previously shown to modulate differentiation of MSCs [95], was not affected by nanotopography and is not relevant for this mechanotransduction.

Polymeric substrates in the form of nanofibers have been widely explored as scaffolds for TE applications. Nanofiber meshes fabricated by electrospun display the required flexibility in terms structure and composition which can be modulated to orchestrate specific cellular responses [158]. Polymeric nanofibers fabricated by electrospun have also been investigated in the context of BTE.

Examples include the development of PCL-based nanofibrous scaffolds incorporating HA-NPs or β-TCP, which demonstrated the ability to direct the osteogenic differentiation of MSCs without the need for chemical osteogenic inducers [159]. In addition, incorporation of HA-NPs into poly(3-hydroxybutyrate-*co*-3-hydroxyvalerate) (PHVB) nanofibrous electrospun scaffolds have also been considered [160]. PHVB is a biodegradable, non-antigenic and biocompatible polyester and together with HA-NPs supported the commitment of MSCs towards the osteogenic lineage. Furthermore, the implantation of the PHVB/HA-NPs scaffolds into critical-sized rabbit radius defects, resulted in complete reparation as shown by the radiographic results [161]. PHBV/HA-NPs scaffolds with aligned and random-oriented nanofibers have also been prepared and compared. While no differences could be observed on bone growth following implantation into critical-sized rabbit bone defects depending on the fiber orientation, the mechanical properties of the repaired bones were different depending on the orientation of the fibers. Specifically, sixteen weeks after implantation, the rabbit's radii were taken out and the mechanical properties were evaluated conducting stress-strain curves. The elasticity modulus of the repaired bone yielded values closer to the control group (i.e., normal radii of rabbits) for the scaffolds with aligned fibers. Thus, these results suggested that the PHBV/HA scaffolds with aligned nanofibers are more suitable for BTE applications as compared to the random-oriented counterparts [161]. The same group also conducted studies to assess the underlying mechanism of the spontaneous osteogenic differentiation of MSCs seeded onto PHVB/HA-NPs scaffolds [162]. By immunofluorescence staining after 3 days and 7 days of cell seeding, it was possible to demonstrate that Wnt/β-catenin signaling pathway, BMP- small mothers against decapentaplegic (Smad) signaling pathway and MAPK were involved in the osteogenic differentiation of MSCs. These results are not surprising since the Wnt/β-catenin signaling pathway, which has a central role in bone cell function, it is known to be involved in the cell responses to implant surfaces [163,164]. The BMP-Smad signaling pathway has also been reported to mediate the biological effects of the implant surfaces [165], while MAPKs are known to be key mediators of the cellular responses to a variety of extracellular stimuli [162,166]. Mechanistic investigations were also conducted by the Soleimani group [167]. The authors evaluated the effect of fiber orientation on the osteogenic differentiation of MSCs using a poly-L-lactide acid (PLLA) scaffold. The different nanotopographical cues (i.e., aligned versus random) had an effect on the osteogenic commitment of MSCs through the cross-regulation between microRNAs (miRNAs) and long noncoding RNAs. Additionally, by using a combination of computational and experimental approaches, the authors could identify a set of miRNAs that can efficiently regulate BMP signaling pathway through multi-target genes as accelerators of osteogenic differentiation.

Scaffolds constituted by nanofiber meshes fabricated by electrospinning are of particular interest for tendon and ligament (T/L) TE. T/L are constituted by a dense regular connective tissue composed primarily by COLI and collagen type III (COLIII) fibers [168]. These fibers consist of triple helix tropocollagen molecules which are hierarchically arranged in a multilevel creating parallel collagenous fibrils. This hierarchical architecture provides tendons and ligaments with exceptional axial tensile load-bearing capacity. However, following injury, tendons and ligaments have difficulties to heal. This is a consequence of a lack of vasculature in these tissues, which results in a suboptimal supply of essential reparative factors at the injury site [169]. When T/L heal by themselves, the initially formed healing matrix does not display the arranged hierarchical architecture but, instead, is constituted by randomly arranged collagen mainly composed by the smaller and less organized COLIII [170]. As a result of this marked decrease in the COLI and COLIII ratio, the tissue will experience a reparative phase, whereby the structural integrity between the torn ends will be restored. However, this reconstruction does not result in tissue regeneration, where the COLI to COLIII ratio is similar to that of the native tissues. Gradual remodeling of this initially formed healing matrix will take place over the course of a year, where the tissue will experience enhanced fibril alignment along the tissue length, as well as an increase in the COLI and COLIII ratio. However, despite the gradual alignment of the collagen fibrils over time, minor disorganization and abnormalities persist and the tissue never regains its original properties and functionality [169,171]. Approaches to repair T/L tears include suture, however this usually results in scar tissue and the subsequent shortening of the tendon. This tendon shortening permanently limits the motion range and, furthermore, it also leads to a high rate of injury reoccurrence [172]. A different strategy is to employ autografts or allografts to replace the injured tissue. However, such an approach involves the risk of donor site morbidity (for autografts) and disease transmission or rejection (for allografts). Alternatively, synthetic or natural ECM-derived grafts can also be employed as replacements. However, these grafts fail in replicating the mechanical properties of the native tissues [173]. Thus, there is a need for tissue-engineered T/L that better mimic the features of native tissue for improved repair. Topographical cell guidance has been investigated to create scaffolds with the ability to promote aligned cellular orientation for T/L applications [169].

First studies demonstrating the beneficial effects of culturing cells on aligned electrospun fibers for T/L TE include the work by Lee et al. [162]. The authors fabricated aligned polyurethane fibers and observed how human ligament fibroblasts cultured onto the fibers displayed a spindle-shaped morphology and oriented in the direction of the fibers. These aligned cells were able to secrete much more collagen as compared to randomly-oriented fibroblasts. The ligaments' tensile strength has also been shown to improve when fibroblasts are aligned along the length [169,174]. Aligned fibers were also fabricated by the Ouyang group and demonstrated that, when aligned, PLLA nanofibers were able to regulate the orientation of MSCs and promote their differentiation into the teno-lineage [175]. Collagen threads have also been aligned electrochemically and evaluated in the context of tendon regeneration [176]. Pre-seeded MSCs oriented in a parallel manner to the aligned fibers and it was shown how this anisotropic orientation induced the tenogenic differentiation even in the absence of biochemical factors. Tong et al. replicated not only the physical topography but also the elasticity of the tendons' ECM using nanoimprinting [177]. MSCs cultured on such scaffolds displayed shape and alignment similar to that of MSCs seeded onto the "longitudinally-cut" section of tendons. While, in this first study, no significantly enhanced tenomodulin (TNMD) expression could be detected, the results were different upon coating this nanoimprint topography with COLI and enhanced TNMD expression could be observed [178]. Thus, these results highlight the need for an additional signal to complete the full differentiation process towards a tenogenic lineage. Coatings with COLI-mimetic peptides have also been considered [158]. Using peptides mimicking a specific domain has several advantages over using natural adhesive proteins, such as COLI, fibronectin, or laminin. Limitations of natural proteins include a potential immunogenic response by the host, protein purification issues or the fact that proteins display multiple adhesion domains that may induce conflicting intracellular signals [179]. PCL meshes with an aligned topography were fabricated by electrospinning and coated with the glycine-phenylalanine-hydroxyproline-glycine-glutamate-arginine (GFOGER) motif of COLI [158]. Such a peptide was chosen since this domain (which corresponds to the residues 502–507 of the α_1(I) chain of COLI) interacts with $\alpha_2\beta_1$ integrin, thus mediating the adhesion and differentiation of OBs, as well as the mineralization of the matrix [158,180,181]. In this study, the in vitro migration and differentiation of MSCs as a result of both the functionalization with the peptide motif and orientation of the fibers was evaluated. The results showed how coating with the GFOGER peptide resulted in enhanced migration, proliferation, and osteogenic differentiation of MSCs. A further enhancement on cell migration was observed on the aligned fiber meshes as compared to the randomly-oriented ones, and the migration took place along the direction of the fiber orientation. However, no enhanced osteogenic differentiation could be detected as a result of the fiber alignment.

In addition, in the context of T/L TE, the Goh group fabricated a 3D scaffold consisting of aligned SF electrospun fibers (SFEFs) [169,182]. In a first study, it was demonstrated that, following seeding with MSCs, the aligned SFEFs promoted both cellular and ECM alignment [182]. Additionally, enhanced differentiation of MSCs towards the ligament fibroblast lineage was also demonstrated by the expression and production of ligament-related proteins (i.e., COLI, COLIII, tenascin-C, and TNMD). Due to the relevance of mechanical stimulation during the healing of tendons and ligaments [170], on a follow-up study, the effects of mechanical conditioning on the 3D aligned scaffolds were evaluated [169].

The authors hypothesized that mechanical loading was as important as the cellular alignment for the maintenance of an appropriate in vivo cellular behavior and matrix remodeling in the context of T/L regeneration [169]. As such, the biophysical effect of topography and mechanical stimuli on the tenogenesis of MSCs was investigated. To do so, mechanical stimulation of the scaffolds was achieved by rolling them up and loading them into the chamber vessels of a bioreactor where the ends of the scaffolds were used for anchorage. Next, both a 5% translational strain and a 90° rotational strain mimicking the multidimensional straining environment of the in vivo situation, were applied. Interestingly, the results, as shown by the enhanced production of T/L-related proteins, revealed that the introduction of mechanical forces accelerated and intensified the tenogenic differentiation of the pre-seeded MSCs. These findings are in agreement with previous studies showing an increase in the levels of intracellular Ca^{2+} in both tendon and ligament cells following mechanical stimulation [169,183,184]. Other reports have also shown how tendon cells respond to mechanical loads by releasing adenosine triphosphate [169,185]. Additionally, applying a cyclic strain intensifies the expression of ECM proteins while also inducing the assembly of ECM structures [169,186]. However, to the best of our knowledge, this is the first report on the combined and synergistic effect of aligned nanopatterns and mechanical stimuli in the context of T/L TE. Overall, these results suggest accelerated matrix deposition and remodeling process for the treatment group which will, in turn, improve the tensile properties.

Three-dimensional scaffolds constituted by braided polymeric microfibers have also shown promising mechanical properties for T/L TE applications [187]. Braided scaffolds are an interesting material since their mechanical properties can be fine-tuned by changing certain braiding parameters, such as the number of the braided fiber bundle, fiber diameter, or braiding angle. However, most of the reported braided scaffolds are constituted by fibers with a size in the micrometer range, which results in suboptimal biological activity. In contrast, 3D scaffolds consisting of braided fibers in the nano range will better mimic the dimensionality of native collagen fibrils [188], thus providing a more favorable environment for T/L tissue regeneration [187]. As such, Barber et al. pioneered the fabrication of hierarchical scaffolds for T/L TE based on braided nanosized fibers [187]. The scaffolds were fabricated by braiding 3, 4, or 5 aligned bundles of electrospun PLLA nanofibers and the tenogenic differentiation of pre-seeded MSCs was evaluated. The results showed how MSCs adhered and aligned in parallel to the length of the fibers experiencing also a realignment of the actin cytoskeleton. This is an important fact since, aligned actin stress fibers, have previously shown protection from mechanical loading while also mediating appropriate mechanotransduction signals within tendon fibroblasts [187,189]. When cultured in the growth medium, the different braided scaffolds supported the proliferation of MSCs enhancing their pluripotency as shown by the upregulation of key pluripotency genes (i.e., Oct3/4 and Sox2). In contrast, when the MSCs seeded scaffolds were cultured in differentiation medium containing fibroblast growth factor 2 (FGF-2), growth differentiation factor 5 (GDF-5), and BMP-2 as growth factors, downregulation of the three embryonic stem cell markers was detected. However, the differentiation medium influenced the commitment of the MSCs towards osteogenesis and away from the tenogenic lineage. This was shown by the upregulation of RUNX2 and the downregulation of Scleraxis (SCX). Interestingly, such a trend could be reversed following mechanical stimulation. Upon applying a cyclic tensile strain onto the MSCs-seeded scaffolds, significant upregulation of SCX gene expression was observed. Additionally, the MSCS-seeded scaffolds were able to produce tissue-specific ECM in the presence of both mechanical stimuli and differentiation medium. Altogether, these results highlight the importance of a suitable mechanobiological environment for scaffolds to be used in T/L TE applications.

In a follow-up study by the same group, the effect of the braiding angle onto the mechanical properties of the resulting braided electrospun nanofibers was evaluated [173]. The purpose of the study was to assess the effect of the mechanical properties of the resulting braided scaffold on the tenogenic commitment of MSCs. Interestingly, the authors employed iPSC-MSCs. iPSCs can be obtained by reprogramming many adult somatic tissues and, thus, they are an interesting cell source when MSCs are less available (e.g., from donors with increased age) [190]. In addition, iPSC-MSCs will render patient-specific cell lines for TE applications. PLLA and PCL fibers were prepared with different stitches per inch (SPI) (i.e., 8, 12, 16, 20, and 24) (Figure 4A,B) [173]. The different number of SPI resulted in braided nanofibers with different braid angles (ranging from 47 to 67°) (Figure 4C). The resulting fibers displayed different mechanical properties depending on the braiding angle, as shown by conducting stress-strain curves. Braided nanofibers with fewer SPI showed enhanced modulus and strength than those with more SPI. In addition, the mechanical evaluation demonstrated that PCL braided nanofibers were stronger and stiffer than their PLLA counterparts. The cell experiments were conducted by applying a cyclic strain stimulation, due to the relevance of mechanical loading in the in vivo cellular behavior of tissues and ligaments. It was demonstrated that PLLA was better than PCL in supporting cell adhesion, encouraging spindle-like morphology and inducing the characteristic tenogenic gene expression. These results are likely due to PCL's hydrophobicity which will lead to less protein adhesion and the subsequent reduced cell attachment, proliferation, and differentiation [191]. Interestingly, while the nanofibers chemistry had an effect on the adhesion of iPSC-MSCs, the braiding angle played a major role in regulating their differentiation. Thus, braided nano fibers with fewer SPI promoted upregulation of tenogenic markers (i.e., SCX, TNMD, COLI to COLIII expression ratio) while downregulating the expression of the osteogenic ones (i.e., RUNX2, OCN) (Figure 4D). The authors speculated that, following application of a cyclic strain, the different mechanical properties of the scaffolds translated into varying mechanobiological forces onto the pre-seeded MSCs. This, in turn, by means of mechanotransduction, had an effect on the tenogenic differentiation of the cells. In a more recent study, aligned PCL- and PLLA-nanofibers with either a stacked or a braided architecture were fabricated and evaluated [192]. While both stacked [193] and braided [194] scaffolds have been fabricated both as tendon and ligament grafts, a direct comparison still remained to be conducted. The authors evaluated the effect of the macro-architecture on the mechanical properties and the subsequent tenogenic differentiation of MSCs. The results showed that while both types of scaffolds supported the tenogenic differentiation of MSCs, the braided constructs enhanced the upregulation tenogenesis to a greater degree than their stacked counterparts. As such, these results indicate that macro-architecture is also an important parameter that needs to be taken into account in T/L tissue engineering applications.

While the aforementioned reports have shown the ability to control the fate of MSCs in vitro by modifying the physical properties of the substrates, studies assessing whether the topography-induced effects observed in vitro can be also recapitulated in vivo, remain scarce. One of these few studies was reported by the Ouyang group [195]. The authors assessed whether the in vitro diverging differentiation pathways of MSCs promoted by aligned and randomly-oriented PLLA fibrous scaffolds could also be observed in vivo. SEM micrographs showed how MSCs were well attached to both types of scaffolds but displaying different morphologies. While MSCs cultured onto the randomly-oriented scaffolds were spread out showing a polygonal phenotype, when seeded onto the aligned scaffold, they displayed an elongated and spindle-shape morphology and were furthermore oriented parallel to the aligned fibers. The randomly-oriented fibers triggered the differentiation of the MSCs to the osteogenic lineage, as shown by the enhanced expression of osteogenic markers and the positive ALP staining. In contrast, increased tenogenic differentiation was observed for MSCs on the aligned substrate, as indicated by the enhanced expression of the tenogenic transcription factor gene SCX. Additionally, differences in the distribution of FA complexes, as well as the cytoskeletal organization of MSCs were observed depending on the alignments of the fibrous scaffolds. To get an in-depth understanding of the topographic induced differentiation of MSCs, the authors also conducted mechanistic studies. Specifically, the effects of cytochalasin D (cyto D) and the Rho kinase (ROCK) inhibitor Y-27632 on cytoskeletal reorganization and mechanotransduction were evaluated [195]. Addition of cyto D, a potent inhibitor of actin polymerization, onto the cell medium resulted in the cells to become rounded and no notable differences in morphology depending on the scaffold (i.e., aligned versus randomly-oriented) could be observed. Additionally, cyto D treatment downregulated the expression levels of both osteogenic RUNX2 and tenogenic SCX markers. Thus, cyto D could attenuate both the orientation-induced osteogenesis and alignment-induced tenogenesis. The addition to the cell media of Y-27632, which is an inhibitor of the myosin-generated cytoskeletal tension, eliminated cell elongation for MSCs on the aligned scaffold and reduced the projected surface area of MSCs in the randomly-oriented one. Thus, treatment with both cyto D and Y-27632, resulted in significant changes in cell morphology followed by loss of lineage commitment. These results suggest that the actin cytoskeleton is of high relevance for the lineage commitment of MSCs. Within the same study, both scaffolds were also evaluated in vivo on a rat Achilles tendon repair model. Following 4 weeks post-implantation, histological examination and immunohistochemical analysis showed that the aligned fibrous scaffolds promoted tendon-like tissue formation and yielded much more mature tendon as compared to the randomly-oriented ones. In contrast, for the randomly-oriented fibrous scaffold group, substantial chondrogenesis followed by tissue ossification and bone formation was observed at the implantation site. Thus, importantly, this study highlights the relationship between the topographic features of the scaffold and tissue formation and reveals possible risks of ectopic ossification induced by the design of the scaffolds in tendon-tissue repair applications.

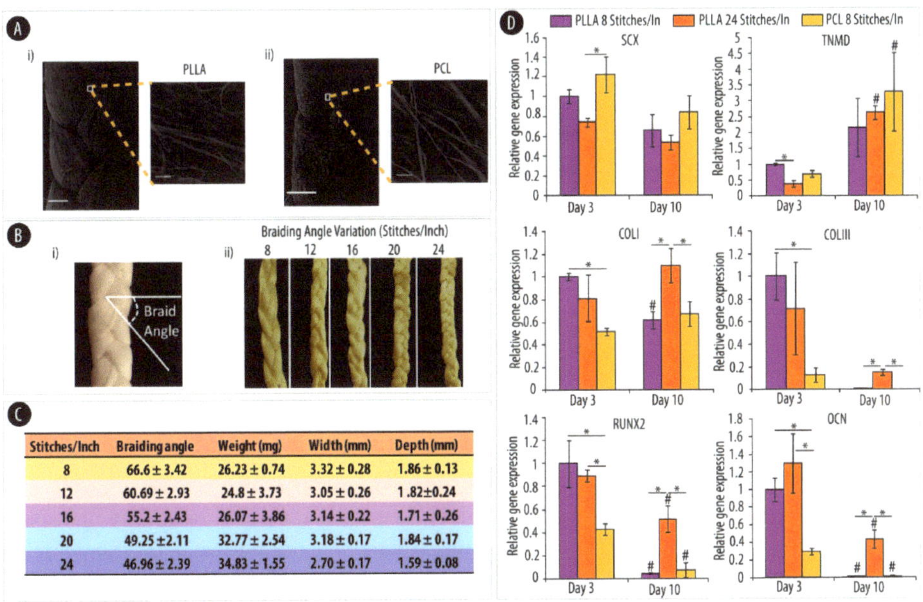

Figure 4. Properties of braided submicron fibrous scaffolds (BSMFSs) direct the tenogenic differentiation of human-induced pluripotent stem cell-derived mesenchymal stem cells (hiPSC-MSCs). (**A**) Representative SEM images of poly-L-lactide acid (PLLA) (i) and polycaprolactone (PCL) (ii) BSMFSs with insets showing a high magnification view of the fiber alignment. (scale bar: 500 μm, 10 μm (insets)). (**B**) Demonstration of braiding angle in a BSMFS (i) and macrographs of PLLA BSMFSs fabricated with different braiding angles by altering the number of bundle stitches per inch (8, 12, 16, 20, and 24) (ii) (scale bar: 5 μm). (**C**) BSMFSs with different braiding angles. (**D**) Quantitative RT-PCR results of mRNA expression of the tendon and ligament (T/L)-associated markers, Scleraxis (SCX), tenomodulin (TNMD), collagen type I (COLI), and collagen type III (COLIII), and the bone-associated markers, runt-related transcription factor 2 (RUNX2) and osteocalcin (OCN), at days 3 and 10. * denotes significant differences ($p < 0.05$; $n = 3$) between conditions at the same time point and # denotes significant differences ($p < 0.05$; $n = 3$) within the same condition between days 10 and 3. Adapted from [173] with permission from Elsevier, 2014.

Nanopatterned substrates have also been evaluated in the context of cartilage TE. Despite its simple appearance, treating adult cartilage tissue following injury remains a significant clinical challenge. This poor ability for self-regeneration is a consequence of several features: cartilage is an avascular tissue, has a low supply of progenitor cells and it also is a scarcely cellularized tissue with a single constituent cell type (chondrocytes) with a very poor proliferation ability [3,196]. However, articular cartilage is frequently damaged mainly due to trauma, inflammatory causes or degenerative joint diseases, such as osteoarthritis [197]. To date, the major treatment strategies for cartilage repair include microfracture (marrow stimulation) [198], autografts [199], and autologous chondrocyte implantation [196,200]. However, these methods have important limitations, which include lack of integration, graft instability, and calcification, as well as donor-site morbidity [196,197,201–203]. Therefore, TE approaches have emerged as a promising strategy to restore injured cartilage [197]. Due to the limited source of healthy donor cartilage, as well as the reduced proliferative ability of isolated autologous chondrocytes, the use of SCs has emerged as an alternative for generating the required cell numbers for viable cartilage repair [3,204]. MSCs offer an attractive cell source since they have the ability to differentiate into chondrocytes, have a high proliferation capacity and can also be easily extracted from donor BM [3,205,206].

The Mauck group was one of the firsts in exploring the potential of MSCs for cartilage tissue engineering [207,208]. MSCs were used in the context of fibrocartilaginous tissues, which are tissues sharing the characteristics of both articular cartilage and fibrous tissues, such as tendon and ligament. Fibrocartilaginous tissues, such as the annulus fibrosus of the intervertebral disc and the knee meniscus, conduct mechanical roles that are essential for healthy joint function. In a first study, Baker and Mauk demonstrated the utility of MSCs for meniscus tissue engineering by showing how MSCs were able to align with the predominant nanofiber direction of a PCL scaffold and deposit fibrocartilaginous ECM [208]. Enhanced ECM deposition and improvement in tensile properties were observed for the aligned nanofibers as compared to the random-oriented ones [208]. The same group also reported a strategy for the engineering of annulus fibrosus, which is a multi-lamellar fibrocartilage of the intervertebral disk [207]. The study shows the fabrication of PCL-based scaffolds capturing the multi-scale structural architecture of the annulus fibrosus, which is a requirement to achieve the functionality of their native counterparts. Anisotropic nanofibrous laminates were created by electrospinning and, following MSCs seeding, they directed the deposition of a collagen-rich ECM mimicking the multi-lamellar hierarchical structure of the native annulus fibrosus. Additionally, such an architecture displayed similar mechanical properties than the native tissue after 10 weeks of in vitro culture. While this prior work was conducted making use of bovine MSCs, more recently, the same group employed human MSCs and evaluated their potential in fibrocartilage engineering [209]. Specifically, MSCs were harvested from the BM of patients undergoing total knee arthroplasty and, as a control, fibrochondrocytes were isolated from the meniscus of the same donors. Both cell types were cultured on aligned nanofibrous scaffolds and their ability to create a mechanically functional, fibrocartilaginous matrix was assessed over 9 weeks. Surprisingly, and in contrast to the results obtained using bovine MSCs, little matrix formation was observed for MSCs as compared to fibrochondrocytes taken from the same donor. These varying results could be attributed to the lack of a spatially-defined topography rendered by the electrospinning technique. Only by spatially controlling the patterned nanofeatures, it will be possible to conduct systematic studies on the chondrogenesis of MSCs. Thus, nanoimprinting has been also used to create specific nano-topographies and evaluate their ability to regulate the formation of cartilage tissue [3,210]. Nano-grill, nano-hole and nano-pillar patterns were chosen due to their resemblance to the topographical features of the cartilage collagen fibril (Figure 5A) [3,211]. According to the authors, approximately, a nano-grill topography resembles the parallel collagen fibers of the superficial and deep zone cartilage, nano-holes better mimic the pores among the randomly oriented collagen fibers while the nano-pillars represent the intersections of the random collagen fibers. PCL was also employed as the polymeric material to create a nanopatterned film. However, in this study, a chondroitin sulfate (CS) coating was deposited onto the substrate to provide chondro-inductive biochemical cues [3,212,213]. The results, as shown by quantifying the expression level COLI and collagen type II (COLII) by enzyme linked immunosorbent assays (ELISA) and the cartilaginous mRNA markers (COLI and COLII), demonstrated the ability of nano-pillar and nano-hole topography to enhance the differentiation of MSCs towards the chondrogenic lineage (Figure 5B). Additionally, the nano-pillar and nano-hole topographies facilitated the formation of hyaline cartilage as compared to the non-patterned substrate. In contrast, on a nano-grill topography, the MSCs experienced delayed chondrogenesis and were induced to fibro/superficial zone cartilage formation (Figure 5C) [3]. In a follow-up study by the same group, the combined effect of nano-topography and mechanical stiffness of the substrate was evaluated [210]. To do so, nano-grating and nano-pillar patterns were fabricated employing polyesters of varying stiffness [210]. PCL, polylactide (PLA) and polyglycolide (PGA) which have different mechanical stiffness (PCL<PLA<PGA) were thermally imprinted and coated with CS to obtain an even chondro-inductive surface across all polymeric substrates. The results demonstrated that substratum stiffness and topographical cues affected the phenotypic development at the earlier stage of chondrogenic differentiation. Additionally, different types of cartilage-like tissue were obtained depending on the topography and stiffness of the substrate. This is an important fact since, while previous studies of cartilage tissue engineering using MSCs resulted in homogeneous cartilage tissue structures [214], this has very little resemblance to the in vivo situation. In our body, articular cartilage can be classified into different zones (i.e., the superficial, middle, and deep zones), where both the chondrocytes morphology, as well as the ECM composition and structural arrangement, are greatly different. Additionally, the amount of proteoglycans, the alignment and complex interactions among collagen fibers and other matrix components, result in different mechanical properties that inevitably influence the overall function of articular cartilage [3,214,215]. Within this study, hyaline-like cartilage with middle/deep zone cartilage characteristics was obtained on softer surfaces with nano-pillar topographies. This was shown by the upregulation of aggrecan, CILP, COMP, and COLIX, which are middle/deep zone cartilage markers. In contrast, on stiffer and also nano-pillar substrates, MSCs generated constituents of hyaline/fibro/hypertrophic cartilage as shown by the high expression of COLII which was accompanied by low levels of COLI. Fibro/superficial zone-like cartilage could be obtained following culture on nano-grating surfaces of softer stiffness. This was demonstrated by the analysis of PRG4, which is a specific superficial zone cartilage marker. In contrast, stiffer nano-grating did not promote chondrogenesis.

4.3. Porosity

Scaffolds in tissue engineering are meant to act as a temporary ECM and, as such, they are expected to provide supports for cells and guide cell differentiation. Thus, since bioactive materials are designed to replicate the properties of in vivo environment, having an appropriate porosity and pore size are essential to control biological functions [74].

Thanks to recent developments in fabrication technologies, it is now possible to construct biomaterials with a well-defined pore structure that can be sensed by the cells of the host tissue (mainly SCs) and thus regulate cell fates during differentiation. The scaffolds pore structure, in terms of both porosity and pore size, provides space for the cell attachment and growth but also ensures the transport of nutrients, metabolites and waste products [74].

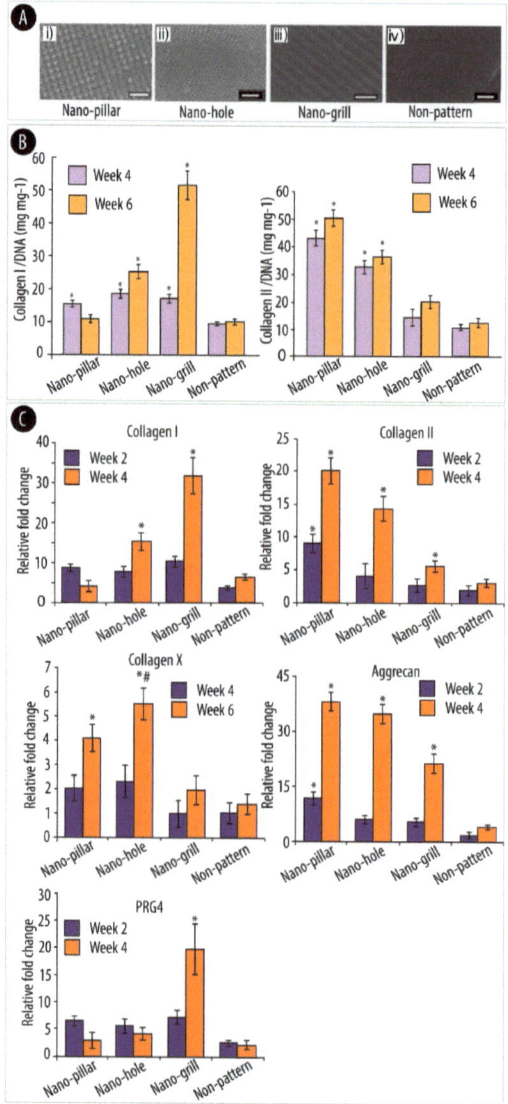

Figure 5. Substrate topography determines the fate of chondrogenesis from human mesenchymal stem cells (MSCs) resulting in specific cartilage phenotype formation. (**A**) SEM images showing the surface morphology of non-patterned and nano-patterned polycaprolactone (PCL) films with a chondroitin sulfate (CS) coating (scale bar: 1 μm). (**B**) Collagen I and II expressed by MSCs cultured on the different-patterned surface. The results were analyzed by enzyme linked immunosorbent assays (ELISA) and normalized to DNA at week 4 and week 6 of chondrogenic differentiation. * denotes $p < 0.05$, significant increase compared to non-patterned surface. (**C**) Real-time PCR analysis of mRNA expression levels of cartilaginous genes at weeks 2, 4, or 6 of chondrogenic differentiation normalized to their respective glyceraldehyde-3-phosphate dehydrogenase (GAPDH) expression and expressed as fold changes relative to undifferentiated MSCs. * denotes $p < 0.05$, significant increase compared to non-patterned surface. # denotes $p < 0.05$, significant different between nano-pillar and nano-hole surface at the same time point. Adapted from [3] with permission from Elsevier, 2014.

An appropriate porous structure of the scaffold is central to achieve an optimal osteogenic effect. This is not surprising given the porous structure of native bone. Specifically, there are two main types of bone: while the cancellous (or trabecular) bone shows a porosity ranging from 40–95%, the denser cortical (or compact) bone has a lower porosity of 5–25% [216]. Therefore, scaffolds for BTE require a porous architecture that will allow for the ingrowth of bone tissue and blood vessels while also ensuring bone oxygenation [74,217–219]. In particular, pore sizes of about 100 μm are considered beneficial for cell migration and nutrient transport [218]. However, prior reports have also found that pores larger than 200 μm were also able to promote vascularization and the formation of new bone [217]. Several studies have also found that smaller pores of around 50 to 100 μm were better for inducing endochondral ossification (i.e., osteochondral formation before osteogenesis) while, large pores (100–300 μm) promoted vascularization and were also beneficial for nutrient supply and waste removal [220–222]. Additionally, the porosity and connectivity among the pores are two important parameters. While high porosities of ~80% are regarded as optimal for the regeneration of new bone, cell ingrowth is also highly dependent on the degree of connectivity among the pores [220–224]. Additionally, recent studies are also indicating that interconnected pores are necessary to allow for the circulation of nutrients, waste and to promote the migration of cells to the center of the implant [225,226]. Thus, biomaterials scaffolds for BTE should exhibit both suitable pore sizes and porosity, as well as good interconnectivity among the pores [74,227–229].

The first studies assessing the effect of porosity on ceramic scaffolds date back to the early 1970s [230]. Hulbert et al. were able to demonstrate that pore sizes between 100 and 300 μm initiated bone formation, while a pore size range of 10 to 100 μm conditioned fibrous tissue or unmineralized osteoids [230,231]. Porous polymeric scaffolds were also developed in the 70s showing how bone ingrowth predominated when the scaffolds displayed a pore size around 450 μm [232]. Since these early studies, a wide range of pore sizes (from 20 to 1500 μm) have been investigated, however, the optimal pore size for OB activity is still controversial [233–236]. For example, one study demonstrated that OBs proliferation and bone formation on PLGA scaffolds was optimal for pore sizes ranging from 150 to 300 μm [237]. A different group also employing PLGA-based scaffolds showed how pore sizes in the range of 150 to 710 μm had no impact on the OBs attachment and proliferation [230,238]. Smaller pore sizes have also been investigated. For example, Akay et al. employed a highly porous polymeric foam (the so-called polyHIPE polymer) and, following the evaluation of different pore sizes, enhanced proliferation of OBs was demonstrated for small pores of 40 μm [239]. In contrast, scaffolds with larger pores (of about 100 μm) facilitated cell migration. However, no significant differences in the extent of mineralization or cell penetration depth could be observed depending on the pore size [239]. Collagen-based scaffolds have also been widely explored to identify an optimal pore size for BTE applications. In particular, the effect of the pore size on the preosteoblastic cell line MC3T3-E1 was evaluated employing collagen–glycosaminoglycan scaffolds [233]. In this case, optimal cell proliferation and infiltration were found for scaffolds with mean pore sizes larger than 300 μm. What is more, the ability of larger pores to facilitate cell infiltration was shown to counteract the favorable effect of the enhanced initial cell attachment surface areas provided by smaller pores. Thus, this study supports the hypothesis of previous reports highlighting the important role of pore sizes greater than 300 μm to promote osteogenesis [240]. The porosity and pore sizes in polymeric scaffolds have also been evaluated in the context of cartilage TE. Scaffolds consisting of biodegradable estane polymers displaying macropores in the range from 150 to 355 μm which are highly interconnected to micropores (<50 μm), have been shown to promote the ingrowth into fibrocartilaginous tissue preventing degeneration of the articular cartilage [230,241]. Studies evaluating the lineage commitment of MSCs depending on the scaffolds pore size have been conducted with polymeric substrates. Specifically, collagen-hyaluronic acid scaffolds with pore sizes of 94, 130, and 300 μm in diameter were evaluated [242]. The results showed that the largest (300 μm) pores promoted the best outcome in terms of chondrogenic differentiation. PLGA scaffolds with pore sizes in the range from 90 to 180 μm with a fibrin coating enhanced both the deposition of cartilage-specific ECM, as well as the chondrogenic differentiation, of pre-seeded MSCs [243]. iPSCs have also been studied in the context of regeneration and restoration of cartilage defects on 3D nanofibrous scaffolds with similar pore sizes [230,244].

Metals and alloys have been widely used as bone implants [245–247]. Thus, methods developed so far to fabricate porous metals include liquid state processing (e.g., direct foaming and spray foaming), solid-state processing, such as powder metallurgy and sintering of powders and fibers, electro-deposition, and vapor deposition [248]. While these methods offer control over both the shape and size of the pores by adjusting several parameters of the fabrication process, only a randomly organized porous structure can be obtained. In contrast, more sophisticated fabrication techniques, such as selective laser melting and electron beam melting, which are computer-controlled fabrication methods based on layer-wise manufacturing principles, render porous metals with a predefined external shape and internal architecture [248]. Similarly, as with ceramics or polymeric scaffolds, the literature evaluating the optimal pore size within metallic scaffolds still remains open to discussion. For example, Itala and co-workers, fabricated triangle-shaped Ti implants with pore sizes ranging from 50 to 125 μm [249]. Following implantation into rabbit femur, they found no clear lower limit of pore sizes for consistent bone ingrowth. The possible effect of microporosity, that is, small pores in the range of 0.5 to 10 μm on osseointegration, has also been evaluated [250]. Braem et al. fabricated microporous Ti implants that were placed in the cortical bone of rabbit tibiae. Following 2 and 4 weeks post-implantation, bone ingrowth could be observed throughout the porous Ti implant, highlighting that osseointegration of microporosities (<10 μm) could be achieved. Interestingly, when surface modifications inside the porous structure further reduced the interconnective pore size to a submicrometer level, the ingrowth of bone tissue was impaired. Extra-large pores in the range from 300 to 900 μm have also been investigated. Porous Ti scaffolds with pore sizes of 300, 600, and 900 μm were fabricated and implanted into rabbit tibias, and the results showed the highest bone ingrowth for the 600 and 900 μm-pore scaffolds [251].

Apart from the pore size, the level of porosity is an important parameter in bone regeneration since its envisioned that scaffolds with high porosity will facilitate the in vivo ingrowth of bone cells [230]. For example, Chan et al. evaluated the impact of different levels of porosity on the osteoinductive potential employing an ectopic bovine model [252]. Silicate substituted calcium phosphate scaffolds with porosities of 22.5, 32.0, and 46.0%, respectively, were implanted into the ovine parapsinalis muscle for 12 weeks. Histomorphometry analyses showed the largest amount of bone formation for the 46.0% porosity scaffolds with a 6.17 % increase in bone formation. In contrast, the lowest porosity level (22.2%) rendered only a 1.33 % increase in bone formation. Interestingly, bone formation was also observed in pores with a size smaller than 10 μm. Campion et al. also conducted studies with the same type of scaffolds in ovine critical-sized defect sites [253]. Their results also showed increased bone formation for the most porous scaffold but also more advanced neovascularization. Thus, these studies highlight the clinical importance of porosity for silicate-substituted calcium phosphate bone grafts. The effect of the level of porosity on the osteogenic differentiation of SCs has also been evaluated. Two groups of HA/TCP scaffolds with different porosity (i.e., 45 and 30 pores per inch, respectively) but controlled pore size were fabricated to evaluate the proliferation and osteogenic differentiation of skeletal SCs (Figure 6A) [254]. Following implantation of the seeded scaffolds into the rear flank of immunodeficient mice, the scaffold displaying 45 pores per inch, demonstrated the greatest increase in density (Figure 6B). While both scaffolds supported the proliferation of the skeletal SCs, the scaffold with the highest porosity was better in promoting the differentiation towards the osteogenic lineage (Figure 6C).

While a high level of porosity is envisioned to facilitate recruitment and penetration of cells from the surrounding bone tissue and its vascularization, thus benefiting bone ingrowth [248], it will also have a profound effect on the implants' mechanical properties. This is of special interest for metallic implants. While metals and alloys have been widely used as bone implants, they are usually much stiffer than native bones. While cortical bones have elastic moduli ranging from 3 to 30 GPa, the elastic moduli of cancellous bones is only about 0.02–2 GPa [248]. In contrast, most current metallic implants used for bone regeneration have a much higher moduli. For example, a modulus of around 110 GPa and

210 GPa has been reported for Ti6Al4V and CoCrMo alloys, respectively [245–247]. Such a difference in stiffness between natural bone and metallic implants results in stress shielding which is a major source for bone resorption subsequently leading to the eventual failure of the implants [245,248]. An efficient method to circumvent stress shielding at the bone-implant interface, is to introduce controlled porosity in the materials [248,255]. Examples of porous metallic scaffolds to be used in BTE applications include cylindrical dental implants of cancellous structured Ti which were placed into a canine model. Following 8 weeks after the operation, more bone ingrowth could be detected for the implants with the highest level of porosity [256]. While a high level of porosity may be beneficial to reduce the elastic moduli of metallic implants, it might be detrimental for other less stiff materials. For BTE applications, a porous scaffold should also display appropriate mechanical properties (e.g., stiffness, strength, and toughness). Only scaffolds with satisfactory mechanical properties will be able to keep the shape following implantation in the body, while providing sufficient structural support during bone formation [225]. Therefore, the porosity of the scaffolds must be carefully tailored to achieve appropriate mechanical properties and biological performance [74]. As an example, Eqtesadi et al. evaluated the effect of porosity on the mechanical properties of 13–93 BG scaffolds. By immersing the scaffolds into simulated body fluid to mimic the in vivo degradation behavior, they demonstrated that the formed HA crystal on the porous scaffold surface was larger and more numerous than that on the dense scaffold surface. However, the presence of the porosity significantly diminished the scaffolds mechanical strength [257].

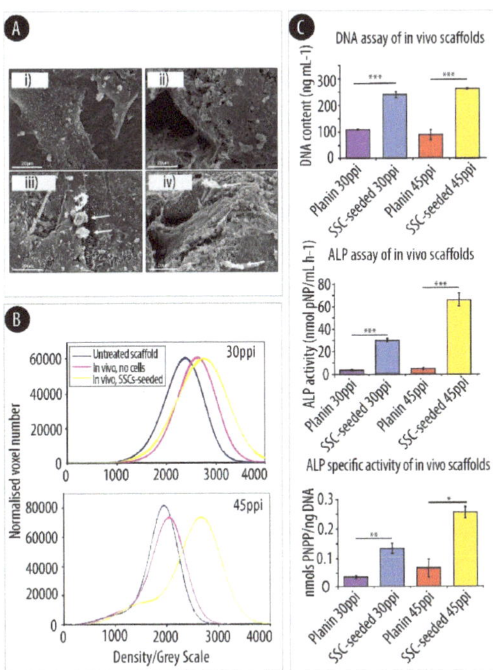

Figure 6. The effect of porosity of a biphasic ceramic scaffold on human skeletal stem cell (SSC) growth and differentiation in vivo. (**A**) SEM images of ex vivo scaffold with 30 pores per inch (30 ppi) (i), scaffold with 45 pores per inch (45 ppi) (ii), SSCs-seeded 30 ppi scaffold (iii), and SSCs-seeded 45 ppi scaffold (iv). Thick arrows indicate the SSC-like cells. (scale bar: 20 μm). (**B**) Density plots of scaffolds following 4 weeks in vivo growth with and without prior SSC-seeding. (**C**) Biochemical analysis of cellular material on the scaffolds following 4 weeks in vivo growth. * denotes $p < 0.05$, ** denotes $p < 0.01$, *** denotes $p < 0.001$ between the marked groups. Adapted from [254] with permission from Wiley, 2013.

5. Conclusions and Outlook

The magic ability to regenerate injured or diseased human body parts to restore missing or lost function, has long been a dream of humanity. Biomaterials-based scaffolds in combination with MSCs represent a promising tool in regenerative medicine. Such an approach will instruct cell fate to regain lost tissue volume and function, thus improving patients' quality of life.

Human MSCs have the ability of multi-lineage differentiation. Additionally, they possess the capability to self-renew over long periods of time, and they can be easily isolated with minimally invasive procedures. Although it has been over 50 years since their discovery, recent progress in extraction, culture and differentiation methods have paved the way for MSCs to become closer to clinical applications for tissue reconstruction.

During the course of tissue regeneration, the formation of new tissue is largely influenced by the physico-chemical cues of the microenvironment. Tissue cells reside in a complex milieu where they can communicate with each other while also responding to multiple stimuli provided by the surrounding matrix. As such, biomaterials-based scaffolds can provide multiple physico-chemical cues to modulate the growth and differentiation of MSCs and the subsequent formation of new tissue.

Traditionally, biochemical soluble compounds (i.e., growth factors) have been employed to regulate the differentiation of stem cells into a specific cell lineage. However, due to the limitations of these soluble factors (e.g., high costs, off-target effects, or challenges in obtaining optimal release kinetics), the focus has shifted towards adjusting the physical aspects of the stem cells environment. As such, biomaterials-based scaffolds with a well-defined topography have emerged as a powerful strategy. In this review, we have aimed to demonstrate how the different topographical features (i.e., surface roughness, patterns, and porosity) are an efficient method to control different aspects of cell behavior.

Despite the promising advancements presented herein, much work still remains to be conducted to successfully apply a biomaterials-mediated approach to guide MSCs for clinical use. While the studies conducted so far show the potential of the different topographical features to substantially modulate several aspects of MSCs, including proliferation, migration, alignment, differentiation, and even cell structure (e.g., cytoskeleton remodeling); such results have been observed on a case-by-case basis. Therefore, a universal correlation between the different topographical features and the subsequent cell response still needs to be established. To do so, future studies should focus on the interactions of cells and substrates comprehensively inspecting the biofunctions of the scaffold biophysical cues to unravel the underlying mechanisms. This is an important, but complicated, aspect due to the difficulty in reproducing the tissue structure in micro- and nanodimension. This translates into the need of developing new fabrication methods. In general, tissue formation takes place at multiple length scales. This hierarchical structure is responsible for many functional aspects of the tissue. Thus, it is envisioned that new or multiple combined fabrication techniques will be needed to engineer the scaffolds structure at molecular, nanoscale, microscale, and macroscale dimensions. Addressing these challenges will allow us to fabricate advanced implants with proper structural features able to provide the optimal cues to achieve the desired cell responses in vivo.

Author Contributions: The manuscript was written through contributions of all authors. All authors have read and agreed to the published version of the manuscript.

Funding: This work has been supported by the Novo Nordisk Foundation, Denmark (Grant No. NNF19OC0058413).

Conflicts of Interest: The authors declare no conflict of interest.

References

1. Curtis, A.S.G.; Varde, M. Control of cell behavior: Topological factors 2. *J. Natl. Cancer Inst.* **1964**, *33*, 15–26. [CrossRef] [PubMed]
2. Stevens, M.M. Exploring and engineering the cell surface interface. *Science* **2005**, *310*, 1135–1138. [CrossRef] [PubMed]
3. Wu, Y.-N.; Law, J.B.K.; He, A.Y.; Low, H.Y.; Hui, J.H.; Lim, C.T.; Yang, Z.; Lee, E.H. Substrate topography determines the fate of chondrogenesis from human mesenchymal stem cells resulting in specific cartilage phenotype formation. *Nanomed. Nanotechnol. Biol. Med.* **2014**, *10*, 1507–1516. [CrossRef] [PubMed]
4. Metavarayuth, K.; Sitasuwan, P.; Zhao, X.; Lin, Y.; Wang, Q. Influence of surface topographical cues on the differentiation of mesenchymal stem cells in vitro. *ACS Biomater. Sci. Eng.* **2016**, *2*, 142–151. [CrossRef]
5. Brafman, D.A. Constructing stem cell microenvironments using bioengineering approaches. *Physiol. Genom.* **2013**, *45*, 1123–1135. [CrossRef]
6. Lee, J.-H.; Park, H.-K.; Kim, K.S. Intrinsic and extrinsic mechanical properties related to the differentiation of mesenchymal stem cells. *Biochem. Biophys. Res. Commun.* **2016**, *473*, 752–757. [CrossRef]
7. Moghaddam, M.M.; Bonakdar, S.; Shariatpanahi, M.R.; Shokrgozar, M.A.; Faghihi, S. The effect of physical cues on the stem cell differentiation. *Curr. Stem Cell Res. Ther.* **2019**, *14*, 268–277. [CrossRef]
8. Curtis, A.; Wilkinson, C. New depths in cell behaviour: Reactions of cells to nanotopography. *Biochem. Soc. Symp.* **1999**, *65*, 15–26.
9. Norman, J.J.; Desai, T.A. Methods for fabrication of nanoscale topography for tissue engineering scaffolds. *Ann. Biomed. Eng.* **2006**, *34*, 89–101. [CrossRef]
10. Nie, Z.; Kumacheva, E. Patterning surfaces with functional polymers. *Nat. Mater.* **2008**, *7*, 277–290. [CrossRef]
11. Dalby, M.J.; Gadegaard, N.; Herzyk, P.; Agheli, H.; Sutherland, D.S.; Wilkinson, C.D. Group analysis of regulation of fibroblast genome on low-adhesion nanostructures. *Biomaterials* **2007**, *28*, 1761–1769. [CrossRef] [PubMed]
12. Nguyen, K.T.; Shukla, K.P.; Moctezuma, M.; Tang, L. Cellular and molecular responses of smooth muscle cells to surface nanotopography. *J. Nanosci. Nanotechnol.* **2007**, *7*, 2823–2832. [CrossRef] [PubMed]
13. Dalby, M.J.; McCloy, D.; Robertson, M.; Agheli, H.; Sutherland, D.S.; Affrossman, S.; Oreffo, R.O. Osteoprogenitor response to semi-ordered and random nanotopographies. *Biomaterials* **2006**, *27*, 2980–2987. [CrossRef] [PubMed]
14. Oh, S.; Brammer, K.S.; Li, Y.S.J.; Teng, D.; Engler, A.J.; Chien, S.; Jin, S. Stem cell fate dictated solely by altered nanotube dimension. *Proc. Natl. Acad. Sci. USA* **2009**, *106*, 2130–2135. [CrossRef]
15. Yim, E.K.; Darling, E.M.; Kulangara, K.; Guilak, F.; Leong, K.W. Nanotopography-induced changes in focal adhesions, cytoskeletal organization, and mechanical properties of human mesenchymal stem cells. *Biomaterials* **2010**, *31*, 1299–1306. [CrossRef]
16. Hoch, A.I.; Mittal, V.; Mitra, D.; Vollmer, N.; Zikry, C.A.; Leach, J.K. Cell-secreted matrices perpetuate the bone-forming phenotype of differentiated mesenchymal stem cells. *Biomaterials* **2016**, *74*, 178–187. [CrossRef]
17. Leach, J.K.; Whitehead, J. Materials-directed differentiation of mesenchymal stem cells for tissue engineering and regeneration. *ACS Biomater. Sci. Eng.* **2017**, *4*, 1115–1127. [CrossRef]
18. Dobbenga, S.; Fratila-Apachitei, L.E.; Zadpoor, A.A. Nanopattern-induced osteogenic differentiation of stem cells-A systematic review. *Acta Biomater.* **2016**, *46*, 3–14. [CrossRef]
19. Mashinchian, O.; Turner, L.-A.; Dalby, M.J.; Laurent, S.; Shokrgozar, M.A.; Bonakdar, S.; Imani, M.; Mahmoudi, M. Regulation of stem cell fate by nanomaterial substrates. *Nanomedicine* **2015**, *10*, 829–847. [CrossRef]
20. Engler, A.J.; Sweeney, H.L.; Discher, D.E.; Schwarzbauer, J.E. Extracellular matrix elasticity directs stem cell differentiation. *J. Musculoskelet. Neuronal Interact.* **2007**, *7*, 335.
21. Engler, A.J.; Sen, S.; Sweeney, H.L.; Discher, D.E. Matrix elasticity directs stem cell lineage specification. *Cell* **2006**, *126*, 677–689. [CrossRef]
22. Kang, E.-S.; Kim, D.-S.; Suhito, I.R.; Lee, W.; Song, I.; Kim, T.-H. Two-dimensional material-based bionano platforms to control mesenchymal stem cell differentiation. *Biomater. Res.* **2018**, *22*, 10. [CrossRef] [PubMed]
23. Wang, N.; Tytell, J.D.; Ingber, D.E. Mechanotransduction at a distance: Mechanically coupling the extracellular matrix with the nucleus. *Nat. Rev. Mol. Cell Biol.* **2009**, *10*, 75–82. [CrossRef] [PubMed]
24. Kim, Y.J.; Hwang, S.J.; Bae, Y.C.; Jung, J.S. miR-21 Regulates adipogenic differentiation Through the modulation of TGF-β signaling in mesenchymal stem cells derived from human adipose tissue. *STEM CELLS* **2009**, *27*, 3093–3102. [CrossRef]

25. Nieden, N.I.Z.; Kempka, G.; Rancourt, D.E.; Ahr, H.-J. Induction of chondro-, osteo- and adipogenesis in embryonic stem cells by bone morphogenetic protein-2: Effect of cofactors on differentiating lineages. *BMC Dev. Biol.* **2005**, *5*, 1. [CrossRef]
26. Derfoul, A.; Perkins, G.L.; Hall, D.J.; Tuan, R.S. Glucocorticoids promote chondrogenic differentiation of adult human mesenchymal stem cells by enhancing expression of cartilage extracellular matrix genes. *STEM CELLS* **2006**, *24*, 1487–1495. [CrossRef] [PubMed]
27. Wang, J.; Chen, Z.; Sun, M.; Xu, H.; Gao, Y.; Liu, J.; Li, M. Characterization and therapeutic applications of mesenchymal stem cells for regenerative medicine. *Tissue Cell* **2020**, *64*, 101330. [CrossRef] [PubMed]
28. Friedenstein, A.J.; Piatetzky-Shapiro, I.I.; Petrakova, K.V. Osteogenesis in transplants of bone marrow cells. *Development* **1966**, *16*, 381–390.
29. Caplan, A.I. Mesenchymal stem cells. *J. Orthop. Res.* **1991**, *9*, 641–650. [CrossRef]
30. Caplan, A.I. Mesenchymal Stem Cells: Time to Change the Name! *STEM CELLS Transl. Med.* **2017**, *6*, 1445–1451. [CrossRef]
31. Andrzejewska, A.; Lukomska, B.; Janowski, M. Concise review: Mesenchymal stem cells: From roots to boost. *STEM CELLS* **2019**, *37*, 855–864. [CrossRef] [PubMed]
32. Bianco, P.; Robey, P.G.; Simmons, P.J. Mesenchymal stem cells: Revisiting history, concepts, and assays. *Cell Stem Cell* **2008**, *2*, 313–319. [CrossRef] [PubMed]
33. Han, Y.; Li, X.; Zhang, Y.; Han, Y.; Chang, F.; Ding, J. Mesenchymal stem cells for regenerative medicine. *Cells* **2019**, *8*, 886. [CrossRef] [PubMed]
34. Peng, Y.; Ke, M.; Xu, L.; Liu, L.; Chen, X.; Xia, W.; Li, X.; Chen, Z.; Ma, J.; Liao, D.; et al. Donor-derived mesenchymal stem cells combined with low-dose tacrolimus prevent acute rejection after renal transplantation. *Transplantation* **2013**, *95*, 161–168. [CrossRef]
35. Marędziak, M.; Marycz, K.; Tomaszewski, K.A.; Kornicka, K.; Henry, B.M. The Influence of aging on the regenerative potential of human adipose derived mesenchymal stem cells. *Stem Cells Int.* **2016**, *2016*, 1–15. [CrossRef]
36. Golpanian, S.; El-Khorazaty, J.; Mendizabal, A.; Difede, D.L.; Suncion, V.Y.; Karantalis, V.; Fishman, J.E.; Ghersin, E.; Balkan, W.; Hare, J.M. Effect of aging on human mesenchymal stem cell therapy in ischemic cardiomyopathy patients. *J. Am. Coll. Cardiol.* **2015**, *65*, 125–132. [CrossRef]
37. Dominici, M.; Le Blanc, K.; Mueller, I.; Slaper-Cortenbach, I.; Marini, F.; Krause, D.; Deans, R.; Keating, A.; Prockop, D.; Horwitz, E. Minimal criteria for defining multipotent mesenchymal stromal cells. The International Society for Cellular Therapy position statement. *Cytotherapy* **2006**, *8*, 315–317. [CrossRef]
38. Soleimani, M.; Nadri, S. A protocol for isolation and culture of mesenchymal stem cells from mouse bone marrow. *Nat. Protoc.* **2009**, *4*, 102–106. [CrossRef]
39. Kim, M.J.; Shin, K.S.; Jeon, J.H.; Lee, D.R.; Shim, S.H.; Kim, J.K.; Cha, D.-H.; Yoon, T.K.; Kim, G.J. Human chorionic-plate-derived mesenchymal stem cells and Wharton's jelly-derived mesenchymal stem cells: A comparative analysis of their potential as placenta-derived stem cells. *Cell Tissue Res.* **2011**, *346*, 53–64. [CrossRef]
40. Zuk, P.A.; Zhu, M.; Ashjian, P.; De Ugarte, D.A.; Huang, J.I.; Mizuno, H.; Alfonso, Z.C.; Fraser, J.K.; Benhaim, P.; Hedrick, M.H. Human adipose tissue is a source of multipotent stem cells. *Mol. Biol. Cell* **2002**, *13*, 4279–4295. [CrossRef]
41. Sibov, T.T.; Tobo, P.R.; Marti, L.C.; Pavon, L.F.; Oliveira, D.M.; Tobo, P.R.; Campos, A.H.; Paes, A.T.; Amaro, E.; Gamarra, L.F.; et al. Mesenchymal stem cells from umbilical cord blood: Parameters for isolation, characterization and adipogenic differentiation. *Cytotechnology* **2012**, *64*, 511–521. [CrossRef] [PubMed]
42. Gao, W.-X.; Sun, Y.; Shi, J.; Li, C.-L.; Fang, S.-B.; Wang, D.; Deng, X.-Q.; Wen, W.; Fu, Q.-L. Effects of mesenchymal stem cells from human induced pluripotent stem cells on differentiation, maturation, and function of dendritic cells. *Stem Cell Res. Ther.* **2017**, *8*, 48. [CrossRef] [PubMed]
43. Lin, H.; Sohn, J.; Shen, H.; Langhans, M.T.; Tuan, R.S. Bone marrow mesenchymal stem cells: Aging and tissue engineering applications to enhance bone healing. *Biomaterials* **2019**, *203*, 96–110. [CrossRef]
44. Mochizuki, T.; Muneta, T.; Sakaguchi, Y.; Nimura, A.; Yokoyama, A.; Koga, H.; Sekiya, I. Higher chondrogenic potential of fibrous synovium-and adipose synovium–derived cells compared with subcutaneous fat–derived cells: Distinguishing properties of mesenchymal stem cells in humans. *Arthritis Rheum.* **2006**, *54*, 843–853. [CrossRef] [PubMed]

45. Jin, H.J.; Bae, Y.K.; Kim, M.; Kwon, S.; Jeon, H.B.; Choi, S.J.; Kim, S.W.; Yang, Y.S.; Oh, W.; Chang, J.W. Comparative analysis of human mesenchymal stem cells from bone marrow, adipose tissue, and umbilical cord blood as sources of cell therapy. *Int. J. Mol. Sci.* **2013**, *14*, 17986–18001. [CrossRef] [PubMed]
46. Kargozar, S.; Mozafari, M.; Hashemian, S.J.; Milan, P.B.; Hamzehlou, S.; Soleimani, M.; Joghataei, M.T.; Gholipourmalekabadi, M.; Korourian, A.; Mousavizadeh, K.; et al. Osteogenic potential of stem cells-seeded bioactive nanocomposite scaffolds: A comparative study between human mesenchymal stem cells derived from bone, umbilical cord Wharton's jelly, and adipose tissue. *J. Biomed. Mater. Res. Part B Appl. Biomater.* **2016**, *106*, 61–72. [CrossRef]
47. Burrow, K.L.; Hoyland, J.A.; Richardson, S.M. Human adipose-derived stem cells exhibit enhanced proliferative capacity and retain multipotency longer than donor-matched bone marrow mesenchymal stem cells during expansion in vitro. *Stem Cells Int.* **2017**, *2017*, 1–15. [CrossRef]
48. Oreffo, R.O.; Tare, R.S.; Yang, L.-Y.; Williams, D.F.; Ou, K.-L.; Oreffo, R.O. Biofabrication of bone tissue: Approaches, challenges and translation for bone regeneration. *Biomaterials* **2016**, *83*, 363–382. [CrossRef]
49. Khojasteh, A.; Fahimipour, F.; Jafarian, M.; Sharifi, D.; Jahangir, S.; Khayyatan, F.; Eslaminejad, M.B. Bone engineering in dog mandible: Coculturing mesenchymal stem cells with endothelial progenitor cells in a composite scaffold containing vascular endothelial growth factor. *J. Biomed. Mater. Res. Part B Appl. Biomater.* **2016**, *105*, 1767–1777. [CrossRef]
50. Katagiri, W.; Watanabe, J.; Toyama, N.; Osugi, M.; Sakaguchi, K.; Hibi, H. Clinical study of bone regeneration by conditioned medium from mesenchymal stem cells after maxillary sinus floor elevation. *Implant. Dent.* **2017**, *26*, 607–612. [CrossRef]
51. Gjerde, C.; Mustafa, K.; Hellem, S.; Rojewski, M.; Gjengedal, H.; Yassin, M.A.; Feng, X.; Skaale, S.; Berge, T.; Rosen, A.; et al. Cell therapy induced regeneration of severely atrophied mandibular bone in a clinical trial. *Stem Cell Res. Ther.* **2018**, *9*, 213. [CrossRef] [PubMed]
52. Liebergall, M.; Schroeder, J.; Mosheiff, R.; Gazit, Z.; Yoram, Z.; Rasooly, L.; Daskal, A.; Khoury, A.; Weil, Y.; Beyth, S. Stem cell–based therapy for prevention of delayed fracture union: A randomized and prospective preliminary study. *Mol. Ther.* **2013**, *21*, 1631–1638. [CrossRef] [PubMed]
53. Fernandes, T.L.; Kimura, H.A.; Pinheiro, C.C.G.; Shimomura, K.; Nakamura, N.; Ferreira, J.R.; Gomoll, A.H.; Hernandez, A.J.; Bueno, D.F. Human synovial mesenchymal stem cells good manufacturing practices for articular cartilage regeneration. *Tissue Eng. Part C Methods* **2018**, *24*, 709–716. [CrossRef] [PubMed]
54. Bougioukli, S.; Sugiyama, O.; Pannell, W.; Ortega, B.; Tan, M.H.; Tang, A.H.; Yoho, R.; Oakes, D.A.; Lieberman, J.R. Gene therapy for bone repair using human cells: Superior osteogenic potential of bone morphogenetic protein 2–transduced mesenchymal stem cells derived from adipose tissue compared to bone marrow. *Hum. Gene Ther.* **2018**, *29*, 507–519. [CrossRef]
55. Wakitani, S.; Goto, T.; Pineda, S.J.; Young, R.G.; Mansour, J.M.; Caplan, A.I.; Goldberg, V.M. Mesenchymal cell-based repair of large, full-thickness defects of articular cartilage. *J. Bone Jt. Surg. Am. Vol.* **1994**, *76*, 579–592. [CrossRef]
56. Orozco, L.; Munar, A.; Soler, R.; Alberca, M.; Soler, F.; Huguet, M.; Sentís, J.; Sánchez, A.; García-Sancho, J. Treatment of knee osteoarthritis with autologous mesenchymal stem cells. *Transplantation* **2013**, *95*, 1535–1541. [CrossRef]
57. De Windt, T.S.; Vonk, L.A.; Slaper-Cortenbach, I.C.M.; Broek, M.P.H.V.D.; Nizak, R.; Van Rijen, M.H.P.; De Weger, R.A.; Dhert, W.J.A.; Saris, D.B. Allogeneic mesenchymal stem cells stimulate cartilage regeneration and are safe for single-stage cartilage repair in humans upon mixture with recycled autologous chondrons. *STEM CELLS* **2016**, *35*, 256–264. [CrossRef]
58. Orozco, L.; Soler, R.; Morera, C.; Alberca, M.; Sánchez, A.; García-Sancho, J. Intervertebral disc repair by autologous mesenchymal bone marrow cells: A pilot study. *Transplant.* **2011**, *92*, 822–828. [CrossRef]
59. Pettine, K.A.; Suzuki, R.K.; Sand, T.T.; Murphy, M.B. Autologous bone marrow concentrate intradiscal injection for the treatment of degenerative disc disease with three-year follow-up. *Int. Orthop.* **2017**, *41*, 2097–2103. [CrossRef]
60. Shapiro, S.A.; Kazmerchak, S.E.; Heckman, M.G.; Zubair, A.C.; O'Connor, M.I. A Prospective, single-blind, placebo-controlled trial of bone marrow aspirate concentrate for knee osteoarthritis. *Am. J. Sports Med.* **2016**, *45*, 82–90. [CrossRef]

61. Chen, M.; Guo, W.; Gao, S.; Hao, C.; Shen, S.; Zhang, Z.; Wang, Z.; Li, X.; Jing, X.; Zhang, X.; et al. Biomechanical stimulus based strategies for meniscus tissue engineering and regeneration. *Tissue Eng. Part B Rev.* **2018**, *24*, 392–402. [CrossRef] [PubMed]
62. Gülecyüz, M.F.; Macha, K.; Pietschmann, M.F.; Ficklscherer, A.; Sievers, B.; Roßbach, B.P.; Jansson, V.; Müller, P.E. Allogenic myocytes and mesenchymal stem cells partially improve fatty rotator cuff degeneration in a rat model. *Stem Cell Rev. Rep.* **2018**, *14*, 847–859. [CrossRef] [PubMed]
63. Bae, K.S.; Park, J.B.; Kim, H.S.; Kim, D.S.; Park, D.J.; Kang, S.J. Neuron-like differentiation of bone marrow-derived mesenchymal stem cells. *Yonsei Med. J.* **2011**, *52*, 401–412. [CrossRef] [PubMed]
64. Oh, S.K.; Choi, K.H.; Yoo, J.Y.; Kim, D.Y.; Kim, S.J.; Jeon, S.R. A Phase III clinical trial showing limited efficacy of autologous mesenchymal stem cell therapy for spinal cord injury. *Neurosurgery* **2015**, *78*, 436–447. [CrossRef]
65. Wang, S.; Cheng, H.; Dai, G.; Wang, X.; Hua, R.; Liu, X.; Wang, P.; Chen, G.; Yue, W.; An, Y. Umbilical cord mesenchymal stem cell transplantation significantly improves neurological function in patients with sequelae of traumatic brain injury. *Brain Res.* **2013**, *1532*, 76–84. [CrossRef]
66. Harris, V.K.; Stark, J.; Vyshkina, T.; Blackshear, L.; Joo, G.; Stefanova, V.; Sara, G.; Sadiq, S.A. Phase I trial of intrathecal mesenchymal stem cell-derived neural progenitors in progressive multiple sclerosis. *EBioMedicine* **2018**, *29*, 23–30. [CrossRef] [PubMed]
67. Shichinohe, H.; Kawabori, M.; Iijima, H.; Teramoto, T.; Abumiya, T.; Nakayama, N.; Kazumata, K.; Terasaka, S.; Arato, T.; Houkin, K. Research on advanced intervention using novel bone marrOW stem cell (RAINBOW): A study protocol for a phase I, open-label, uncontrolled, dose-response trial of autologous bone marrow stromal cell transplantation in patients with acute ischemic stroke. *BMC Neurol.* **2017**, *17*, 1–8. [CrossRef]
68. Venkatesh, K.; Sen, D. mesenchymal stem cells as a source of dopaminergic neurons: A potential cell based therapy for Parkinson's disease. *Curr. Stem Cell Res. Ther.* **2017**, *12*, 326–347. [CrossRef]
69. Tse, H.-F.; Kwong, Y.-L.; Chan, J.K.F.; Lo, G.; Ho, C.-L.; Lau, C.-P. Angiogenesis in ischaemic myocardium by intramyocardial autologous bone marrow mononuclear cell implantation. *Lancet* **2003**, *361*, 47–49. [CrossRef]
70. Aggarwal, S.; Pittenger, M.F. Human mesenchymal stem cells modulate allogeneic immune cell responses. *Blood* **2005**, *105*, 1815–1822. [CrossRef]
71. Petersen, B.E. Bone marrow as a potential source of hepatic oval cells. *Science* **1999**, *284*, 1168–1170. [CrossRef] [PubMed]
72. Von Der Mark, K.; Park, J. Engineering biocompatible implant surfaces. *Prog. Mater. Sci.* **2013**, *58*, 327–381. [CrossRef]
73. Lehnert, D.; Wehrle-Haller, B.; David, C.; Weiland, U.; Ballestrem, C.; Imhof, B.A.; Bastmeyer, M. Cell behaviour on micropatterned substrata: Limits of extracellular matrix geometry for spreading and adhesion. *J. Cell Sci.* **2004**, *117*, 41–52. [CrossRef] [PubMed]
74. Chen, X.; Fan, H.; Deng, X.; Wu, L.; Yi, T.; Gu, L.; Zhou, C.; Fan, Y.; Zhang, X. Scaffold structural microenvironmental cues to guide tissue regeneration in bone tissue applications. *Nanomaterials* **2018**, *8*, 960. [CrossRef] [PubMed]
75. Hynes, R.O. Integrins. *Cell* **2002**, *110*, 673–687. [CrossRef]
76. Arima, Y.; Iwata, H. Effect of wettability and surface functional groups on protein adsorption and cell adhesion using well-defined mixed self-assembled monolayers. *Biomaterials* **2007**, *28*, 3074–3082. [CrossRef]
77. Zhu, X.; Fan, H.; Xiao, Y.; Li, D.; Zhang, H.; Luxbacher, T.; Zhang, X. Effect of surface structure on protein adsorption to biphasic calcium-phosphate ceramics in vitro and in vivo. *Acta Biomater.* **2009**, *5*, 1311–1318. [CrossRef]
78. Higgins, A.M.; Brown, J.L. Osteoinductive biomaterial geometries for bone regenerative engineering. *Curr. Pharm. Des.* **2013**, *19*, 3446–3455. [CrossRef]
79. Vogel, V.; Sheetz, M. Local force and geometry sensing regulate cell functions. *Nat. Rev. Mol. Cell Biol.* **2006**, *7*, 265–275. [CrossRef] [PubMed]
80. Cukierman, E.; Pankov, R.; Stevens, D.R.; Yamada, K.M. Taking cell-matrix adhesions to the third dimension. *Science* **2001**, *294*, 1708–1712. [CrossRef] [PubMed]
81. McBeath, R.; Pirone, D.M.; Nelson, C.M.; Bhadriraju, K.; Chen, C.S. Cell Shape, cytoskeletal tension, and RhoA regulate stem cell lineage commitment. *Dev. Cell* **2004**, *6*, 483–495. [CrossRef]
82. Lavenus, S.; Ricquier, J.-C.; Louarn, G.; Layrolle, P. Cell interaction with nanopatterned surface of implants. *Nanomedicine* **2010**, *5*, 937–947. [CrossRef] [PubMed]

83. Zhu, X.; Chen, J.; Scheideler, L.; Altebaeumer, T.; Geis-Gerstorfer, J.; Kern, D. Cellular reactions of osteoblasts to micron-and submicron-scale porous structures of titanium surfaces. *Cells Tissues Organs* **2004**, *178*, 13–22. [CrossRef] [PubMed]
84. Limat, A.; Hunziker, T.; Breitkreutz, D.; Fusenig, N.E.; Braathen, L.R. Organotypic cocultures as models to study cell-cell and cell-matrix interactions of human hair follicle cells. *Ski. Pharmacol. Physiol.* **1994**, *7*, 47–54. [CrossRef]
85. Giancotti, F.G. Integrin signaling. *Science* **1999**, *285*, 1028–1033. [CrossRef] [PubMed]
86. McNamara, L.E.; McMurray, R.J.; Biggs, M.J.P.; Kantawong, F.; Oreffo, R.O.; Dalby, M.J. Nanotopographical control of stem cell differentiation. *J. Tissue Eng.* **2010**, *2010*, 120623. [CrossRef]
87. Harrison, R.G. On the stereotropism of embryonic cells. *Science* **1911**, *34*, 279–281. [CrossRef]
88. Lozano-Calderón, S.A.; Swaim, S.O.; Federico, A.; Anderson, M.E.; Gebhardt, M.C. Predictors of soft-tissue complications and deep infection in allograft reconstruction of the proximal tibia. *J. Surg. Oncol.* **2016**, *113*, 811–817. [CrossRef]
89. Jiang, Y.; Jahagirdar, B.N.; Reinhardt, R.L.; Schwartz, R.E.; Keene, C.D.; Ortiz-Gonzalez, X.R.; Reyes, M.; Lenvik, T.; Lund, T.; Blackstad, M.; et al. Pluripotency of mesenchymal stem cells derived from adult marrow. *Nat. Cell Biol.* **2002**, *418*, 41–49. [CrossRef]
90. Yang, W.; Han, W.; He, W.; Li, J.; Wang, J.; Feng, H.; Qian, Y. Surface topography of hydroxyapatite promotes osteogenic differentiation of human bone marrow mesenchymal stem cells. *Mater. Sci. Eng. C* **2016**, *60*, 45–53. [CrossRef]
91. Matsuzaka, K.; Yoshinari, M.; Shimono, M.; Inoue, T. Effects of multigrooved surfaces on osteoblast-like cellsin vitro: Scanning electron microscopic observation and mRNA expression of osteopontin and osteocalcin. *J. Biomed. Mater. Res.* **2003**, *68*, 227–234. [CrossRef]
92. Faia-Torres, A.B.; Guimond-Lischer, S.; Rottmar, M.; Charnley, M.; Goren, T.; Maniura-Weber, K.; Spencer, N.D.; Reis, R.L.; Textor, M.; Neves, N.M. Differential regulation of osteogenic differentiation of stem cells on surface roughness gradients. *Biomaterials* **2014**, *35*, 9023–9032. [CrossRef] [PubMed]
93. Hefti, T.; Frischherz, M.; Spencer, N.D.; Hall, H.; Schlottig, F. A comparison of osteoclast resorption pits on bone with titanium and zirconia surfaces. *Biomaterials* **2010**, *31*, 7321–7331. [CrossRef]
94. Faia-Torres, A.B.; Charnley, M.; Goren, T.; Guimond-Lischer, S.; Rottmar, M.; Maniura-Weber, K.; Spencer, N.D.; Reis, R.L.; Textor, M.; Neves, N.M. Osteogenic differentiation of human mesenchymal stem cells in the absence of osteogenic supplements: A surface-roughness gradient study. *Acta Biomater.* **2015**, *28*, 64–75. [CrossRef] [PubMed]
95. Hong, J.-H.; Wilson, W.H.; Schroeder, D.C.; Allen, M.J.; Holden, M.T.G.; Parkhill, J.; Barrell, B.G.; Churcher, C.; Hamlin, N.; Mungall, K.; et al. TAZ, a Transcriptional modulator of mesenchymal stem cell differentiation. *Science* **2005**, *309*, 1074–1078. [CrossRef]
96. Dupont, S.; Morsut, L.; Aragona, M.; Enzo, E.; Giulitti, S.; Cordenonsi, M.; Zanconato, F.; Le Digabel, J.; Forcato, M.; Bicciato, S.; et al. Role of YAP/TAZ in mechanotransduction. *Nat. Cell Biol.* **2011**, *474*, 179–183. [CrossRef] [PubMed]
97. Schantz, J.-T.; Lim, T.C.; Ning, C.; Teoh, S.H.; Tan, K.C.; Wang, S.C.; Hutmacher, D.W. Cranioplasty after trephination using a novel biodegradable burr hole cover: Technical case report. *Oper. Neurosurg.* **2006**, *58*. [CrossRef] [PubMed]
98. Hou, Y.; Xie, W.; Yu, L.; Camacho, L.C.; Nie, C.; Zhang, M.; Haag, R.; Wei, Q. Surface roughness gradients reveal topography-specific mechanosensitive responses in human mesenchymal stem cells. *Small* **2020**, *16*, e1905422. [CrossRef] [PubMed]
99. Greer, A.I.; Goriainov, V.; Kanczler, J.; Black, C.R.; Turner, L.-A.; Meek, R.M.; Burgess, K.; MacLaren, I.; Dalby, M.J.; Oreffo, R.O.; et al. Nanopatterned titanium implants accelerate bone formation in vivo. *ACS Appl. Mater. Interfaces* **2020**, *12*, 33541–33549. [CrossRef]
100. Sjöström, T.; McNamara, L.E.; Dalby, M.J.; Meek, R.M.D.; Su, B. 2D and 3D Nanopatterning of titanium for enhancing osteoinduction of stem cells at implant surfaces. *Adv. Heal. Mater.* **2013**, *2*, 1285–1293. [CrossRef] [PubMed]
101. Chen, W.; Xu, K.; Tao, B.; Dai, L.; Yu, Y.; Mu, C.; Shen, X.; Hu, Y.; He, Y.; Cai, K. Multilayered coating of titanium implants promotes coupled osteogenesis and angiogenesis in vitro and in vivo. *Acta Biomater.* **2018**, *74*, 489–504. [CrossRef] [PubMed]

102. Nelson, K.L.; Cox, M.D.; Richter, G.T.; Dornhoffer, J.L. A comparative review of osseointegration failure between osseointegrated bone conduction device models in pediatric patients. *Otol. Neurotol.* **2016**, *37*, 276–280. [CrossRef] [PubMed]
103. Sjöström, T.; Dalby, M.J.; Hart, A.; Tare, R.; Oreffo, R.O.; Su, B. Fabrication of pillar-like titania nanostructures on titanium and their interactions with human skeletal stem cells. *Acta Biomater.* **2009**, *5*, 1433–1441. [CrossRef] [PubMed]
104. Khang, D.; Choi, J.; Im, Y.-M.; Kim, Y.-J.; Jang, J.-H.; Kang, S.S.; Nam, T.-H.; Song, J.; Park, J.-W. Role of subnano-, nano- and submicron-surface features on osteoblast differentiation of bone marrow mesenchymal stem cells. *Biomaterials* **2012**, *33*, 5997–6007. [CrossRef] [PubMed]
105. Cai, K.; Bossert, J.; Jandt, K.D. Does the nanometre scale topography of titanium influence protein adsorption and cell proliferation? *Colloids Surf. B Biointerfaces* **2006**, *49*, 136–144. [CrossRef]
106. Xu, K.; Chen, W.; Hu, Y.; Shen, X.; Xu, G.; Ran, Q.; Yu, Y.; Mu, C.; Cai, K. Influence of strontium ions incorporated into nanosheet-pore topographical titanium substrates on osteogenic differentiation of mesenchymal stem cells in vitro and on osseointegration in vivo. *J. Mater. Chem. B* **2016**, *4*, 4549–4564. [CrossRef]
107. Xu, K.; Shen, X.; Chen, W.; Mu, C.; Jiang, C.; Zhao, Y.; Cai, K. Nanosheet-pore topographical titanium substrates: A biophysical regulator of the fate of mesenchymal stem cells. *J. Mater. Chem. B* **2016**, *4*, 1797–1810. [CrossRef]
108. Zhao, L.; Wang, H.; Huo, K.; Zhang, X.; Wang, W.; Zhang, Y.; Wu, Z.; Chu, P.K. The osteogenic activity of strontium loaded titania nanotube arrays on titanium substrates. *Biomaterials* **2013**, *34*, 19–29. [CrossRef]
109. Huang, Y.; Zha, G.; Luo, Q.; Zhang, J.; Zhang, F.; Li, X.; Zhao, S.; Zhu, W.; Li, X. The construction of hierarchical structure on Ti substrate with superior osteogenic activity and intrinsic antibacterial capability. *Sci. Rep.* **2014**, *4*, 6172. [CrossRef]
110. Heitz-Mayfield, L.J. Peri-implant diseases: Diagnosis and risk indicators. *J. Clin. Periodontol.* **2008**, *35*, 292–304. [CrossRef]
111. Lotz, E.M.; Berger, M.B.; Schwartz, Z.; Boyan, B.D. Regulation of osteoclasts by osteoblast lineage cells depends on titanium implant surface properties. *Acta Biomater.* **2018**, *68*, 296–307. [CrossRef] [PubMed]
112. Zhao, C.; Wang, X.; Gao, L.; Jing, L.; Zhou, Q.; Chang, J. The role of the micro-pattern and nano-topography of hydroxyapatite bioceramics on stimulating osteogenic differentiation of mesenchymal stem cells. *Acta Biomater.* **2018**, *73*, 509–521. [CrossRef] [PubMed]
113. Hu, C.; Ashok, D.; Nisbet, D.R.; Gautam, V. Bioinspired surface modification of orthopedic implants for bone tissue engineering. *Biomaterials* **2019**, *219*, 119366. [CrossRef]
114. Xia, L.; Lin, K.; Jiang, X.; Xu, Y.; Zhang, M.; Chang, J.; Zhang, Z. Enhanced osteogenesis through nano-structured surface design of macroporous hydroxyapatite bioceramic scaffolds via activation of ERK and p38 MAPK signaling pathways. *J. Mater. Chem. B* **2013**, *1*, 5403–5416. [CrossRef]
115. Yuan, H.; Fernandes, H.; Habibovic, P.; De Boer, J.; Barradas, A.M.C.; De Ruiter, A.; Walsh, W.R.; Van Blitterswijk, C.A.; De Bruijn, J.D. Osteoinductive ceramics as a synthetic alternative to autologous bone grafting. *Proc. Natl. Acad. Sci. USA* **2010**, *107*, 13614–13619. [CrossRef] [PubMed]
116. Wang, C.; Fan, Z.; Han, Y. Formation and osteoblast behavior of HA nano-rod/fiber patterned coatings on tantalum in porous and compact forms. *J. Mater. Chem. B* **2015**, *3*, 5442–5454. [CrossRef] [PubMed]
117. Zhou, J.; Li, B.; Lu, S.; Zhang, L.; Han, Y. Regulation of osteoblast proliferation and differentiation by interrod spacing of Sr-HA nanorods on microporous titania coatings. *ACS Appl. Mater. Interf.* **2013**, *5*, 5358–5365. [CrossRef] [PubMed]
118. Zhou, J.; Li, B.; Han, Y.; Zhao, L. The osteogenic capacity of biomimetic hierarchical micropore/nanorod-patterned Sr-HA coatings with different interrod spacings. *Nanomed. Nanotechnol. Biol. Med.* **2016**, *12*, 1161–1173. [CrossRef] [PubMed]
119. Zhou, J.; Zhao, L.; Li, B.; Han, Y. Nanorod diameter modulated osteogenic activity of hierarchical micropore/nanorod-patterned coatings via a Wnt/β-catenin pathway. *Nanomed. Nanotechnol. Biol. Med.* **2018**, *14*, 1719–1731. [CrossRef]
120. Loye, A.M.; Kinser, E.R.; Bensouda, S.; Shayan, M.; Davis, R.; Wang, R.; Chen, Z.; Schwarz, U.D.; Schroers, J.; Kyriakides, T.R. Regulation of mesenchymal stem cell differentiation by nanopatterning of bulk metallic glass. *Sci. Rep.* **2018**, *8*, 8758. [CrossRef]
121. Kumar, G.; Tang, H.X.; Schroers, J. Nanomoulding with amorphous metals. *Nat. Cell Biol.* **2009**, *457*, 868–872. [CrossRef] [PubMed]

122. Kumar, G.; Staffier, P.A.; Blawzdziewicz, J.; Schwarz, U.D.; Schroers, J. Atomically smooth surfaces through thermoplastic forming of metallic glass. *Appl. Phys. Lett.* **2010**, *97*, 101907. [CrossRef]
123. Kumar, G.; Blawzdziewicz, J.; Schroers, J. Controllable nanoimprinting of metallic glasses: Effect of pressure and interfacial properties. *Nanotechnology* **2013**, *24*, 105301. [CrossRef] [PubMed]
124. Wang, W.; Dong, C.; Shek, C. Bulk metallic glasses. *Mater. Sci. Eng. R Rep.* **2004**, *44*, 45–89. [CrossRef]
125. Lee, D.-K.; Yi, T.; Park, K.-E.; Lee, H.-J.; Cho, Y.-K.; Lee, S.J.; Lee, J.; Park, J.H.; Lee, M.-Y.; Song, S.U.; et al. Non-invasive characterization of the adipogenic differentiation of human bone marrow-derived mesenchymal stromal cells by HS-SPME/GC-MS. *Sci. Rep.* **2014**, *4*, 6550. [CrossRef]
126. Lin, K.; Xia, L.; Zhai, D.; Zhang, N.; Liu, J.; Fang, B.; Chang, J.; Lin, K. Designing ordered micropatterned hydroxyapatite bioceramics to promote the growth and osteogenic differentiation of bone marrow stromal cells. *J. Mater. Chem. B* **2015**, *3*, 968–976. [CrossRef]
127. Yang, C.; Zhao, C.; Wang, X.; Shi, M.; Zhu, Y.; Jing, L.; Wu, C.; Chang, J. Stimulation of osteogenesis and angiogenesis by micro/nano hierarchical hydroxyapatite via macrophage immunomodulation. *Nanoscale* **2019**, *11*, 17699–17708. [CrossRef]
128. Xia, L.; Lin, K.; Jiang, X.; Fang, B.; Xu, Y.; Liu, J.; Zeng, D.; Zhang, M.; Zhang, X.; Chang, J.; et al. Effect of nano-structured bioceramic surface on osteogenic differentiation of adipose derived stem cells. *Biomaterials* **2014**, *35*, 8514–8527. [CrossRef]
129. Lin, K.; Xia, L.; Gan, J.; Zhang, Z.; Chen, H.; Jiang, X.; Chang, J. Tailoring the nanostructured surfaces of hydroxyapatite bioceramics to promote protein adsorption, osteoblast growth, and osteogenic differentiation. *ACS Appl. Mater. Interf.* **2013**, *5*, 8008–8017. [CrossRef]
130. Spiller, K.L.; Koh, T.J. Macrophage-based therapeutic strategies in regenerative medicine. *Adv. Drug Deliv. Rev.* **2017**, *122*, 74–83. [CrossRef]
131. Sridharan, R.; Cameron, A.R.; Kelly, D.J.; Kearney, C.J.; O'Brien, F.J. Biomaterial based modulation of macrophage polarization: A review and suggested design principles. *Mater. Today* **2015**, *18*, 313–325. [CrossRef]
132. Lee, C.Z.W.; Kozaki, T.; Ginhoux, F. Studying tissue macrophages in vitro: Are iPSC-derived cells the answer? *Nat. Rev. Immunol.* **2018**, *18*, 716–725. [CrossRef] [PubMed]
133. Alaarg, A.; Pérez-Medina, C.; Metselaar, J.M.; Nahrendorf, M.; Fayad, Z.A.; Storm, G.; Mulder, W.J.M. Applying nanomedicine in maladaptive inflammation and angiogenesis. *Adv. Drug Deliv. Rev.* **2017**, *119*, 143–158. [CrossRef] [PubMed]
134. Zhang, Q.; Hwang, J.W.; Oh, J.-H.; Park, C.H.; Chung, S.H.; Lee, Y.-S.; Baek, J.-H.; Ryoo, H.-M.; Woo, K.M. Effects of the fibrous topography-mediated macrophage phenotype transition on the recruitment of mesenchymal stem cells: An in vivo study. *Biomaterials* **2017**, *149*, 77–87. [CrossRef]
135. Mukherjee, S.; Darzi, S.; Rosamilia, A.; Kadam, V.; Truong, Y.B.; Werkmeister, J.A.; Gargett, C.E. Blended nanostructured degradable mesh with endometrial mesenchymal stem cells promotes tissue integration and anti-inflammatory response in vivo for pelvic floor application. *Biomacromolecules* **2018**, *20*, 454–468. [CrossRef]
136. Li, K.; Hu, D.; Xie, Y.; Huang, L.; Zheng, X. Sr-doped nanowire modification of Ca–Si-based coatings for improved osteogenic activities and reduced inflammatory reactions. *Nanotechnology* **2018**, *29*, 084001. [CrossRef]
137. Zhang, Q.; Ma, L.; Zheng, S.; Wang, Y.; Feng, M.; Shuai, Y.; Duan, B.; Fan, X.; Yang, M.; Mao, C. Air-plasma treatment promotes bone-like nano-hydroxylapatite formation on protein films for enhanced in vivo osteogenesis. *Biomater. Sci.* **2019**, *7*, 2326–2334. [CrossRef]
138. Lin, K.-F.; He, S.; Song, Y.; Wang, C.-M.; Gao, Y.; Li, J.-Q.; Tang, P.; Wang, Z.; Bi, L.; Pei, G.-X. Low-temperature additive manufacturing of biomimic three-dimensional hydroxyapatite/collagen scaffolds for bone regeneration. *ACS Appl. Mater. Interf.* **2016**, *8*, 6905–6916. [CrossRef]
139. Wagener, V.; Boccaccini, A.R.; Virtanen, S. Protein-adsorption and Ca-phosphate formation on chitosan-bioactive glass composite coatings. *Appl. Surf. Sci.* **2017**, *416*, 454–460. [CrossRef]
140. Cardoso, G.B.C.; Maniglio, D.; Volpato, F.Z.F.Z.; Tondon, A.; Migliaresi, C.; Kaunas, R.; Zavaglia, C.A.C.C.A.C. Oleic acid surfactant in polycaprolactone/hydroxyapatite-composites for bone tissue engineering. *J. Biomed. Mater. Res. Part B Appl. Biomater.* **2015**, *104*, 1076–1082. [CrossRef]
141. Gogoi, D.; Choudhury, A.J.; Chutia, J.; Pal, A.R.; Khan, M.; Choudhury, M.; Pathak, P.; Das, G.; Patil, D.S. Development of advanced antimicrobial and sterilized plasma polypropylene grafted muga (antheraea assama) silk as suture biomaterial. *Biopolymers* **2014**, *101*, 355–365. [CrossRef] [PubMed]

142. Jiang, J.; Hao, W.; Li, Y.; Yao, J.; Zheng, G.; Li, H.; Yang, J.; Chen, S. Erratum to: Hydroxyapatite/regenerated silk fibroin scaffold-enhanced osteoinductivity and osteoconductivity of bone marrow-derived mesenchymal stromal cells. *Biotechnol. Lett.* **2013**, *35*, 1349–1350. [CrossRef]
143. Liu, H.; Xu, G.W.; Wang, Y.F.; Zhao, H.S.; Xiong, S.; Wu, Y.; Heng, B.C.; An, C.R.; Zhu, G.H.; Xie, D.H. Composite scaffolds of nano-hydroxyapatite and silk fibroin enhance mesenchymal stem cell-based bone regeneration via the interleukin 1 alpha autocrine/paracrine signaling loop. *Biomaterials* **2015**, *49*, 103–112. [CrossRef] [PubMed]
144. Wang, N.; Li, H.; Lü, W.; Li, J.; Wang, J.; Zhang, Z.; Liu, Y. Effects of TiO2 nanotubes with different diameters on gene expression and osseointegration of implants in minipigs. *Biomaterials* **2011**, *32*, 6900–6911. [CrossRef]
145. Qiu, J.; Li, J.; Wang, S.; Ma, B.; Zhang, S.; Guo, W.; Zhang, X.; Tang, W.; Sang, Y.; Liu, H. TiO_2 nanorod array constructed nanotopography for regulation of mesenchymal stem cells fate and the realization of location-committed stem cell differentiation. *Small* **2016**, *12*, 1770–1778. [CrossRef] [PubMed]
146. Lv, L.; Liu, Y.; Zhang, P.; Zhang, X.; Liu, J.; Chen, T.; Su, P.; Li, H.; Zhou, Y. The nanoscale geometry of TiO_2 nanotubes influences the osteogenic differentiation of human adipose-derived stem cells by modulating H_3K_4 trimethylation. *Biomaterials* **2015**, *39*, 193–205. [CrossRef] [PubMed]
147. Lavenus, S.; Trichet, V.; Le Chevalier, S.; Hoornaert, A.; Louarn, G.; Layrolle, P. Cell differentiation and osseointegration influenced by nanoscale anodized titanium surfaces. *Nanomedicine* **2012**, *7*, 967–980. [CrossRef]
148. Oh, S.; Daraio, C.; Chen, L.-H.; Pisanic, T.R.; Fiñones, R.R.; Jin, S. Significantly accelerated osteoblast cell growth on aligned TiO2 nanotubes. *J. Biomed. Mater. Res. Part A* **2006**, *78*, 97–103. [CrossRef]
149. Hou, W.; Fu, H.; Liu, X.; Duan, K.; Lu, X.; Lu, M.; Sun, T.; Guo, T.; Weng, J. Cation Channel Transient Receptor Potential Vanilloid 4 Mediates Topography-Induced Osteoblastic Differentiation of Bone Marrow Stem Cells. *ACS Biomater. Sci. Eng.* **2019**, *5*, 6520–6529. [CrossRef]
150. Lee, H.-P.; Stowers, R.; Chaudhuri, O. Volume expansion and TRPV4 activation regulate stem cell fate in three-dimensional microenvironments. *Nat. Commun.* **2019**, *10*, 529. [CrossRef]
151. Jiang, X.; Zhang, W.; Li, Z.; Xu, L.; Li, J.; Jin, Y.; Wang, G.; Liu, X.; Huang, Q. Effects of a hybrid micro/nanorod topography-modified titanium implant on adhesion and osteogenic differentiation in rat bone marrow mesenchymal stem cells. *Int. J. Nanomed.* **2013**, *8*, 257–265. [CrossRef] [PubMed]
152. Dalby, M.J.; Gadegaard, N.; Tare, R.; Andar, A.; Riehle, M.O.; Herzyk, P.; Wilkinson, C.D.W.; Oreffo, R.O.C. The control of human mesenchymal cell differentiation using nanoscale symmetry and disorder. *Nat. Mater.* **2007**, *6*, 997–1003. [CrossRef] [PubMed]
153. McMurray, R.J.; Gadegaard, N.; Tsimbouri, P.M.; Burgess, K.V.; McNamara, L.E.; Tare, R.; Murawski, K.; Kingham, E.; Oreffo, R.O.C.; Dalby, M.J. Nanoscale surfaces for the long-term maintenance of mesenchymal stem cell phenotype and multipotency. *Nat. Mater.* **2011**, *10*, 637–644. [CrossRef] [PubMed]
154. Abagnale, G.; Steger, M.; Nguyen, V.H.; Hersch, N.; Sechi, A.; Joussen, S.; Denecke, B.; Merkel, R.; Hoffmann, B.; Dreser, A.; et al. Surface topography enhances differentiation of mesenchymal stem cells towards osteogenic and adipogenic lineages. *Biomaterials* **2015**, *61*, 316–326. [CrossRef] [PubMed]
155. Rubehn, B.; Stieglitz, T. In vitro evaluation of the long-term stability of polyimide as a material for neural implants. *Biomaterials* **2010**, *31*, 3449–3458. [CrossRef]
156. Sun, Y.; Lacour, S.P.; Brooks, R.A.; Rushton, N.; Fawcett, J.W.; Cameron, R.E. Assessment of the biocompatibility of photosensitive polyimide for implantable medical device use. *J. Biomed. Mater. Res. Part A* **2009**, *90*, 648–655. [CrossRef]
157. Prichard, H.L.; Reichert, W.M.; Klitzman, B. Adult adipose-derived stem cell attachment to biomaterials. *Biomaterials* **2007**, *28*, 936–946. [CrossRef]
158. Kolambkar, Y.M.; Bajin, M.; Wojtowicz, A.; Hutmacher, D.W.; Garcia, A.J.; Guldberg, R.E. Nanofiber orientation and surface functionalization modulate human mesenchymal stem cell behavior in vitro. *Tissue Eng. Part A* **2013**, *20*, 398–409. [CrossRef]
159. Polini, A.; Pisignano, D.; Parodi, M.; Quarto, R.; Scaglione, S. Osteoinduction of human mesenchymal stem cells by bioactive composite scaffolds without supplemental osteogenic growth factors. *PLoS ONE* **2011**, *6*, e26211. [CrossRef]
160. Lü, L.-X.; Wang, Y.-Y.; Mao, X.; Xiao, Z.; Huang, N.-P. The effects of PHBV electrospun fibers with different diameters and orientations on growth behavior of bone-marrow-derived mesenchymal stem cells. *Biomed. Mater.* **2012**, *7*, 015002. [CrossRef]

161. Lü, L.-X.; Zhang, X.-F.; Wang, Y.-Y.; Ortiz, L.; Mao, X.; Jiang, Z.-L.; Xiao, Z.-D.; Huang, N.-P. Effects of hydroxyapatite-containing composite nanofibers on osteogenesis of mesenchymal stem cells in vitro and bone regeneration in vivo. *ACS Appl. Mater. Interf.* **2013**, *5*, 319–330. [CrossRef]
162. Zhang, N.; Xiao, Q.-R.; Man, X.-Y.; Liu, H.-X.; Lü, L.-X.; Huang, N.-P. Spontaneous osteogenic differentiation of mesenchymal stem cells on electrospun nanofibrous scaffolds. *RSC Adv.* **2016**, *6*, 22144–22152. [CrossRef]
163. Liu, F.; Kohlmeier, S.; Wang, C.-Y. Wnt signaling and skeletal development. *Cell. Signal.* **2008**, *20*, 999–1009. [CrossRef] [PubMed]
164. Guan, J.; Zhang, J.; Guo, S.; Zhu, H.; Zhu, Z.; Li, H.; Wang, Y.; Zhang, C.; Chang, J. Human urine-derived stem cells can be induced into osteogenic lineage by silicate bioceramics via activation of the Wnt/β-catenin signaling pathway. *Biomaterials* **2015**, *55*, 1–11. [CrossRef]
165. Nishimura, R.; Hata, K.; Harris, S.; Ikeda, F.; Yoneda, T. Core-binding factor α1 (Cbfa1) induces osteoblastic differentiation of C2C12 cells without interactions with Smad1 and Smad5. *Bone* **2002**, *31*, 303–312. [CrossRef]
166. Wang, W.; Liu, Q.; Zhang, Y.; Zhao, L. Involvement of ILK/ERK1/2 and ILK/p38 pathways in mediating the enhanced osteoblast differentiation by micro/nanotopography. *Acta Biomater.* **2014**, *10*, 3705–3715. [CrossRef] [PubMed]
167. Izadpanahi, M.; Seyedjafari, E.; Arefian, E.; Hamta, A.; Hosseinzadeh, S.; Kehtari, M.; Soleimani, M.; Kehtari, M. Nanotopographical cues of electrospun PLLA efficiently modulate non-coding RNA network to osteogenic differentiation of mesenchymal stem cells during BMP signaling pathway. *Mater. Sci. Eng. C* **2018**, *93*, 686–703. [CrossRef]
168. Kaplan, L.; Fu, F. Clinical Applications of Orthopedic Tissue Engineering. *Orthop. Tissue Eng.* **2004**, 261. [CrossRef]
169. Teh, T.K.; Toh, S.-L.; Goh, J.C. Aligned fibrous scaffolds for enhanced mechanoresponse and tenogenesis of mesenchymal stem cells. *Tissue Eng. Part A* **2013**, *19*, 1360–1372. [CrossRef]
170. Woo, S.L.-Y.; Debski, R.E.; Zeminski, J.; Abramowitch, S.D.; Saw, S.; Fenwick, J.A. Injury and repair of ligaments and tendons. *Annu. Rev. Biomed. Eng.* **2000**, *2*, 83–118. [CrossRef]
171. Woo, S.L.-Y.; Vogrin, T.M.; Abramowitch, S.D. Healing and repair of ligament injuries in the knee. *J. Am. Acad. Orthop. Surg.* **2000**, *8*, 364–372. [CrossRef]
172. Hogan, M.V.; Bagayoko, N.; James, R.; Starnes, T.; Katz, A.; Chhabra, B.A. Tissue engineering solutions for tendon repair. *J. Am. Acad. Orthop. Surg.* **2011**, *19*, 134–142. [CrossRef] [PubMed]
173. Czaplewski, S.K.; Tsai, T.-L.; Duenwald-Kuehl, S.E.; Vanderby, R.; Li, W.-J. Tenogenic differentiation of human induced pluripotent stem cell-derived mesenchymal stem cells dictated by properties of braided submicron fibrous scaffolds. *Biomaterials* **2014**, *35*, 6907–6917. [CrossRef]
174. Ma, Z.; Kotaki, M.; Inai, R.; Ramakrishna, S. Potential of nanofiber matrix as tissue-engineering scaffolds. *Tissue Eng.* **2005**, *11*, 101–109. [CrossRef] [PubMed]
175. Yin, Z.; Chen, X.; Chen, J.L.; Shen, W.-L.; Nguyen, T.M.H.; Gao, L.; Ouyang, H.W. The regulation of tendon stem cell differentiation by the alignment of nanofibers. *Biomaterials* **2010**, *31*, 2163–2175. [CrossRef] [PubMed]
176. Kishore, V.; Bullock, W.; Sun, X.; Van Dyke, W.S.; Akkus, O. Tenogenic differentiation of human MSCs induced by the topography of electrochemically aligned collagen threads. *Biomaterials* **2012**, *33*, 2137–2144. [CrossRef] [PubMed]
177. Tong, W.Y.; Shen, W.; Yeung, C.W.; Zhao, Y.; Cheng, S.H.; Chu, P.K.; Chan, D.; Chan, G.C.; Cheung, K.M.; Yeung, K.W.K.; et al. Functional replication of the tendon tissue microenvironment by a bioimprinted substrate and the support of tenocytic differentiation of mesenchymal stem cells. *Biomaterials* **2012**, *33*, 7686–7698. [CrossRef]
178. Ker, E.D.; Nain, A.S.; Weiss, L.E.; Wang, J.; Suhan, J.; Amon, C.H.; Campbell, P.G. Bioprinting of growth factors onto aligned sub-micron fibrous scaffolds for simultaneous control of cell differentiation and alignment. *Biomaterials* **2011**, *32*, 8097–8107. [CrossRef]
179. Lutolf, M.P.; Hubbell, J.A. Synthetic biomaterials as instructive extracellular microenvironments for morphogenesis in tissue engineering. *Nat. Biotechnol.* **2005**, *23*, 47–55. [CrossRef]
180. Mizuno, M.; Fujisawa, R.; Kuboki, Y. Type I collagen-induced osteoblastic differentiation of bone-marrow cells mediated by collagen-α2β1 integrin interaction. *J. Cell. Physiol.* **2000**, *184*, 207–213. [CrossRef]
181. Mizuno, M.; Kuboki, Y. Osteoblast-related gene expression of bone marrow cells during the osteoblastic differentiation induced by type I collagen. *J. Biochem.* **2001**, *129*, 133–138. [CrossRef] [PubMed]
182. Teh, T.K.; Toh, S.-L.; Goh, J.C. Aligned hybrid silk scaffold for enhanced differentiation of mesenchymal stem cells into ligament fibroblasts. *Tissue Eng. Part C Methods* **2011**, *17*, 687–703. [CrossRef]

183. Banes, A.J.; Lee, G.; Graff, R.; Otey, C.; Archambault, J.; Tsuzaki, M.; Elfervig, M.; Qi, J. Mechanical forces and signaling in connective tissue cells: Cellular mechanisms of detection, transduction, and responses to mechanical deformation. *Curr. Opin. Orthop.* **2001**, *12*, 389–396. [CrossRef]
184. Hung, C.T.; Allen, F.D.; Pollack, S.R.; Attia, E.T.; Hannafin, J.A.; Torzilli, P.A. Intracellular calcium response of ACL and MCL ligament fibroblasts to fluid-induced shear stress. *Cell. Signal.* **1997**, *9*, 587–594. [CrossRef]
185. Tsuzaki, M.; Bynum, D.; Almekinders, L.; Yang, X.; Faber, J.; Banes, A. ATP modulates load-inducible IL-1, COX 2, and MMP-3 gene expression in human tendon cells. *J. Cell. Biochem.* **2003**, *89*, 556–562. [CrossRef] [PubMed]
186. Chiquet, M.; Gelman, L.; Lutz, R.; Maier, S. From mechanotransduction to extracellular matrix gene expression in fibroblasts. *Biochim. Biophys. Acta (BBA) Mol. Cell Res.* **2009**, *1793*, 911–920. [CrossRef] [PubMed]
187. Barber, J.G.; Handorf, A.M.; Allee, T.J.; Li, W.-J. Braided nanofibrous scaffold for tendon and ligament tissue engineering. *Tissue Eng. Part A* **2013**, *19*, 1265–1274. [CrossRef]
188. Li, W.-J.; Jiang, Y.J.; Tuan, R.S. Chondrocyte phenotype in engineered fibrous matrix is regulated by fiber size. *Tissue Eng.* **2006**, *12*, 1775–1785. [CrossRef]
189. Ralphs, J.; Waggett, A.D.; Benjamin, M. Actin stress fibres and cell-cell adhesion molecules in tendons: Organisation in vivo and response to mechanical loading of tendon cells in vitro. *Matrix Biol.* **2002**, *21*, 67–74. [CrossRef]
190. Diekman, B.O.; Christoforou, N.; Willard, V.P.; Sun, H.; Sanchez-Adams, J.; Leong, K.W.; Guilak, F. Cartilage tissue engineering using differentiated and purified induced pluripotent stem cells. *Proc. Natl. Acad. Sci. USA* **2012**, *109*, 19172–19177. [CrossRef]
191. Li, W.-J.; Cooper, J.A.; Mauck, R.L.; Tuan, R.S. Fabrication and characterization of six electrospun poly(α-hydroxy ester)-based fibrous scaffolds for tissue engineering applications. *Acta Biomater.* **2006**, *2*, 377–385. [CrossRef] [PubMed]
192. Rothrauff, B.B.; Lauro, B.B.; Yang, G.; Debski, R.E.; Musahl, V.; Tuan, R.S. Braided and stacked electrospun nanofibrous scaffolds for tendon and ligament tissue engineering. *Tissue Eng. Part A* **2017**, *23*, 378–389. [CrossRef] [PubMed]
193. Orr, S.B.; Chainani, A.; Hippensteel, K.; Kishan, A.; Gilchrist, C.; Garrigues, N.W.; Ruch, D.S.; Guilak, F.; Little, D. Aligned multilayered electrospun scaffolds for rotator cuff tendon tissue engineering. *Acta Biomater.* **2015**, *24*, 117–126. [CrossRef]
194. Vuornos, K.; Björninen, M.; Talvitie, E.; Paakinaho, K.; Kellomäki, M.; Huhtala, H.; Miettinen, S.; Seppänen-Kaijansinkko, R.; Haimi, S. Human adipose stem cells differentiated on braided polylactide scaffolds is a potential approach for tendon tissue engineering. *Tissue Eng. Part A* **2016**, *22*, 513–523. [CrossRef] [PubMed]
195. Yin, Z.; Chen, X.; Song, H.-X.; Hu, J.-J.; Tang, Q.-M.; Zhu, T.; Shen, W.-L.; Chen, J.-L.; Liu, H.; Heng, B.C.; et al. Electrospun scaffolds for multiple tissues regeneration in vivo through topography dependent induction of lineage specific differentiation. *Biomaterials* **2015**, *44*, 173–185. [CrossRef]
196. Liu, Y.; Zhou, G.; Cao, Y. Recent Progress in Cartilage Tissue Engineering—Our Experience and Future Directions. *Engineering* **2017**, *3*, 28–35. [CrossRef]
197. Vinatier, C.; Guicheux, J. Cartilage tissue engineering: From biomaterials and stem cells to osteoarthritis treatments. *Ann. Phys. Rehabil. Med.* **2016**, *59*, 139–144. [CrossRef]
198. Steadman, J.R.; Rodkey, W.G.; Rodrigo, J.J. Microfracture: Surgical Technique and Rehabilitation to Treat Chondral Defects. *Clin. Orthop. Relat. Res.* **2001**, *391*, S362–S369. [CrossRef]
199. Hangody, L.; Füles, P. Autologous osteochondral mosaicplasty for the treatment of full-thickness defects of weight-bearing joints. *J. Bone Jt. Surg. Am. Vol.* **2003**, *85*, 25–32. [CrossRef]
200. Brittberg, M.; Lindahl, A.; Nilsson, A.; Ohlsson, C.; Isaksson, O.; Peterson, L. Treatment of deep cartilage defects in the knee with autologous chondrocyte transplantation. *New Engl. J. Med.* **1994**, *331*, 889–895. [CrossRef]
201. Spiller, K.L.; Maher, S.A.; Lowman, A.M. Hydrogels for the repair of articular cartilage defects. *Tissue Eng. Part B Rev.* **2011**, *17*, 281–299. [CrossRef] [PubMed]
202. Moran, C.J.; Pascual-Garrido, C.; Chubinskaya, S.; Potter, H.G.; Warren, R.F.; Cole, B.J.; Rodeo, S.A. Restoration of articular cartilage. *J. Bone Jt. Surg. Am. Vol.* **2014**, *96*, 336–344. [CrossRef]
203. Huey, D.J.; Hu, J.C.; Athanasiou, K.A. Unlike bone, cartilage regeneration remains elusive. *Science* **2012**, *338*, 917–921. [CrossRef] [PubMed]

204. Van Osch, G.J.V.M.; Brittberg, M.; Dennis, J.; Bastiaansen-Jenniskens, Y.; Erben, R.; Konttinen, Y.; Luyten, F. Cartilage repair: Past and future-lessons for regenerative medicine. *J. Cell. Mol. Med.* **2009**, *13*, 792–810. [CrossRef] [PubMed]
205. Jorgensen, C.; Gordeladze, J.; Noel, D. Tissue engineering through autologous mesenchymal stem cells. *Curr. Opin. Biotechnol.* **2004**, *15*, 406–410. [CrossRef]
206. Caplan, A.I. Review: Mesenchymal stem cells: Cell-based reconstructive therapy in orthopedics. *Tissue Eng.* **2005**, *11*, 1198–1211. [CrossRef] [PubMed]
207. Nerurkar, N.L.; Baker, B.M.; Sen, S.; Wible, E.E.; Elliott, D.M.; Mauck, R.L. Nanofibrous biologic laminates replicate the form and function of the annulus fibrosus. *Nat. Mater.* **2009**, *8*, 986–992. [CrossRef]
208. Baker, B.M.; Mauck, R.L. The effect of nanofiber alignment on the maturation of engineered meniscus constructs. *Biomaterials* **2007**, *28*, 1967–1977. [CrossRef]
209. Baker, B.M.; Nathan, A.S.; Gee, A.O.; Mauck, R.L. The influence of an aligned nanofibrous topography on human mesenchymal stem cell fibrochondrogenesis. *Biomaterials* **2010**, *31*, 6190–6200. [CrossRef]
210. Wu, Y.; Yang, Z.; Law, J.B.K.; He, A.Y.; Abbas, A.A.; Denslin, V.; Kamarul, T.; Hui, J.; Lee, E.H. The Combined effect of substrate stiffness and surface topography on chondrogenic differentiation of mesenchymal stem cells. *Tissue Eng. Part A* **2017**, *23*, 43–54. [CrossRef]
211. Jurvelin, J.; Müller, D.J.; Wong, M.; Studer, D.; Engel, A.; Hunziker, E. Surface and subsurface morphology of bovine humeral articular cartilage as assessed by atomic force and transmission electron microscopy. *J. Struct. Biol.* **1996**, *117*, 45–54. [CrossRef] [PubMed]
212. Wu, Y.-N.; Yang, Z.; Hui, J.H.; Ouyang, H.-W.; Lee, E.H. Cartilaginous ECM component-modification of the micro-bead culture system for chondrogenic differentiation of mesenchymal stem cells. *Biomaterials* **2007**, *28*, 4056–4067. [CrossRef] [PubMed]
213. Varghese, S.; Hwang, N.S.; Canver, A.C.; Theprungsirikul, P.; Lin, D.W.; Elisseeff, J. Chondroitin sulfate based niches for chondrogenic differentiation of mesenchymal stem cells. *Matrix Biol.* **2008**, *27*, 12–21. [CrossRef] [PubMed]
214. Klein, T.J.; Malda, J.; Sah, R.L.; Hutmacher, D.W. Tissue engineering of articular cartilage with biomimetic zones. *Tissue Eng. Part B Rev.* **2009**, *15*, 143–157. [CrossRef]
215. Hayes, A.J.; Hall, A.; Brown, L.; Tubo, R.; Caterson, B. Macromolecular organization and in vitro growth characteristics of scaffold-free neocartilage grafts. *J. Histochem. Cytochem.* **2007**, *55*, 853–866. [CrossRef]
216. McMahon, R.E.; Wang, L.; Skoracki, R.; Mathur, A.B. Development of nanomaterials for bone repair and regeneration. *J. Biomed. Mater. Res. Part B Appl. Biomater.* **2012**, *101*, 387–397. [CrossRef]
217. Walpole, A.R.; Xia, Z.; Wilson, C.W.; Triffitt, J.T.; Wilshaw, P.R. A novel nano-porous alumina biomaterial with potential for loading with bioactive materials. *J. Biomed. Mater. Res. Part A* **2009**, *90*, 46–54. [CrossRef]
218. Zhang, B.; Pei, X.; Song, P.; Sun, H.; Li, H.; Fan, Y.; Jiang, Q.; Zhou, C.; Zhang, X. Porous bioceramics produced by inkjet 3D printing: Effect of printing ink formulation on the ceramic macro and micro porous architectures control. *Compos. Part B Eng.* **2018**, *155*, 112–121. [CrossRef]
219. Schumacher, M.; Deisinger, U.; Detsch, R.; Ziegler, G. Indirect rapid prototyping of biphasic calcium phosphate scaffolds as bone substitutes: Influence of phase composition, macroporosity and pore geometry on mechanical properties. *J. Mater. Sci. Mater. Electron.* **2010**, *21*, 3119–3127. [CrossRef]
220. Wu, C.; Ramaswamy, Y.; Zreiqat, H. Porous diopside ($CaMgSi_2O_6$) scaffold: A promising bioactive material for bone tissue engineering. *Acta Biomater.* **2010**, *6*, 2237–2245. [CrossRef]
221. Pei, X.; Ma, L.; Zhang, B.; Sun, J.; Sun, Y.; Fan, Y.; Gou, Z.; Zhou, C.; Zhang, X. Creating hierarchical porosity hydroxyapatite scaffolds with osteoinduction by three-dimensional printing and microwave sintering. *Biofabrication* **2017**, *9*, 045008. [CrossRef]
222. Daugela, P.; Pranskunas, M.; Juodzbalys, G.; Liesiene, J.; Baniukaitiene, O.; Afonso, A.; Gomes, P.S. Novel cellulose/hydroxyapatite scaffolds for bone tissue regeneration: In vitro and in vivo study. *J. Tissue Eng. Regen. Med.* **2018**, *12*, 1195–1208. [CrossRef]
223. Godoy-Gallardo, M.; Portolés-Gil, N.; López-Periago, A.M.; Domingo, C.; Hosta-Rigau, L. Multi-layered polydopamine coatings for the immobilization of growth factors onto highly-interconnected and bimodal PCL/HA-based scaffolds. *Mater. Sci. Eng. C* **2020**, *117*, 111245. [CrossRef] [PubMed]

224. Godoy-Gallardo, M.; Portolés-Gil, N.; López-Periago, A.M.; Domingo, C.; Hosta-Rigau, L. Immobilization of BMP-2 and VEGF within multilayered polydopamine-coated scaffolds and the resulting osteogenic and angiogenic synergy of co-cultured human mesenchymal stem cells and human endothelial progenitor cells. *Int. J. Mol. Sci.* **2020**, *21*, 6418. [CrossRef] [PubMed]
225. Zhang, K.; Fan, Y.; Dunne, N.; Li, X. Effect of microporosity on scaffolds for bone tissue engineering. *Regen. Biomater.* **2018**, *5*, 115–124. [CrossRef] [PubMed]
226. Lord, M.S.; Foss, M.; Besenbacher, F. Influence of nanoscale surface topography on protein adsorption and cellular response. *Nano Today* **2010**, *5*, 66–78. [CrossRef]
227. Zhao, F.; Mc Garrigle, M.J.; Vaughan, T.J.; McNamara, L.M. In silico study of bone tissue regeneration in an idealised porous hydrogel scaffold using a mechano-regulation algorithm. *Biomech. Model. Mechanobiol.* **2017**, *17*, 5–18. [CrossRef]
228. Raafat, A.I.; Abd-Allah, W.M. In vitro apatite forming ability and ketoprofen release of radiation synthesized (gelatin-polyvinyl alcohol)/bioglass composite scaffolds for bone tissue regeneration. *Polym. Compos.* **2016**, *39*, 606–615. [CrossRef]
229. Shim, K.-S.; Kim, H.J.; Yun, Y.-P.; Jeon, D.I.; Kim, H.J.; Park, K.; Song, H.-R. Surface immobilization of biphasic calcium phosphate nanoparticles on 3D printed poly (caprolactone) scaffolds enhances osteogenesis and bone tissue regeneration. *J. Ind. Eng. Chem.* **2017**, *55*, 101–109. [CrossRef]
230. Bružauskaitė, I.; Bironaitė, D.; Bagdonas, E.; Bernotienė, E. Scaffolds and cells for tissue regeneration: Different scaffold pore sizes-different cell effects. *Cytotechnology* **2015**, *68*, 355–369. [CrossRef]
231. Hulbert, S.F.; Young, F.A.; Mathews, R.S.; Klawitter, J.J.; Talbert, C.D.; Stelling, F.H. Potential of ceramic materials as permanently implantable skeletal prostheses. *J. Biomed. Mater. Res.* **1970**, *4*, 433–456. [CrossRef] [PubMed]
232. Ashman, A.; Moss, M.L. Implantation of porous polymethylmethacrylate resin for tooth and bone replacement. *J. Prosthet. Dent.* **1977**, *37*, 657–665. [CrossRef]
233. Murphy, C.M.; Haugh, M.G.; O'Brien, F.J. The effect of mean pore size on cell attachment, proliferation and migration in collagen–glycosaminoglycan scaffolds for bone tissue engineering. *Biomaterials* **2010**, *31*, 461–466. [CrossRef] [PubMed]
234. Nehrer, S.; Breinan, H.A.; Ramappa, A.; Young, G.; Shortkroff, S.; Louie, L.K.; Sledge, C.B.; Yannas, I.V.; Spector, M. Matrix collagen type and pore size influence behaviour of seeded canine chondrocytes. *Biomaterials* **1997**, *18*, 769–776. [CrossRef]
235. Lee, S.J.; Lee, I.W.; Lee, Y.M.; Lee, H.B.; Khang, G. Macroporous biodegradable natural/synthetic hybrid scaffolds as small intestine submucosa impregnated poly(D,L-lactide-co-glycolide) for tissue-engineered bone. *J. Biomater. Sci. Polym. Ed.* **2004**, *15*, 1003–1017. [CrossRef] [PubMed]
236. Baksh, D.; Davies, J.E.; Kim, S. Three-dimensional matrices of calcium polyphosphates support bone growth in vitro and in vivo. *J. Mater. Sci. Mater. Electron.* **1998**, *9*, 743–748. [CrossRef] [PubMed]
237. Peter, S.J.; Miller, M.J.; Yasko, A.W.; Yaszemski, M.J.; Mikos, A.G. Polymer concepts in tissue engineering. *J. Biomed. Mater. Res.* **1998**, *43*. [CrossRef]
238. Ishaug, S.L.; Crane, G.M.; Miller, M.J.; Yasko, A.W.; Yaszemski, M.J.; Mikos, A.G. Bone formation by three-dimensional stromal osteoblast culture in biodegradable polymer scaffolds. *J. Biomed. Mater. Res.* **1997**, *36*. [CrossRef]
239. Akay, G.; Birch, M.; Bokhari, M. Microcellular polyHIPE polymer supports osteoblast growth and bone formation in vitro. *Biomaterials* **2004**, *25*, 3991–4000. [CrossRef]
240. Kuboki, Y.; Jin, Q.; Takita, H. Delivery Systems for the BMPs Geometry of carriers controlling phenotypic expression in BMP-induced osteogenesis and chondrogenesis. *BMPS* **2001**, *83*, 105–115.
241. Van Tienen, T.G.; Heijkants, R.G.J.C.; De Groot, J.H.; Pennings, A.J.; Schouten, A.J.; Veth, R.P.H.; Buma, P. Replacement of the Knee Meniscus by a Porous Polymer Implant. *Am. J. Sports Med.* **2006**, *34*, 64–71. [CrossRef] [PubMed]
242. Matsiko, A.; Gleeson, J.P.; O'Brien, F.J. Scaffold mean pore size influences mesenchymal stem cell chondrogenic differentiation and matrix deposition. *Tissue Eng. Part A* **2015**, *21*, 486–497. [CrossRef] [PubMed]
243. Sha'Ban, M.; Kim, S.H.; Ruszymah, B.; Khang, G. Fibrin and poly(lactic-co-glycolic acid) hybrid scaffold promotes early chondrogenesis of articular chondrocytes: An in vitro study. *J. Orthop. Surg. Res.* **2008**, *3*, 17. [CrossRef] [PubMed]

244. Liu, J.; Nie, H.; Xu, Z.; Niu, X.; Guo, S.; Yin, J.; Guo, F.; Li, G.; Wang, Y.; Zhang, C. The effect of 3D nanofibrous scaffolds on the chondrogenesis of induced pluripotent stem cells and their application in restoration of cartilage defects. *PLoS ONE* **2014**, *9*, e111566. [CrossRef] [PubMed]
245. Long, M.; Rack, H. Titanium alloys in total joint replacement-A materials science perspective. *Biomaterials* **1998**, *19*, 1621–1639. [CrossRef]
246. Niinomi, M. Recent research and development in titanium alloys for biomedical applications and healthcare goods. *Sci. Technol. Adv. Mater.* **2003**, *4*, 445–454. [CrossRef]
247. Liu, X.; Chu, P.K.; Ding, C. Surface modification of titanium, titanium alloys, and related materials for biomedical applications. *Mater. Sci. Eng. R Rep.* **2004**, *47*, 49–121. [CrossRef]
248. Wang, X.; Xu, S.; Zhou, S.; Xu, W.; Leary, M.; Choong, P.; Qian, M.; Brandt, M.; Xie, Y.M. Topological design and additive manufacturing of porous metals for bone scaffolds and orthopaedic implants: A review. *Biomaterials* **2016**, *83*, 127–141. [CrossRef]
249. Itälä, A.I.; Ylänen, H.O.; Ekholm, C.; Karlsson, K.H.; Aro, H.T. Pore diameter of more than 100 μm is not requisite for bone ingrowth in rabbits. *J. Biomed. Mater. Res.* **2001**, *58*, 679–683. [CrossRef]
250. Braem, A.; Chaudhari, A.; Cardoso, M.V.; Schrooten, J.; Duyck, J.; Vleugels, J. Peri- and intra-implant bone response to microporous Ti coatings with surface modification. *Acta Biomater.* **2014**, *10*, 986–995. [CrossRef]
251. Taniguchi, N.; Fujibayashi, S.; Takemoto, M.; Sasaki, K.; Otsuki, B.; Nakamura, T.; Matsushita, T.; Kokubo, T.; Matsuda, S. Effect of pore size on bone ingrowth into porous titanium implants fabricated by additive manufacturing: An in vivo experiment. *Mater. Sci. Eng. C* **2016**, *59*, 690–701. [CrossRef]
252. Chan, O.; Coathup, M.; Nesbitt, A.; Ho, C.-Y.; Hing, K.; Buckland, T.; Campion, C.; Blunn, G.W. The effects of microporosity on osteoinduction of calcium phosphate bone graft substitute biomaterials. *Acta Biomater.* **2012**, *8*, 2788–2794. [CrossRef] [PubMed]
253. Campion, C.R.; Chander, C.; Buckland, T.; Hing, K. Increasing strut porosity in silicate-substituted calcium-phosphate bone graft substitutes enhances osteogenesis. *J. Biomed. Mater. Res. Part B Appl. Biomater.* **2011**, *97*, 245–254. [CrossRef] [PubMed]
254. Aarvold, A.; Smith, J.O.; Tayton, E.R.; Lanham, S.; Chaudhuri, J.B.; Turner, I.G.; Oreffo, R.O. The effect of porosity of a biphasic ceramic scaffold on human skeletal stem cell growth and differentiationin vivo. *J. Biomed. Mater. Res. Part A* **2013**, *101*, 3431–3437. [CrossRef]
255. Gibson, L.J.; Ashby, M.F. *Cellular Solids: Structure and Properties*; Cambridge University Press: Cambridge, UK, 1999.
256. Story, B.J.; Wagner, W.R.; Gaisser, D.M.; Cook, S.D.; Rust-Dawicki, A.M. In vivo performance of a modified CSTi dental implant coating. *Int. J. Oral Maxillofac. Implant.* **1998**, *13*, 749–757.
257. Eqtesadi, S.; Motealleh, A.; Pajares, A.; Miranda, P. Effect of milling media on processing and performance of 13-93 bioactive glass scaffolds fabricated by robocasting. *Ceram. Int.* **2015**, *41*, 1379–1389. [CrossRef]

Publisher's Note: MDPI stays neutral with regard to jurisdictional claims in published maps and institutional affiliations.

© 2020 by the authors. Licensee MDPI, Basel, Switzerland. This article is an open access article distributed under the terms and conditions of the Creative Commons Attribution (CC BY) license (http://creativecommons.org/licenses/by/4.0/).

nanomaterials

MDPI

Review

Harnessing Inorganic Nanoparticles to Direct Macrophage Polarization for Skeletal Muscle Regeneration

Francesca Corsi [1,2,†], Felicia Carotenuto [1,2,3,†], Paolo Di Nardo [2,3,4,‡] and Laura Teodori [1,3,*,‡]

1 Department of Fusion and Technologies for Nuclear Safety and Security, Diagnostic and Metrology (FSN-TECFIS-DIM), ENEA, 00044 Frascati, Italy; francesca.corsi@uniroma2.it (F.C.); carotenuto@med.uniroma2.it (F.C.)
2 Department of Clinical Science and Translational Medicine, University of Rome "Tor Vergata", 00133 Rome, Italy; dinardo@uniroma2.it
3 Interdepartmental Center of Regenerative Medicine (CIMER), University of Rome "Tor Vergata", 00133 Rome, Italy
4 L.L. Levshin Institute of Cluster Oncology, I. M. Sechenov First Medical University, 119991 Moscow, Russia
* Correspondence: laura.teodori@enea.it; Tel.: +39-06-9400-5642
† These authors contributed equally to this work.
‡ These authors share last authorship as senior authors.

Received: 23 July 2020; Accepted: 29 September 2020; Published: 2 October 2020

Abstract: Modulation of macrophage plasticity is emerging as a successful strategy in tissue engineering (TE) to control the immune response elicited by the implanted material. Indeed, one major determinant of success in regenerating tissues and organs is to achieve the correct balance between immune pro-inflammatory and pro-resolution players. In recent years, nanoparticle-mediated macrophage polarization towards the pro- or anti-inflammatory subtypes is gaining increasing interest in the biomedical field. In TE, despite significant progress in the use of nanomaterials, the full potential of nanoparticles as effective immunomodulators has not yet been completely realized. This work discusses the contribution that nanotechnology gives to TE applications, helping native or synthetic scaffolds to direct macrophage polarization; here, three bioactive metallic and ceramic nanoparticles (gold, titanium oxide, and cerium oxide nanoparticles) are proposed as potential valuable tools to trigger skeletal muscle regeneration.

Keywords: nanotechnology; tissue engineering; immunomodulation; macrophage plasticity; skeletal muscle regeneration

1. Introduction

Tissue engineering (TE) is a multidisciplinary field including bio-medicine, material science, and engineering, aimed at manufacturing functional biological tissues to be implanted in organs damaged by otherwise incurable diseases or severe casualties [1,2]. Engineered tissues are also fundamental to set up in vitro models of human physiological systems, replacing animal models for drug development, toxicology studies, etc. [3].

A successful TE treatment depends on the immune response of the recipient tissue. In fact, the engineered tissue must initially challenge the inflammatory microenvironment of the damaged tissue to establish an efficient integration and, after the implantation, the inflammatory reaction characterizes a possible rejection process. The traditional strategy adopted after tissue and organ transplantation aims at minimizing the host immune response through anti-inflammatory and immunosuppressive therapies. However, if, on one hand, the immune system is often the cause of implants rejection, on the other, several immune components positively affect tissue regeneration and healing [4]. Therefore, controlling the balance between immune pro-inflammatory and pro-resolution players represents a better strategy to ensure implant tolerance than suppressing the immune response [5].

Among the immune cells involved in the foreign body response, macrophages play a central role. Macrophages are effector immune cells simplistically divided into two classes, named M1 and M2. To ensure regeneration, a balance between M1 and M2 activities (pro- and anti-inflammatory, respectively) shifting over time is required [6]; hence, scaffold-based approaches to direct macrophage polarization are gaining much interest in TE for regenerative medicine.

Nanomaterials are emerging as effective agents able to target macrophages, perturbing their polarization and thus their activity [7]. As a consequence, in recent years, optimally designed-nanoparticles (NPs) for the modulation of macrophage plasticity have been studied for treating diseases characterized by hyper-tolerant or inflammatory immune microenvironment, such as cancer [8] and inflammatory diseases [9], respectively. As far as TE is concerned, the full potential of nanoparticles as macrophage regulators has not yet been fully realized. This review aims at discussing nanotechnology impact on TE in terms of directing macrophage polarization and reprogramming, focusing on bioactive inorganic nanoparticles prospects for skeletal muscle regeneration.

2. The Immune Response to Implanted Materials: The Driving Role of Macrophages

The immune response against implanted materials originates as an acute inflammatory response to the tissue injury and the foreign material itself introduced with the implant [4]. Innate immunity is the first to be involved, providing an initial barrier against potential pathogens invading the damaged tissue. In the absence of pathogens, the innate response triggers a so-called sterile inflammation, put in motion by the danger signals released from the injured tissue, strongly affecting the outcome of regeneration. The innate response is then followed by the activation of the adaptive immunity. Although it was originally considered less important in the regeneration process, the adaptive immune response to implanted materials plays a pivotal role in tissue regeneration, mainly due to T cells activity [10]. In the following paragraphs, we will address the role of the innate and adaptive immune responses, focusing on macrophage intervention. The main steps are graphically represented in Figure 1.

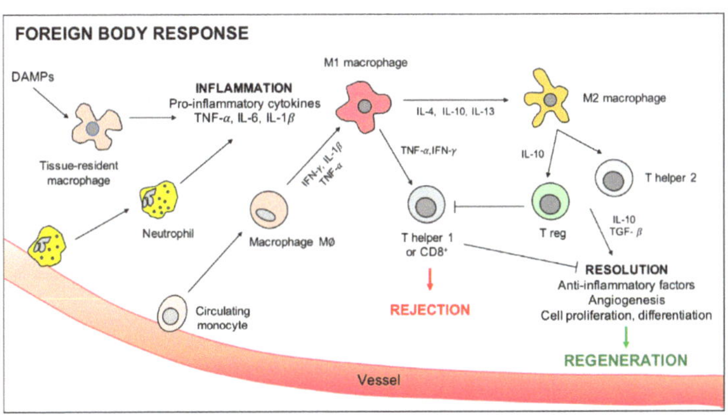

Figure 1. Graphical representation of macrophage intervention in scaffold-induced foreign body response. From the left, danger associated molecular patterns (DAMPs) are released after tissue injury, activating tissue-resident macrophages to promote inflammation and neutrophils recruitment. Inflammation is then sustained by pro-inflammatory M1 macrophages, differentiated by circulating monocytes, before being eventually resolved by macrophage polarization towards the M2 anti-inflammatory phenotype. Macrophage-driven stimulation of T cells is also shown.

2.1. Macrophages Orchestrate Innate Immunity

Tissue injury activates a local inflammatory response by the damage-associated molecular patterns (DAMPs) released from necrotic or stressed cells and damaged extracellular matrix (ECM). Danger signals typically include monosodium urate, extracellular ATP, nucleic acids, heat shock proteins (HSP), and fragments from ECM components, e.g., portions of collagen, hyaluronic acid, elastin, fibronectin, laminin, etc. Released DAMPs are then mainly recognized by Toll-like receptors (TLRs), initiating the inflammatory process via activation of the nuclear transcription factor NF-kB or interferon-regulatory factors. Upon TLR activation, tissue-resident macrophages are awakened to release chemoattractants for recruiting neutrophils, monocytes, and macrophages to the injury site, and to produce pro-inflammatory cytokines such as tumor necrosis factor-α (TNF-α), interleukin-6 (IL-6), and IL-1β [10].

Once crossed the endothelium and recruited to the implantation site, neutrophils produce antimicrobial substances and proteases to kill and degrade pathogens that have potentially invaded the damaged tissue. Furthermore, they also produce cytokines (e.g., IL-17) to recruit and activate more neutrophils, monocytes and macrophages, and growth factors (e.g., vascular endothelial growth factor, VEGF-A) to promote angiogenesis and proliferation of fibroblasts, epithelial cells, etc. [11]. Although being inflammatory cells, neutrophils also exhibit anti-inflammatory activity; indeed, recruiting macrophages, they promote their own removal, preserving the ECM from excessive degradation [6].

The recruited macrophages are differentiated from circulating monocytes and they often greatly exceed the population of tissue-resident macrophages involved in the very early immune response. Macrophages are key actors of tissue regeneration; indeed, they are responsible for the production of several cytokines, proteases, growth factors, soluble mediators, and ECM components, regulating tissue microenvironment [12]. Macrophages modulate the microenvironment and the latter in turn regulates macrophages activity; indeed, they may undergo several phenotypic and functional changes in response to different environmental stimuli [12]. Polarized macrophages can be simplistically classified in two main groups: classically activated macrophages (M1) activated by IFN-γ and lipopolysaccharides (LPS) and alternatively activated macrophages (M2), further divided in M2a (if activated by IL-4 or IL-13), M2b (by IL-1β or LPS) and M2c (by IL-10, TGF-β or glucocorticoids) [13]. This macrophage categorization based on specific activating factors *in vitro*, is, however, very simplistic, ignoring the source and the plurality of the stimuli coexisting in the in vivo environment [14]. In this regard, recent studies employing transcriptomic and proteomic analyzes have revealed a broader spectrum of interrelated macrophage activation states resulting from different sets of stimuli [15]. Coherently, macrophage polarization is presently the subject of debate. Nevertheless, here, for the sake of clarity, we use a bipolar nomenclature (M1 pro-inflammatory vs. M2 anti-inflammatory) to describe the two opposite sides of the spectrum of macrophage activation states. M1 macrophages exhibit pro-inflammatory and microbicidal activities, whereas M2 macrophages are anti-inflammatory immune cells, generally considered pro-resolution players, since they produce several ECM components and growth factors (e.g., VEGF-A) [12]. The kinetic of M1 and M2 macrophages is therefore crucial for tissue regeneration. In skeletal muscle regeneration, M1-related cytokines actively participate in skeletal muscle stem cells (satellite cells) activation. Afterwards, the M2-secreted cytokines IL-4 and IL-10 promote satellite cells proliferation and differentiation, increasing myogenin expression [16]. The mechanisms driving macrophage polarization towards the pro-resolution phenotype are still unclear; besides cytokines, microRNAs (miRs) are surely key regulators of macrophage polarization, via controlling the translation and degradation of messenger RNAs of cytokines and transcription factors [17,18]. MiR-146a, miR146b, miR222-3p, and miR181a, for example, are of particular interest for tissue regeneration, supporting M2 polarization in humans [19]. However, miRs-mediated macrophage regulation is out of the focus of this paper and a further study is under way from the same group.

2.2. Macrophage Role in Adaptive Immunity

Macrophages are not only part of innate immunity: they also prepare adaptive immunity to respond, generating costimulatory molecules for lymphocytes T and B activation by non-self-antigens, such as those occurring in materials carrying allogenic-cells or scaffolds not properly decellularized [4].

M2 macrophages trigger Th2-mediated response, generally having beneficial effects on tissue regeneration, whereas M1 macrophages promote strong T helper 1 (Th1) response, triggering pro-inflammatory cytokines production and cytotoxic ($CD8^+$) T cells activation, impairing tissue healing. After bone injury, for example, Th1 and $CD8^+$ T cells activation strongly correlates with impaired osteogenesis [20]. On the contrary, Th2 and regulatory T cells (Tregs) function as critical pro-resolution mediators; indeed, producing anti-inflammatory cytokines such as IL-10 and TGF-β, they promote at least bone [21], cardiac [22], and skeletal muscle [23] regeneration.

If on one hand it is evident that too sustained M1 macrophage activation can exacerbate inflammation increasing tissue injury, on the other, persistent M2 activation may also be counterintuitively detrimental, leading to pathological fibrosis [24]. Immunomodulation is therefore essential to trigger tissue regeneration, maintaining a delicate balance between the pro-inflammatory and the anti-inflammatory response.

3. Immunomodulation by Implanted Scaffolds

As reported in our previous work [4], materials properties are important for tissue regeneration to: (i) provide mechanical support to the implant, (ii) regulate scaffold adsorption/degradation time, (iii) promote stem cell recruitment and differentiation, and (iv) elicit a balanced immune response. In the following two paragraphs, we briefly examined the effects of native vs. synthetic scaffolds on macrophage polarization, discussing their impact on tissue regeneration.

3.1. Native Scaffolds: Decellularized ECM

Native scaffolds are highly bio-mimetic platforms, providing an ideal microenvironment for cell attachment, proliferation, and stem cell differentiation, having thus demonstrated to be the most appealing tool for TE applications [4]. They are obtained from several tissues and organs of different species by decellularization protocols. The resulting material is free of cells and cellular debris, but enriched with ECM components, such as collagen, fibrin, proteoglycans, and cryptic peptides, conferring it the positional and functional information required to re-build tissues and organs [5,25].

The immunomodulatory effect of decellularized ECM (dECM) mainly depends on scaffold structure and composition. Huleihel and colleagues, for example, recently demonstrated that matrix-bound vesicles (MBVs) present in urinary bladder matrix are responsible for macrophage polarization towards the anti-inflammatory phenotype, via three miRs (miR125b-5p, miR143-3p, and miR145-5p) carried within the vesicles [26]. The authors hypothesized that, during degradation of the ECM-based scaffold, the released MBVs are internalized by macrophages, eliciting their polarization.

The source tissue where the ECM is harvested is another factor determining the immune response to the scaffold. In this context, a recent study comparing macrophage activation after exposure to dECM derived from different tissues, highlighted heterogeneity in macrophage response due to distinct ECM degradation products. In particular, the exposure to urinary bladder matrix or small intestinal submucosa ECM differently affected murine bone marrow-derived macrophages, generating distinct gene expression profiles of several commonly investigated macrophage surface and activation markers [27]. Scaffolds deriving from the same tissue may also behave differently; dECM from skeletal muscle, for example, has been shown to trigger macrophage polarization towards the M2 anti-inflammatory phenotype in an in vivo model of skeletal tissue injury [28], but other studies reported no impact of skeletal muscle-derived ECM on macrophage plasticity [10]. Such a heterogeneity, even among matrices from the same tissue, may derive from the applied decellularization protocol. Indeed, even the detergent choice and treatment with specific enzymes to cleave immunogenic antigens may profoundly modify the immune response [29]. A recent work by Wu [30] showed that the application of a digestion-neutralization protocol to transform bone-derived filler particles into gel bio-scaffolds enhanced tissue regeneration in rat periodontal defect model [30]. Particularly, the applied approach shifted macrophage polarization towards the M2 phenotype, instead of the M1 pro-inflammatory phenotype generally triggered by the commonly used bone-derived ECM particles, enhancing the expression of pro-resolution mediators (e.g., IL-10) and decreasing the pro-inflammatory ones (TNF-α, IL-1β).

3.2. Synthetic Scaffolds

Synthetic scaffolds can be assembled from both naturally and synthetically-derived biocompatible polymers, in order to mimic the architecture and composition of several tissues and organs. Their molecular and structural characteristics can be finely controlled, with greater precision respect to their natural counterparts (i.e., dECM) [25,31]. Each material has different physico-chemical and biological properties, often contributing to a great extent in eliciting the foreign body response, which therefore becomes a critical issue for successful scaffold implantation [2].

As previously mentioned, different materials trigger diverse immune responses. In a recent work by Badylak's group, the hydrogelating self-assembling fiber (hSAF) system, formed from two complementary de novo designed α-helical peptides, strongly modulated macrophage plasticity, increasing the expression of anti-inflammatory markers such as Arginase1, IL-10, and CD206, finally contributing to remodeling and myogenic differentiation [32]. Chitosan also polarized human macrophages towards the M2 phenotype, downregulating the expression of pro-inflammatory TNF-α, while increasing the levels of the anti-inflammatory IL-10 and TGF-β [33], whereas poly-ethylen-glycol (PEG)-based hydrogels enhanced pro-inflammatory cytokines expression in murine RAW 264.7 macrophages [34].

Material-induced immune response is even more complicated, as the same material can divergently stimulate specific immune cell populations. This is the strange case of chitosan, that, on the one hand, triggered macrophage polarization towards the anti-inflammatory subtype, and, on the other, polarized dendritic cells (DCs) towards a pro-inflammatory type [33].

Besides composition, other crucial factors regulating the immune response concern the structural properties of the material (e.g., porosity, pore distribution), surface topography, stiffness, etc. [25]. Regarding the impact mediated by material porosity, Tylek and colleagues recently exploited advanced melt electrowriting (MEW) to fabricate poly-ε-caprolactone (PCL)-based scaffold with box-shaped pores with various inter-fiber spacing, from 100 μm to 40 μm. The resulted scaffolds facilitated human primary macrophages elongation and polarization towards the M2 phenotype, with the smallest pore sized as the most effective [35]. Scaffold surface topography is widely recognized as modulator of the immune players; in the last decade, indeed, many studies highlighted the possibility to attenuate the inflammatory process and drive macropages towards the pro-healing phenotype [36], by controlling scaffold surface grooves and gratings—for example, via 3D lithographic methods. These approaches allow for finely regulating a scaffold micro- and nano-pattern, thus strongly affecting macrophages elongation and differentiation [37].

Scaffold stiffness, instead, is generally considered less important in driving macrophage polarization, especially if compared to scaffold architecture [25]; however, recent studies strongly suggest that substrate stiffness can make the difference. Sridharan and colleagues, for example, varying collagen-coated polyacrylamide gels stiffness, were able to alter human THP-1-derived macrophages behavior in terms of migration mode, polarization state, and, thus, a functional role. Particularly, the highest stiffness (323 kPa) primed macrophages towards the M1 phenotype, whereas the softest (11 kPa) towards the M2 phenotype [38]. The same authors further demonstrated that the variability in macrophage polarization in response to differential stiffnesses may be dependent on the crosslinking agents used to modulate collagen stiffness [39].

It is therefore evident that the macrophage response induced by scaffolds (both natural and synthetic) is highly variable, mainly because there are many factors involved (biomaterial, composition, topography, porosity, etc.). Furthermore, it is arduous to recreate the complexity of the microenvironmental stimuli naturally polarizing macrophages; thus, novel approaches such as microfluidic systems are trying to recreate in vitro, the in vivo stimulation [40].

Altogether, there is a need to find new strategies more finely controlling macrophage plasticity, to help scaffolds addressing the immune response towards tissue regeneration. Nanotechnology may represent the solution.

4. Nanoparticles to Direct Macrophage Polarization

Nanoparticles (NPs) have been initially developed to overcome problems of bioavailability, body retention, solubility, stability, and selectivity of pharmaceutical agents, protecting the carried drug until reaching the desired body district. However, nanomaterials at the nanoscale (1–100 nm) acquired peculiar properties, due to the increased reactive surface/bulk ratio with respect to micro- and macro-structures [41], making them attractive for TE.

Indeed, in the 2000s, a key role has been recognized for nanomaterials in TE, as nanocomposite polymers, both in the form of electrospun fibers and hydrogels, often provide superior mechanical, functional, and electrical properties [42]. In the context of skeletal muscle regeneration, for example, aligned nanofibrous scaffolds (e.g., PCL/collagen) favored cell alignment and myotube formation, thus promoting muscle regeneration [43].

It is noteworthy that the role of NPs as modulators of macrophage plasticity is strongly emerging; indeed, due to their particulate (instead of molecular) nature, nanoparticles preferentially target professional phagocytes, such as macrophages [44]. This property is of a paramount importance for the success of TE procedures. Therefore, several research works focused on the use of organic nanovesicles (liposomes, polysaccharides, capsules, etc.) carrying encapsulated bioactive molecules [45], such as flavonoids [46], miRs, and cytokines [24], in order to regulate macrophage activity. In this study, instead, we investigate three major inorganic NPs (gold, titanium oxide, and cerium oxide NPs, later denoted as AuNPs, TiO_2 NPs, and CeO_2 NPs), not only being exploitable as carriers for drugs and molecules, but also being characterized by intrinsic bioactivity, which renders them promising candidates for TE therapies (Figure 2, Table 1).

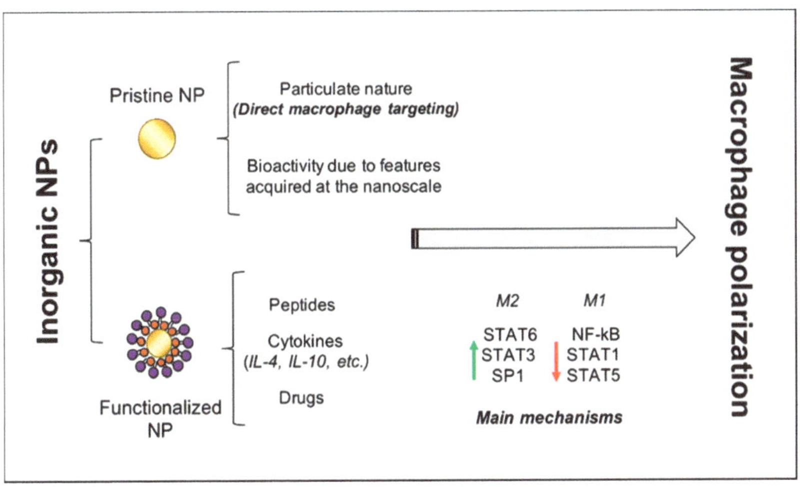

Figure 2. Inorganic NP-mediated macrophage polarization towards regeneration. The schematic representation of inorganic NPs double role is reported. NPs may act as pristine NPs, characterized by intrinsic bioactivity, and as carriers for peptides, cytokines, and drugs. Principal transcription factors involved in macrophage activation and polarization are also included. NP = nanoparticle; IL = interleukin; STAT = signal transducer and activator of transcription; NF-kB = nuclear factor kappa-light-chain-enhancer of activated B cells.

Table 1. Nanoparticle-induced macrophage polarization towards M2 phenotype.

NPs	Shape	Size (nm)	Surface Modification	Initial Phenotype	Polarized Phenotype	Model	Used Markers	Clinical Aim	Ref.
Au	rods	ND	PEGylation + RGD	MØ	M2	Mouse	Arg-1, IL-4, TNF-α, Retnla	Acute hepatitis	[47]
	spheres	13	hexapeptides	M2 M1	M2 M2 M2	Mouse BMDMs ALI Mouse model	IL-12, IL-6, IL-10, iNOS, Arg-1, YM1 CD80, CD206	Acute lung injury	[48]
	spheres	30	PEGylation + IL-4	MØ	M2	Human THP-1 cell line	CD206, CD163, IL-4	Muscle recovery	[49]
		30	PEGylation + IL-4	MØ	M2	Mouse	CD206, CD80		
		100	None	MØ	M1	Human THP-1 cell line	CD206, CD163, IL-4		
TiO_2	tubes	110	None	MØ MØ	M1 M2	Mouse RAW 264.7 cell line Mouse RAW 264.7 cell line + MSCs	CCR7, IL-6, IL-10, VEGF, BMP2, TGF-β1	Osteogenesis	[50]
	disks	10 *	Ca^{2+}, Sr^{2+}	MØ	M2	Mouse RAW 264.7 cell line	Arg-1, CD163, CD86, iNOS	Osteogenesis	[51]
	tubes	65 92 142	None	MØ	M2	Human THP-1 cell line	Arg-1, CD206, IL-10, VEGF	Heart valve replacement	[52]
CeO_2	thin layer	ND	None	MØ	M2	Mouse RAW 264.7 cell line	CD163, CD206, IL-6, TNF-α, IL-10, TGF-β1	Osteogenesis	[53]
	thin layer	ND ND	None None	MØ	M2	Mouse RAW 264.7 cell line + BMSCs Rat	CCR7, CD206, TNF-α, IL-10	Osteogenesis	[54]

NPs = nanoparticles; RGD = arginine-glycine-aspartic acid; M = macrophage; Arg-1 = arginase-1; IL = interleukin; TNF-α = tumor necrosis factor-α; MSCs = mesenchymal stem cells; BMDM = bone marrow derived macrophages; CCR7 = C-C chemokine receptor 7; VEGF = vascular endothelial growth factor; BMP2 = bone morphogenetic protein 2; TGF-β1 = transforming growth factor β1; NOS = nitric oxide synthase; BMSCs = bone mesenchymal stem cells; * Average surface roughness; ND = not determined.

4.1. Harnessing Gold Nanoparticles to Trigger Skeletal Muscle Regeneration

AuNPs have been extensively studied [55], particularly for cardiac tissue regeneration, due to their ability to enhance scaffold conductivity [42]. Furthermore, they were recently found to improve osteogenic differentiation of human bone marrow-derived mesenchymal stem cells [56]. As regards their immunomodulatory activity, a recent work by Ni highlighted the ability of AuNPs to generate a microenvironment with constraint inflammatory and reparative cytokines [57]. In particular, when facing inflammation upon stimulation with LPS, murine macrophages exposed to AuNPs showed decreased pro-inflammatory and enhanced anti-inflammatory markers, with respect to control cells.

The surface chemistry of AuNPs also allows rapid functionalization with bioactive molecules, making them suitable for several purposes. In the context of macrophage polarization, AuNPs functionalized with the peptide arginine-glycine-aspartic acid (RGD), a motif responsible for cell adhesion to ECM, were observed to dramatically decrease M1 markers expression, while increasing the M2 ones, in macrophages isolated from mouse livers [47]. Furthermore, when coated with the hexapeptides Cys-Leu-Pro-Phe-Phe-Asp, able to inhibit LPS-induced activation of TLR4, AuNPs drove mouse bone marrow-derived macrophages towards the M2 phenotype, both in vitro and in vivo [48]. Cytokines may also be conjugated on nanoparticles surface; a recent work by Raimondo and Mooney, reported that IL-4-conjugated AuNPs recovered injured murine skeletal muscle via NP-directed M2 macrophage polarization [49]; indeed, the authors finely demonstrated that macrophage depletion abrogated the therapeutic effect triggered by the functionalized nanoparticles. The potential positive contribution of AuNPs in skeletal muscle regeneration is confirmed by Ge and colleagues. Indeed, they reported that AuNPs significantly enhanced myogenic differentiation of myoblasts, via activation of the p38 mitogen-activated protein kinase (p38 MAPK) signaling pathway, finally promoting in vivo regeneration in muscle defect models of rats [58].

4.2. Titanium Oxide NPs: From Osteogenesis to Muscle Regeneration?

TiO_2 NPs are of significant concern in TE, especially for hard tissue regeneration, because titanium resists corrosion from the body. However, nanometer-thin TiO_2 deposition has also been applied for muscle regeneration, significantly enhancing rat skeletal muscle cell attachment and proliferation [59]. Unlike other ceramic NPs, which induced an inflammatory phenotype in primary macrophages, titanium oxide NPs were reported not to foster inflammation [60]. A recent work demonstrated that TiO_2 nanotubes promoted osteogenesis via mediating the crosstalk between macrophages and mesenchymal stem cells (MSCs) under oxidative stress, reducing early inflammation by accelerating macrophages M1 to M2 transition [50]. A previous work by Lee, also reported increased M2 subtypes markers (Arg1, MR and CD163) on calcium and strontium (Ca and Sr, respectively)-modified titanium implants, leading to enhanced osteogenesis [51]. Furthermore, TiO_2 nanotubes, inducing macrophage elongation and polarization towards the M2 anti-inflammatory state, also promoted endothelialization [52]. In particular, THP-1 cells, when exposed to TiO_2 nanotubes, polarized to M2 macrophages and produced high levels of VEGF, which promoted human umbilical vein endothelial cells (HUVEC) proliferation via two main proliferative pathways: the phosphatidylinositol 3-kinase/AKT and the extracellular signal-regulated kinase 1/2 pathways. As vasculature is needed to facilitate construct engraftment and provide nutrient and waste exchange to the regenerating tissue, TiO_2 nanoparticles may be able to strongly affect the whole foreign body response, addressing it towards regeneration.

4.3. Cerium Oxide NPs: Not Just Antioxidants

CeO_2 NPs are bioactive nanoparticles attracting great interest in the biomedical field, mainly for their strong antioxidant activity, due to the presence on their surface of cerium atoms in both of the oxidation states: Ce^{3+} and Ce^{4+} [41]. Besides their valence in cancer research [61], they have recently been emerging as promoters of tissue regeneration. In 2013, topical application of CeO_2

NPs was reported to accelerate the healing of dermal wounds in mice, enhancing proliferation and migration of fibroblasts, keratinocytes, and vascular endothelial cells [62]. More recently, CeO_2 NPs incorporated into hydroxyapatite coating, via the plasma spraying technique, induced polarization of murine RAW264.7 macrophages towards the M2 phenotype, as demonstrated by the upregulation of the M2 surface markers CD163 and CD206, promoting an osteogenic behavior of bone mesenchymal stem cells (BMSCs) [53]. The improved osteogenic differentiation ability of BMSCs was further found to be proportional to the surface Ce^{4+}/Ce^{3+} ratio; indeed, the higher ratio achieved better osseointegration in vivo, and enhanced murine macrophages polarization towards the M2 phenotype, increasing the levels of anti-inflammatory cytokines [54].

It is noteworthy that CeO_2 NPs demonstrated reliable biocompatibility [63]. This is of a paramount importance; indeed, in general, pharmacological administration of inorganic nanoparticles causes safety issues to arise, rendering nanotoxicology [64] a key discipline for nanoparticles use in biomedicine and TE.

5. Conclusions

Despite the significant progress in TE applications, nanotechnology (both in terms of nanocomposite polymers and/or nanopatterned scaffolds) has been mainly applied to provide scaffolds with superior mechanical, functional, and electrical properties, and to favor muscle cell alignment and differentiation. Nevertheless, as thoroughly discussed in this review, particular attention should be paid to the induced immune response, especially to macrophage polarization. Indeed, a correct balance between pro- and anti-inflammatory factors is crucial for scaffold engraftment and tissue regeneration. In this context, surprisingly, there is still too little effort to investigate the contribution that nanotechnology may give to direct macrophage polarization, while, in other areas of biomedical research, such as inflammatory diseases and tumors, several bioactive inorganic nanoparticles have been recognized as valuable modulators of macrophage plasticity.

The analysis of the literature performed in this review unveiled the unpresented potential of Au, TiO_2, and CeO_2 nanoparticles to effectively modulate macrophage polarization and strongly support TE for regenerative medicine. While AuNPs have already been studied in the field of muscle regeneration, obtaining excellent results both in terms of NP-elicited macrophage polarization and promotion of myogenesis, TiO_2, and CeO_2 NPs, were mainly tested in the field of osteogenesis, but the results summarized here represent good premises for their immediate application also in skeletal muscle regeneration.

Author Contributions: Conceptualization, F.C. (Francesca Corsi), F.C. (Felicia Carotenuto), and L.T.; writing—original draft preparation, F.C. (Francesca Corsi) and F.C. (Felicia Carotenuto); writing—review and editing, F.C. (Francesca Corsi), F.C. (Felicia Carotenuto), P.D.N., and L.T.; supervision, P.D.N. and L.T.; project administration and funding acquisition, L.T. All authors have read and agreed to the published version of the manuscript.

Funding: This research was funded by Regione Lazio through LazioInnova, Progetto 'SMARTIES' (Smart scaffold ingegnerizzati per la rigenerazione dell'apparato muscolo scheletrico), coordinated by ENEA CR Frascati, Grant No.: 85-2017-15004, AVVISO PUBBLICO PROGETTI DI GRUPPI DI RICERCA Conoscenza e cooperazione per un nuovo modello di sviluppo.

Acknowledgments: We want to express the greatest thanks to Martina Di Cesare (LazioInnova) for the excellent management of the SMARTIES funds. We also would like to thank Andrea Reale and Flavia Miele (ENEA CR Frascati) for their precious technical assistance to the SMARTIES research grant and Anna Pagliardini (ENEA CR Frascati) for her administrative valuable support.

Conflicts of Interest: The authors declare no conflict of interest.

References

1. Atala, A. Engineering tissues, organs and cells. *J. Tissue Eng. Regen. Med.* **2007**, *1*, 83–96. [CrossRef] [PubMed]
2. Armstrong, L.; Lako, M.; Buckley, N.; Lappin, T.; Murphy, M.; Nolta, J.; Pittenger, M.; Stojkovic, M. Editorial: Our top 10 developments in stem cell biology over the last 30 years. *Stem Cells* **2011**, *30*, 2–9. [CrossRef] [PubMed]

3. Watson, D.; Hunziker, R.; Wikswo, J. Fitting tissue chips and microphysiological systems into the grand scheme of medicine, biology, pharmacology, and toxicology. *Exp. Biol. Med.* **2017**, *242*, 1559–1572. [CrossRef] [PubMed]
4. Crupi, A.; Costa, A.; Tarnok, A.; Melzer, S.; Teodori, L. Inflammation in tissue engineering: The Janus between engraftment and rejection. *Eur. J. Immunol.* **2015**, *45*, 3222–3236. [CrossRef] [PubMed]
5. Carotenuto, F.; Teodori, L.; Maccari, A.; Delbono, L.; Orlando, G.; Di Nardo, P. Turning regenerative technologies into treatment to repair myocardial injuries. *J. Cell Mol. Med.* **2019**, *24*, 2704–2716. [CrossRef] [PubMed]
6. Koh, T.; DiPietro, L. Inflammation and wound healing: The role of the macrophage. *Expert Rev. Mol. Med.* **2011**, *13*, e23. [CrossRef]
7. Miao, X.; Leng, X.; Zhang, Q. The current state of nanoparticle-induced macrophage polarization and reprogramming research. *Int. J. Mol. Sci.* **2017**, *18*, 336. [CrossRef]
8. Zanganeh, S.; Hutter, G.; Spitler, R.; Lenkov, O.; Mahmoudi, M.; Shaw, A.; Pajarinen, J.; Nejadnik, H.; Goodman, S.; Moseley, M.; et al. Iron oxide nanoparticles inhibit tumour growth by inducing pro-inflammatory macrophage polarization in tumour tissues. *Nat. Nanotechnol.* **2016**, *11*, 986–994. [CrossRef]
9. Jain, S.; Tran, T.; Amiji, M. Macrophage repolarization with targeted alginate nanoparticles containing IL-10 plasmid DNA for the treatment of experimental arthritis. *Biomaterials* **2015**, *61*, 162–177. [CrossRef]
10. Julier, Z.; Park, A.; Briquez, P.; Martino, M. Promoting tissue regeneration by modulating the immune system. *Acta Biomater.* **2017**, *53*, 13–28. [CrossRef]
11. Kasama, T.; Kobayashi, K.; Yajima, N.; Shiozawa, F.; Yoda, Y.; Takeuchi, H.; Mori, Y.; Negishi, M.; Ide, H.; Adachi, M. Expression of vascular endothelial growth factor by synovial fluid neutrophils in rheumatoid arthritis (RA). *Clin. Exp. Immunol.* **2000**, *121*, 533–538. [CrossRef] [PubMed]
12. Zhang, J.; Yang, Y.; Yang, Z.; Li, T.; Chen, F. Snapshot: Targeting macrophages as a candidate for tissue regeneration. *Curr. Issues Mol. Biol* **2018**, *29*, 37–48. [CrossRef] [PubMed]
13. Martinez, F.O.; Sica, A.; Mantovani, A.; Locati, M. Macrophage activation and polarization. *Front. Biosci.* **2008**, *13*, 453–461. [CrossRef] [PubMed]
14. Martinez, F.O.; Gordon, S. The M1 and M2 paradigm of macrophage activation: Time for reassessment. *F1000Prime Rep.* **2014**, *6*, 13. [CrossRef] [PubMed]
15. Xue, J.; Schmidt, S.V.; Sander, J.; Draffehn, A.; Krebs, W.; Quester, I.; Dominic De Nardo, D.; Gohel, T.D.; Emde, M.; Schmidleithner, L.; et al. Transcriptome-based network analysis reveals a spectrum model of human macrophage activation. *Immunity* **2014**, *40*, 274–288. [CrossRef]
16. Chazaud, B.; Sonnet, C.; Lafuste, P.; Bassez, G.; Rimaniol, A.; Poron, F.; Authier, F.; Dreyfus, P.; Gherardi, R. Satellite cells attract monocytes and use macrophages as a support to escape apoptosis and enhance muscle growth. *J. Cell Biol.* **2003**, *163*, 1133–1143. [CrossRef]
17. Teodori, L.; Costa, A.; Campanella, L.; Albertini, M. Skeletal muscle atrophy in simulated microgravity might be triggered by immune-related microRNAs. *Front. Physiol.* **2019**, *9*, 1926. [CrossRef]
18. Carotenuto, F.; Albertini, M.; Coletti, D.; Vilmercati, A.; Campanella, L.; Darzynkiewicz, Z.; Teodori, L. How diet intervention via modulation of DNA damage response through microRNAs may have an effect on cancer prevention and aging, an in silico study. *Int. J. Mol. Sci.* **2016**, *17*, 752. [CrossRef]
19. Li, H.; Jiang, T.; Li, M.; Zheng, X.; Zhao, G. Transcriptional regulation of macrophages polarization by microRNAs. *Front. Immunol.* **2018**, *9*, 1175. [CrossRef]
20. Reinke, S.; Geissler, S.; Taylor, W.; Schmidt-Bleek, K.; Juelke, K.; Schwachmeyer, V.; Dahne, M.; Hartwig, T.; Akyuz, L.; Meisel, C.; et al. Terminally differentiated CD8+ T cells negatively affect bone regeneration in humans. *Sci. Transl. Med.* **2013**, *5*, 177ra36. [CrossRef]
21. Lei, H.; Schmidt-Bleek, K.; Dienelt, A.; Reinke, P.; Volk, H. Regulatory T cell-mediated anti-inflammatory effects promote successful tissue repair in both indirect and direct manners. *Front. Pharmacol.* **2015**, *6*, 184. [CrossRef] [PubMed]
22. Meng, X.; Yang, J.; Dong, M.; Zhang, K.; Tu, E.; Gao, Q.; Chen, W.; Zhang, C.; Zhang, Y. Regulatory T cells in cardiovascular diseases. *Nat. Rev. Cardiol.* **2015**, *13*, 167–179. [CrossRef] [PubMed]
23. Burzyn, D.; Kuswanto, W.; Kolodin, D.; Shadrach, J.; Cerletti, M.; Jang, Y.; Sefik, E.; Tan, T.; Wagers, A.; Benoist, C.; et al. A special population of regulatory T cells potentiates muscle repair. *Cell* **2013**, *155*, 1282–1295. [CrossRef] [PubMed]

24. Wynn, T.; Vannella, K. Macrophages in tissue repair, regeneration, and fibrosis. *Immunity* **2016**, *44*, 450–462. [CrossRef]
25. Smith, T.; Nagalla, R.; Chen, E.; Liu, W. Harnessing macrophage plasticity for tissue regeneration. *Adv. Drug Deliv. Rev.* **2017**, *114*, 193–205. [CrossRef]
26. Huleihel, L.; Bartolacci, J.; Dziki, J.; Vorobyov, T.; Arnold, B.; Scarritt, M.; Pineda Molina, C.; LoPresti, S.; Brown, B.; Naranjo, J.; et al. Matrix-bound nanovesicles recapitulate extracellular matrix effects on macrophage phenotype. *Tissue Eng. Part A* **2017**, *23*, 1283–1294. [CrossRef]
27. Huleihel, L.; Dziki, J.; Bartolacci, J.; Rausch, T.; Scarritt, M.; Cramer, M.; Vorobyov, T.; LoPresti, S.; Swineheart, I.; White, L.; et al. Macrophage phenotype in response to ECM bioscaffolds. *Semin. Immunol.* **2017**, *29*, 2–13. [CrossRef]
28. Fishman, J.; Lowdell, M.; Urbani, L.; Ansari, T.; Burns, A.; Turmaine, M.; North, J.; Sibbons, P.; Seifalian, A.; Wood, K.; et al. Immunomodulatory effect of a decellularized skeletal muscle scaffold in a discordant xenotransplantation model. *Proc. Natl. Acad. Sci. USA* **2013**, *110*, 14360–14365. [CrossRef]
29. Badylak, S.; Gilbert, T. Immune response to biologic scaffold materials. *Semin. Immunol.* **2008**, *20*, 109–116. [CrossRef]
30. Wu, R.; He, X.; Zhu, J.; Yin, Y.; Li, X.; Liu, X.; Chen, F. Modulating macrophage responses to promote tissue regeneration by changing the formulation of bone extracellular matrix from filler particles to gel bioscaffolds. *Mater. Sci. Eng. C* **2019**, *101*, 330–340. [CrossRef]
31. Ciocci, M.; Mochi, F.; Carotenuto, F.; Di Giovanni, E.; Prosposito, P.; Francini, R.; De Matteis, F.; Reshetov, I.; Casalboni, M.; Melino, S.; et al. Scaffold-in-scaffold potential to induce growth and differentiation of cardiac progenitor cells. *Stem Cells Dev.* **2017**, *26*, 1438–1447. [CrossRef] [PubMed]
32. Mehrban, N.; Pineda Molina, C.; Quijano, L.; Bowen, J.; Johnson, S.; Bartolacci, J.; Chang, J.; Scott, D.; Woolfson, D.; Birchall, M.; et al. Host macrophage response to injectable hydrogels derived from ECM and α-helical peptides. *Acta Biomater.* **2020**, *111*, 141–152. [CrossRef] [PubMed]
33. Oliveira, M.; Santos, S.; Oliveira, M.; Torres, A.; Barbosa, M. Chitosan drives anti-inflammatory macrophage polarisation and pro-inflammatory dendritic cell stimulation. *Eur. Cells Mater.* **2012**, *24*, 136–153. [CrossRef]
34. Blakney, A.; Swartzlander, M.; Bryant, S. The effects of substrate stiffness on the in vitro activation of macrophages and in vivo host response to poly(ethylene glycol)-based hydrogels. *J. Biomed. Mater. Res. A* **2012**, *100*, 1375–1386. [CrossRef] [PubMed]
35. Tylek, T.; Blum, C.; Hrynevich, A.; Schlegelmilch, K.; Schilling, T.; Dalton, P.; Groll, J. Precisely defined fiber scaffolds with 40 μm porosity induce elongation driven M2-like polarization of human macrophages. *Biofabrication* **2020**, *12*, 025007. [CrossRef] [PubMed]
36. Karazisis, D.; Petronis, S.; Agheli, H.; Emanuelsson, L.; Norlindh, B.; Johansson, A.; Rasmusson, L.; Thomsen, P.; Omar, O. The influence of controlled surface nanotopography on the early biological events of osseointegration. *Acta Biomater.* **2017**, *53*, 559–571. [CrossRef]
37. Luu, T.U.; Gott, S.C.; Woo, B.W.; Rao, M.P.; Liu, W.F. Micro-and nanopatterned topographical cues for regulating macrophage cell shape and phenotype. *ACS Appl. Mater. Interfaces* **2015**, *7*, 28665–28672. [CrossRef]
38. Sridharan, R.; Cavanagh, B.; Cameron, A.; Kelly, D.; O'Brien, F. Material stiffness influences the polarization state, function and migration mode of macrophages. *Acta Biomater.* **2019**, *89*, 47–59. [CrossRef]
39. Sridharan, R.; Ryan, E.; Kearney, C.; Kelly, D.; O'Brien, F. Macrophage polarization in response to collagen scaffold stiffness is dependent on cross-linking agent used to modulate the stiffness. *ACS Biomater. Sci. Eng.* **2018**, *5*, 544–552. [CrossRef]
40. Jia, Z.; Cheng, Y.; Jiang, X.; Zhang, C.; Wang, G.; Xu, J.; Li, Y.; Peng, Q.; Gao, Y. 3D culture system for liver tissue mimicking hepatic plates for improvement of human hepatocyte (C3A) function and polarity. *BioMed Res. Int.* **2020**, *2020*, 6354183. [CrossRef]
41. Corsi, F.; Caputo, F.; Traversa, E.; Ghibelli, L. Not only redox: The multifaceted activity of cerium oxide nanoparticles in cancer prevention and therapy. *Front. Oncol.* **2018**, *8*, 309. [CrossRef] [PubMed]
42. Hasan, A.; Morshed, M.; Memic, A.; Hassan, S.; Webster, T.; Marei, H. Nanoparticles in tissue engineering: Applications, challenges and prospects. *Int. J. Nanomed.* **2018**, *13*, 5637–5655. [CrossRef] [PubMed]
43. Choi, J.; Lee, S.; Christ, G.; Atala, A.; Yoo, J. The influence of electrospun aligned poly(ε-caprolactone)/collagen nanofiber meshes on the formation of self-aligned skeletal muscle myotubes. *Biomaterials* **2008**, *29*, 2899–2906. [CrossRef] [PubMed]

44. Mehanna, M.M.; Mohyeldin, S.M.; Elgindy, N.A. Respirable nanocarriers as a promising strategy for antitubercular drug delivery. *J. Control. Release* **2014**, *187*, 183–197. [CrossRef] [PubMed]
45. Caso, M.; Carotenuto, F.; Di Nardo, P.; Migliore, A.; Aguilera, A.; Lopez, C.; Venanzi, M.; Cavalieri, F.; Rinaldi, A. Nanoporous microsponge particles (NMP) of polysaccharides as universal carriers for biomolecules delivery. *Nanomaterials* **2020**, *10*, 1075. [CrossRef] [PubMed]
46. Firouzi-Amandi, A.; Dadashpour, M.; Nouri, M.; Zarghami, N.; Serati-Nouri, H.; Jafari-Gharabaghlou, D.; Karzar, B.; Mellatyar, H.; Aghebati-Maleki, L.; Babaloo, Z.; et al. Chrysin-nanoencapsulated PLGA-PEG for macrophage repolarization: Possible application in tissue regeneration. *Biomed. Pharmacother.* **2018**, *105*, 773–780. [CrossRef] [PubMed]
47. Bartneck, M.; Ritz, T.; Keul, H.; Wambach, M.; Bornemann, J.; Gbureck, U.; Ehling, J.; Lammers, T.; Heymann, F.; Gassler, N.; et al. Peptide-functionalized gold nanorods increase liver injury in hepatitis. *ACS Nano* **2012**, *6*, 8767–8777. [CrossRef]
48. Wang, L.; Zhang, H.; Sun, L.; Gao, W.; Xiong, Y.; Ma, A.; Liu, X.; Shen, L.; Li, Q.; Yang, H. Manipulation of macrophage polarization by peptide-coated gold nanoparticles and its protective effects on acute lung injury. *J. Nanobiotechnol.* **2020**, *18*, 38. [CrossRef]
49. Raimondo, T.; Mooney, D. Functional muscle recovery with nanoparticle-directed M2 macrophage polarization in mice. *Proc. Natl. Acad. Sci. USA* **2018**, *115*, 10648–10653. [CrossRef]
50. Shen, X.; Yu, Y.; Ma, P.; Luo, Z.; Hu, Y.; Li, M.; He, Y.; Zhang, Y.; Peng, Z.; Song, G.; et al. Titania nanotubes promote osteogenesis via mediating crosstalk between macrophages and MSCs under oxidative stress. *Colloid Surf. B* **2019**, *180*, 39–48. [CrossRef]
51. Lee, C.; Kim, Y.; Jang, J.; Park, J. Modulating macrophage polarization with divalent cations in nanostructured titanium implant surfaces. *Nanotechnology* **2016**, *27*, 085101. [CrossRef] [PubMed]
52. Xu, W.C.; Dong, X.; Ding, J.L.; Liu, J.C.; Xu, J.J.; Tang, Y.H.; Yi, Y.P.; Lu, C.; Yang, W.; Yang, J.S.; et al. Nanotubular TiO_2 regulates macrophage M2 polarization and increases macrophage secretion of VEGF to accelerate endothelialization via the ERK1/2 and PI3K/AKT pathways. *Int J. Nanomed.* **2019**, *14*, 441–455. [CrossRef] [PubMed]
53. Li, K.; Shen, Q.; Xie, Y.; You, M.; Huang, L.; Zheng, X. Incorporation of cerium oxide into hydroxyapatite coating regulates osteogenic activity of mesenchymal stem cell and macrophage polarization. *J. Biomater. Appl.* **2016**, *31*, 1062–1076. [CrossRef] [PubMed]
54. Li, J.; Wen, J.; Li, B.; Li, W.; Qiao, W.; Shen, J.; Jin, W.; Jiang, X.; Yeung, K.; Chu, P. Valence state manipulation of cerium oxide nanoparticles on a titanium surface for modulating cell fate and bone formation. *Adv. Sci.* **2017**, *5*, 1700678. [CrossRef]
55. Coletti, D.; Scaramuzzo, F.; Montemiglio, L.; Pristerà, A.; Teodori, L.; Adamo, S.; Barteri, M. Culture of skeletal muscle cells in unprecedented proximity to a gold surface. *J. Biomed. Mater. Res. A* **2009**, *91*, 370–377. [CrossRef]
56. Ko, W.; Kim, S.; Heo, D.; Han, I.; Kim, S.; Kwon, I.; Sohn, S. Double layers of gold nanoparticles immobilized titanium implants improve the osseointegration in rabbit models. *Nanomed. Nanotechnol. Biol. Med.* **2020**, *24*, 102129. [CrossRef]
57. Ni, C.; Zhou, J.; Kong, N.; Bian, T.; Zhang, Y.; Huang, X.; Xiao, Y.; Yang, W.; Yan, F. Gold nanoparticles modulate the crosstalk between macrophages and periodontal ligament cells for periodontitis treatment. *Biomaterials* **2019**, *206*, 115–132. [CrossRef]
58. Liu, M.; Li, N.; He, Y.; Ge, Y.; Song, G. Dually emitting gold-silver nanoclusters as viable ratiometric fluorescent probes for cysteine and arginine. *Microchim. Acta* **2018**, *185*, 147. [CrossRef]
59. Ishizaki, K.; Sugita, Y.; Iwasa, F.; Minamikawa, H.; Ueno, T.; Yamada, M.; Suzuki, T.; Ogawa, T. Nanometer-thin TiO_2 enhances skeletal muscle cell phenotype and behavior. *Int. J. Nanomed.* **2011**, *2011*, 2191–2203.
60. Lucarelli, M.; Gatti, A.M.; Savarino, G.; Quattroni, P.; Martinelli, L.; Monari, E.; Boraschi, D. Innate defence functions of macrophages can be biased by nano-sized ceramic and metallic particles. *Eur. Cytokine Netw.* **2004**, *15*, 339–346.
61. Caputo, F.; Giovanetti, A.; Corsi, F.; Maresca, V.; Briganti, S.; Licoccia, S.; Traversa, E.; Ghibelli, L. Cerium oxide nanoparticles re-establish cell integrity checkpoints and apoptosis competence in irradiated HaCaT cells via novel redox-independent activity. *Front. Pharmacol.* **2018**, *9*, 1183. [CrossRef] [PubMed]

62. Chigurupati, S.; Mughal, M.; Okun, E.; Das, S.; Kumar, A.; McCaffery, M.; Seal, S.; Mattson, M. Effects of cerium oxide nanoparticles on the growth of keratinocytes, fibroblasts and vascular endothelial cells in cutaneous wound healing. *Biomaterials* **2013**, *34*, 2194–2201. [CrossRef] [PubMed]
63. Siposova, K.; Huntosova, V.; Shlapa, Y.; Lenkavska, L.; Macajova, M.; Belous, A.; Musatov, A. Advances in the study of cerium oxide nanoparticles: New insights into antiamyloidogenic activity. *ACS Appl. Bio Mater.* **2019**, *2*, 1884–1896. [CrossRef]
64. Walczynska, M.; Jakubowski, W.; Wasiak, T.; Kadziola, K.; Bartoszek, N.; Kotarba, S.; Siatkowska, M.; Komorowski, P.; Walkowiak, B. Toxicity of silver nanoparticles, multiwalled carbon nanotubes, and dendrimers assessed with multicellular organism Caenorhabditis elegans. *Toxicol. Mech. Methods* **2018**, *28*, 432–439. [CrossRef] [PubMed]

© 2020 by the authors. Licensee MDPI, Basel, Switzerland. This article is an open access article distributed under the terms and conditions of the Creative Commons Attribution (CC BY) license (http://creativecommons.org/licenses/by/4.0/).

nanomaterials

MDPI

Review

Smart ECM-Based Electrospun Biomaterials for Skeletal Muscle Regeneration

Sara Politi [1,2,†], Felicia Carotenuto [1,2,3,†], Antonio Rinaldi [4], Paolo Di Nardo [2,3,5], Vittorio Manzari [2], Maria Cristina Albertini [6], Rodolfo Araneo [7], Seeram Ramakrishna [8,‡] and Laura Teodori [1,3,*,‡]

1 Department of Fusion and Technologies for Nuclear Safety and Security, Diagnostic and Metrology (FSN-TECFIS-DIM), ENEA, CR Frascati, 00044 Rome, Italy; sara.politi@uniroma2.it (S.P.); carotenuto@med.uniroma2.it (F.C.)
2 Department of Clinical Science and Translational Medicine, University of Rome "Tor Vergata", 00133 Rome Italy; dinardo@uniroma2.it (P.D.N.); manzari@uniroma2.it (V.M.)
3 Interdepartmental Center for Regenerative Medicine (CIMER), University of Rome "Tor Vergata", 00133 Rome, Italy
4 Department of Sustainability (SSPT), ENEA, 00123 Rome, Italy; antonio.rinaldi@gmail.com
5 L.L. Levshin Institute of Cluster Oncology, I. M. Sechenov First Medical University, Moscow 119991, Russia
6 Department of Biomolecular Sciences, Urbino University "Carlo Bo", 61029 Urbino, Italy; maria.albertini@uniurb.it
7 Department of Astronautics Electrical and Energy Engineering (DIAEE), University of Rome "La Sapienza", 00184 Rome, Italy; rodolfo.araneo@uniroma1.it
8 Centre for Nanofibers and Nanotechnology, Department of Mechanical Engineering, National University of Singapore, Singapore 119260, Singapore; seeram@nus.edu.sg
* Correspondence: laura.teodori@enea.it; Tel.: +39-06-9400-5642
† These authors share the co-first authorship.
‡ These authors shared last authorship as senior authors.

Received: 6 August 2020; Accepted: 5 September 2020; Published: 9 September 2020

Abstract: The development of smart and intelligent regenerative biomaterials for skeletal muscle tissue engineering is an ongoing challenge, owing to the requirement of achieving biomimetic systems able to communicate biological signals and thus promote optimal tissue regeneration. Electrospinning is a well-known technique to produce fibers that mimic the three dimensional microstructural arrangements, down to nanoscale and the properties of the extracellular matrix fibers. Natural and synthetic polymers are used in the electrospinning process; moreover, a blend of them provides composite materials that have demonstrated the potential advantage of supporting cell function and adhesion. Recently, the decellularized extracellular matrix (dECM), which is the noncellular component of tissue that retains relevant biological cues for cells, has been evaluated as a starting biomaterial to realize composite electrospun constructs. The properties of the electrospun systems can be further improved with innovative procedures of functionalization with biomolecules. Among the various approaches, great attention is devoted to the "click" concept in constructing a bioactive system, due to the modularity, orthogonality, and simplicity features of the "click" reactions. In this paper, we first provide an overview of current approaches that can be used to obtain biofunctional composite electrospun biomaterials. Finally, we propose a design of composite electrospun biomaterials suitable for skeletal muscle tissue regeneration.

Keywords: smart biomaterials; electrospinning; biofunctionalization; decellularized extracellular matrix (dECM); skeletal muscle regeneration; click chemistry

1. Introduction

Biomaterials play a prominent role in regenerative medicine and tissue engineering (TE) through the development of functional systems to improve or restore biological functions of damaged tissues. The current research focuses on the design of stimuli-responsive, smart, and intelligent biomaterials systems that are able to modulate their physical, chemical, and mechanical properties in response to external chemical or physical stimuli and changes in the physiological environments adapting their functionality accordingly.

The development of regenerative biomaterials and the progress in their processing represent key factors to generate smart biomimetic scaffolds resembling the structural organization and activity of native tissue in order to guide tissue regeneration [1,2].

Nowadays, researchers aim to develop tissue-specific scaffolds characterized by desired topographical mechanical and physical features. Among them, the skeletal muscle represents a complex and challenging tissue to be generated in vitro for tissue engineering purposes.

Skeletal muscle is a complex system of oriented muscle fibers acting together to produce a contractile force and support body movement. In addition, the proper function of skeletal muscle encompasses breathing, metabolic control, thermoregulation, and energy storage [3]. Skeletal muscle has innate regenerative potential following injury or disease. However, endogenous self-repair is severely impaired due to volume traumatic muscle loss.

Severe traumatic injuries of skeletal muscles are responsible for functional deficits in patients. Owing to the wide prevalence of these injuries and the associated socio-economic implications, muscle regeneration has been a topic of scientific and clinical interests [4]. The potential strategies for designing a suitable biomaterial for skeletal muscle regeneration should take into account important physical, biochemical, and inflammatory cues that effectively affect cell adhesion and proliferation and thus guide muscle regeneration. Among the most important properties that biomaterials should possess for successful skeletal muscle TE include: porosity, aligned architecture, and bioavailability of bioactive molecules to promote the cell activity [3].

Many studies have demonstrated that the advanced manufacturing of biomimetic scaffolds is a crucial aspect of a successful regenerative process [5,6]. Scaffolding strategies aim to create ideal scaffolds with mechanical, chemical, and biological properties that mimic the composition and the structure of extracellular matrix (ECM) of native tissue to encourage cell adhesion and proliferation [7,8]. The ECM represents the major structural component of the human body, and it is composed of a three-dimensional arrangement of natural polymers such as collagen, elastin, and fibrinogen as well as a mixture of macromolecules such as growth factors, peptides, glycoproteins, and proteoglycans. However, the ECM is not only a structural framework for tissues but through interactions with receptors on the surfaces of cells, it plays an important role in both day-to-day cellular activity and in tissue regeneration [9,10]. For these reasons, the engineering of an ECM-mimicking scaffold is extremely challenging.

In the native tissues, most ECM components consist of interwoven fibrous structures in the micronanoscale range and thus, the fabrication of scaffolds mimicking ECM structural organization is an active area of research in TE. To date, phase separation, self-assembly, and electrospinning have been used to make scaffolds with a fibrous network [11]. Among these techniques, electrospinning has continued to be the most commonly used.

Electrospinning is a powerful and scalable production method [12], which allows the fabrication of micro- or nanofibers possessing large surface areas and high porosity, which are favorable for biomedical applications in terms of cellular interactions [13,14].

A typical electrospinning set-up consists of a high-voltage power supply, a syringe with a needle, and an electrically conductive collector. During the process, a high voltage is applied to the needle of a syringe, which contains a polymeric solution, and a collector. Once the electric field reaches a critical value at which the repulsive electric force overcomes the surface tension of polymeric solution, this is ejected from the syringe needle. The fibers are formed during the fast evaporation of the solvent and are deposited on the collector [14].

The advantages of electrospinning for scaffold fabrication are its versatility, flexibility, the possibility to use different materials combinations, and malleability to conform over a wide variety of form factors (e.g., sizes and shapes) [15].

However, during a typical electrospinning process, fibers are densely deposited on a collector forming closed packed fibers that are associated with poor cell infiltration. To date, many interesting approaches are investigated to enhance porosity in electrospun scaffolds and to acquire pores of the suitable size in order to obtain greater cell infiltration [3,16].

In particular, to control the fiber arrangement, topography, morphology, and the overall performance of electrospun polymeric scaffolds for TE applications, different electrospinning processing factors, such as applied voltage, tip of needle-collector distance, solution viscosity, and feed rate need to be optimized [17]. The role of main effects and interactions between process parameters can be effectively captured by regression and design of experiments methods [18,19]. Randomly oriented or aligned fibers can be formed by using a stationary or rotating collector, respectively. The orientation of fibers is an important feature of an ideal and promising scaffold because this aspect greatly influences cell growth and related functions in cells such as nerve and smooth muscle cells [20]. In particular, the creation of polymeric aligned electrospun biomaterials that mimic skeletal muscle allows the efficient organization of muscle cells to form aligned myotubes during muscle regeneration [3].

Over the past two decades, electrospun scaffolds have been obtained using natural polymers, synthetic polymers, or a combination of those [21–23] according to the desired combination of mechanical and chemical properties that ultimately dictate the biological response toward the targeted tissue regeneration application.

Recently, the development of the decellularization processes of organs and tissues has made possible the creation of an electrospun scaffold prepared from processing methods of a native extracellular matrix. The decellularized extracellular matrix (dECM) is an organ or a tissue devoid of its cellular content but which preserves a rather large portion of its original composition including its architecture, structural organization, and biochemical cues [24,25].

This dECM material can be used as a scaffold that maintains its original geometry directly in medical interventions or can be repopulated with cells before use. More recently, dECMs have been further processed to generate dECM products that can be used as a starting material for biofabrication techniques such as electrospinning [26]. Indeed, electrospun dECM scaffolds are typically achieved by adding components of dECM after the electrospinning process or by direct electrospinning dECM components [27]. However, during processing to obtain the dECM materials, some key components such as glycosaminoglycans (GAGs) can be lost. Indeed, GAG concentrations are especially important for their interaction with growth factors and chemokines, influencing cell signaling [28].

To overcome these obstacles, some authors are developing ad hoc decellurization protocols for specific tissues and the optimization of subsequent dECM process methods with the aim of avoiding the loss of some important bioactive components of the native ECM [29,30].

Another effective way to ensure the bioactivity of the electrospun scaffolds is the judicious combination of electrospun fibers with appropriate biomolecules, such as small molecules, growth factors, short peptides, or proteins [31], in order to stimulate a specific cell response.

In this context, the development of innovative and forefront electrospinning techniques and an accurate choice of polymeric materials and biomolecules enables the loading of bioactive components during the manufacturing process. A local and controlled release of biomolecules enhances the functional properties of scaffold and influence surrounding tissue regeneration.

Furthermore, the regeneration process is strongly influenced by the interactions between cells and biomaterial surfaces. For this reason, the postelectrospinning modification of the high surface of fibers with biomolecules represents a suitable method in order to enhance cell adhesion and organization [22,32]. Such postelectrospinning modifications can be achieved by the physical adsorption or chemical bonding of biomolecules to the fibers. In this context, the biofunctionalization of the fiber surface using "click" reactions has drawn significant interest recently, owing to their simple reaction conditions, high reaction rate, and high chemical selectivity [33,34].

In this paper, we discuss current approaches that can be used to obtain functional composite electrospun biomaterials with a focus on the possible strategies to achieve bioactive scaffolds. The main steps discussed concern (i) obtaining composite electrospun biomaterials with particular attention to the use of native ECM in the solution or fusion of electrospinning; (ii) the incorporation of bioactive molecules during electrospinning (bulk functionalization) and the surface modification of electrospun systems through biomolecules; (iii) biofunctionalization postprocessing using "click" reactions, as schematically represented in Figure 1.

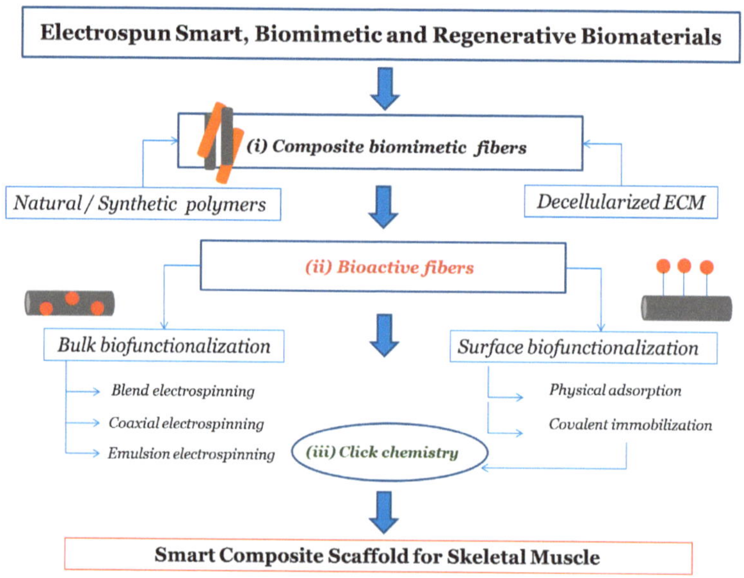

Figure 1. Flow chart for designing process of smart electrospun scaffold for skeletal muscle regeneration. The main steps concerns (i) composite electrospun biomaterials; (ii) the incorporation (bulk biofunctionalization) and surface attachment of bioactive molecules (surface biofunctionalization) during and after electrospinning process, respectively; (iii) surface biofunctionalization postprocessing using "click" reactions.

Finally, we designed a strategy for dECM-based composite bioactive electrospun scaffolds to promote skeletal muscle regeneration.

2. Biomaterials for Electrospinning in Tissue Engineering

2.1. Synthetic and Natural Polymers

The choice of polymeric materials is the first design step for the electrospun scaffold. Natural and synthetic polymers have been widely used to realize scaffolds due to their properties such as biodegradation, mechanical properties, high porosity and surface to volume ratio, and also small pore size [8,35].

In particular, the growing success of synthetic polymers as scaffolding materials is due to their facile synthesis and processing flexibility, enabling good reproducibility and tuneable characteristics. Polyesters (e.g., poly(ε-caprolactone) (PCL), polylactic acid (PLA), polyglycolic acid (PGA), etc.), and polyethers (e.g., polyethylene oxide (PEO), polyurethane (PU), etc.) have been electrospun for TE applications [21,22]. Furthermore, electrospun fibers made of shape memory polymers (SMPs) have also been obtained thanks to the optimization of the experimental parameters [36].

Despite the excellent biodegradability, chemical, and mechanical properties of electrospun nanofibrous scaffolds from synthetic polymers, they often require further modification to their surface and structure to promote their biofunctionality, since synthetic biomaterials lack bioactive functional sites and result in poor biochemical similarities with ECM [37].

Conversely, compared to synthetic polymers, natural polymers offer intrinsic similarity to ECM components and bring with them a biological signature that is advantageous to support cell adhesion and proliferation [8,21,38]. Collagen, gelatin, chitosan, hyaluronic acid, cellulose, and glycans in general have in fact been proposed to produce electrospun fibers in TE because of their biocompatibility and low immunogenicity [21,39]. However, these scaffolds usually lack load-bearing capability due to lower mechanical strength and higher degradation rates than synthetic polymers [21,22].

It is worth noting from a manufacturing standpoint that the electrospinning of natural polymers is very often possible with water-based recipes and with minimal or no recourse to organic solvent, which is a definitive advantage in the ongoing societal transition to green chemistry and circular economy [40] and is only possible with a handful of synthetic counterparts (e.g., PEO, PVOH).

2.2. Composite Polymeric Electrospun Fibers

A blend of polymers can be an attractive option to mix the benefits and overcome the limitations of the individual one-component systems described above. In particular, the combination of natural and synthetic polymers is a means to effectively improve the biocompatibility, mechanical, and structural properties of the scaffold to boost cell adhesion and growth and to promote tissue regeneration [41].

A homogeneous blend composed of a selected mixture of natural and synthetic polymers can be designed and electrospun to fabricate composite scaffolds that are endowed with physicochemical properties of its components, such as hydrophilicity/hydrophobicity, surface charge, and mechanical strength of synthetic polymers as well as the biological features of natural fibers [22,23].

Focusing on binary compound systems, several attempts are found in literature, especially featuring the combination of PCL with another natural component. For example, Zhang et al. [42] fabricated a composite electrospun scaffold based on PCL and gelatin, which exhibited enhanced mechanical properties and wettability (vs. water) compared to those obtained from either PCL or gelatin alone. In addition, cell culture experiments highlighted favorable interactions between the composite scaffold and bone marrow stromal cells (BMSCs), suggesting the potential use of composite gelatin/PCL nanofibers in TE applications.

Other combinations of PCL with collagen [43] and chitosan [44] are reported. The results invariably showed an improvement in the biocompatibility and mechanical properties of the composite systems.

In general, electrospun PCL-based composites are characterized by modulable mechanical properties, namely: flexural strain, tensile strain, stiffness (i.e., Young's modulus and in-plane stiffness at the mesoscale), and thermomechanical strength [18,45,46] for biomedical treatments, particularly important in the implantation of artificial bone and muscle tissues regeneration. In another study, PCL and collagen type I-based systems were investigated for creating implantable engineered muscle tissue that closely mimics the native tissue [47]. The results suggested that aligned PCL/collagen nanofibers facilitate skeletal muscle cell organization and myotube formation as compared to randomly oriented nanofibers. Moreover, the stiffness of the electrospun PCL-based composites makes these scaffolds promising for in-vivo applications [48].

Apart from PCL, other binary synthetic-natural systems are possible. Biomimetic and biocompatible nanofibrous scaffolds of polyamide-6,6 (PA 6,6) blended with chitosan were fabricated for applications in bone TE [49]. In particular, the effects of three different weight percentages of chitosan were investigated. The results showed that the increase in the concentration of chitosan enhances the performances of the composite scaffold in terms of cell growth, adhesion, differentiation, and proliferation.

Additionally, composite electrospun fibers containing gelatin and synthetic polymers exhibited attractive physicochemical, biomechanical, and biocompatibility properties that mimic the important features of natural ECM [50].

Other synthetic polymers such as poly(vinyl alcohol) (PVA) [51] and poly(lactic-co-glycolic acid) (PLGA) [52] were used in combination with natural polymers as electrospun scaffolds, reporting promising results in TE applications.

Recent advances in the field of functional polymeric biomaterials systems using a combination of synthetic and natural polymers enabled the fabrication of composite electrospun fibers characterized by promising and peculiar properties for tissue regeneration process.

2.3. Decellularized Extracellular Matrix (dECM)-Based Electrospun Fibers

In recent years, the research in scaffold engineering has shifted from using natural polymers to using the decellularized extracellular matrix (dECM) to obtain scaffolds mimicking native ECM [53]. Methods of decellularizing tissues have been extensively reviewed [54,55].

In general, decellularization procedures utilize mechanical, enzymatic, and chemical means to remove cellular material, that could induce an unwanted immune response and the rejection of dECM-based biomaterials and retain the intricate mixture of structural and functional elements of ECM.

However, the challenge of each decellularization method is to completely remove the cellular component and DNA content without removing and/or damaging structural components and functional ECM proteins such as glycosaminoglycans and growth factors [29].

Nowadays, experimental decellularization protocols vary widely between tissue and organ types, and optimization of the processes is often required in optimal decellularization of the native extracellular matrix. In addition, further processing of dECM allows the production of powders or hydrogels that can be further treated in order to obtain biomaterials suitable for regenerative applications.

Indeed, the use of dECM products provides an ideal microenvironment for cells, with ample biological and chemical cues necessary to regulate cell behavior [56]. In addition, dECM was able to support the correct phenotypic differentiation of progenitor cells and the maintenance of tissue-specific cell phenotypes.

The limitations of dECM products involve a low reproducibility derived from batch-to-batch variability across different donors, and limited capacity to tune properties, e.g., mechanical stability, porosity, stiffness, and degradability [26]. In order to improve and refine the physical-chemical properties of dECM systems, a viable solution is their combination with natural and/or synthetic polymers to create composite biomaterials.

The mechanical properties, as well as the internal architecture of the scaffold, may play an important role in the regeneration of complex tissue such as skeletal muscle tissue, influencing cell proliferation, alignment, and myogenic differentiation. For this reason, electrospinning may provide a promising method for the fabrication of dECM-based scaffolds allowing for the modulation of architecture and mechanical properties [3].

In particular, dECM from different tissues of origin may be processed in order to obtain a suitable product that can be directly loaded into an electrospinning polymeric solution to fabricate composite scaffolds that combine the versatility of polymeric materials and the biological complexity of natural ECM. Baguiera et al. [57] decellularized rat brains and blended them with gelatin before electrospinning. The obtained fibrous scaffolds provide a suitable microenvironment for mesenchymal stromal cell adhesion, proliferation, and survival. A similar approach was reported to obtain an electrospun scaffold composed of decellularized porcine cardiac tissue and poly(ethylene oxide) [58]. In another work, Gao et al. [59] reported the fabrication and characterization of an electrospun scaffold composed of a decellularized meniscus extracellular matrix and PCL.

Decellularized skeletal muscle has been used in a number of different forms, but few studies have used processed dECM using electrospinning techniques.

For example, Patel et al. [60] investigated the physical and mechanical properties of aligned nanofibers of decellularized muscle tissue and PCL. In vitro testing showed that the electrospun scaffold supports satellite cell growth, myogenic protein expression, and myokine production. This latter result suggests that dECM proteins provide the cues for attachment and growth of satellite cells onto the nanofibers, while the presence of PCL assures structural integrity and elasticity to the scaffold. A study developed a method to fabricate electrospun scaffolds from the decellularized skeletal muscle without the need for a carrier polymer is noteworthy [30]. The resulting scaffolds showed tunable physicochemical properties, including fiber alignment, while important extracellular matrix components for regeneration such as GAGs, were preserved.

3. Bioactivity and Biofunctionalization of Electrospun Scaffolds

In recent years, bioactive scaffolds obtained through the combination of polymeric materials and biomolecules attracted great attention due to the capability to express biological signals and, thus, to support cellular activity and promote tissue regeneration acting as molecules/drug delivery systems. The biomolecules range from growth factors, proteins, short peptides, genes, and enzymes, such that electrospun scaffolds can be functionalized largely and by several methods. Bioactive components can be both internalized (encapsulated) into the fibrous scaffold during the electrospinning and attached on the fiber surface in a postprocessing step. Some strategies are examined.

3.1. Bulk Biofunctionalization

The biomolecules can be introduced directly into fibers during the fabrication process. A simple method is to blend the bioactive molecules directly with the polymeric solution prior to the electrospinning process (blend electrospinning) [31]. In this way, bioactive components are dispersed into the electrospun scaffolds and can provide continued or controlled release of drugs/biomolecules for tissue regeneration. A drawback is the potential loss or alteration of the activity of the incorporated biomolecules caused by voltage or interaction with polymeric solution (e.g., in case of organic solvents such as chloroform, DMF, methanol, dichloromethane used for synthetic polymers such as PCL).

To overcome this inconvenience, loading biomolecules such as proteins or enzymes could require the use of more complex techniques, such as emulsion or coaxial electrospinning. These approaches enable to produce multiphase core-shell fibers, which allow a high loading capacity of bioactive molecules into the core and thus the preservation of their activity. In coaxial electrospinning, two different solutions are coaxially and simultaneous electrospun through different annular channels of a single needle. Furthermore, the peculiar coaxial structure is essential to control the release rate and maintain the mechanical and biological properties of the scaffold [61].

Emulsion electrospinning is a novel and simple technique to fabricate core-shell nanofibers, and either water-in-oil or oil-in-water emulsions can be electrospun to directly encapsulate hydrophilic or hydrophobic bioactive compounds into core-shell fibers, respectively [62].

Table 1 summarized some representative bioactive electrospun scaffolds characteristics in terms of polymeric matrices used, loaded biomolecules and the method of electrospinning used.

Table 1. Bioactive electrospun scaffolds obtained with different methods of preparation, in chronological order.

Polymeric Component	Loaded Biomolecules	Method of Preparation	Reference
PLGA	bFGF	Coaxial electrospinning	[63]
PVA core PCL shell	GF loaded liposomes	Coaxial electrospinning	[64]
PCLC	VEGF	Emulsion electrospinning	[65]
PELCL core PELCL shell	VEGF PDGF	Coaxial electrospinning	[66]
PCL	bFGF	Emulsion electrospinning	[67]
PCL	VEGF	Blend electrospinning	[68]
PVA core PLA shell	CTGF	Coaxial electrospinning	[69]
PLA	PDGF	Coaxial electrospinning	[70]

Abbreviation: PLGA, poly(lactic-co-glycolic acid); bFGF, basic fibroblast growth factor; PVA, poly(vinyl alcohol); PCL, poly(ε-caprolactone); GF, growth factor; PLCL, poly(L-lactic acid-co-ε-caprolactone); VEGF, vascular endothelial growth factor; PELCL poly(ethyleneglycol)-b-poly(L-lactide-co-caprolactone); PDGF, platelet-derived growth factor; PLA, polylacticacid; CTGF, connective tissue growth factor.

3.2. *Surface Biofunctionalization and Click Chemistry*

Surface biofunctionalization strategies aim to immobilize biomolecules such as proteins, peptides and polysaccharides, or bioactive drugs onto the surface of the electrospun fibers to modulate the interactions between biomaterial surfaces and biological systems. In order to promote the interaction between biomolecules and scaffold polymers, often it is necessary to perform surface activation. Relatively sophisticated surface modification can be accomplished via plasma treatment or wet chemical etching, which generate specific functional groups on the surface [11,32]. An ideal surface biofunctionalization process should be selective with easy reaction conditions avoiding the degradation of fiber, and assuring sufficient yields with a defined release profile of biomolecules without loss of their activity. This can be achieved via adsorption techniques, usually following the two fundamental strategies below.

- Physical adsorption is a simple approach that involves incubating the scaffold in a solution containing biomolecules. The biomolecules attach onto the scaffold surface owing to surface interactions, e.g., electrostatic forces, van der Waals forces, and hydrogen bonds.

In this context, electrospun nanofibrous PCL scaffolds combined with the osteogenetic growth factor bone, morphogenetic protein 2 (BMP-2), in association with chitosan by the layer by-layer (LbL) method have been shown to stimulate the regeneration of bone [71]. Layer-by-layer (LbL) is a simple and versatile technique for the realization of multimaterial coatings on the different substrate surfaces, which involves the sequential surface adsorption of alternating layers of oppositely charged polyelectrolytes [72]. Peptides and proteins chemically bound to polyelectrolytes, incorporated in polyelectrolyte multilayer architecture, have been shown to retain their biological activities [73].

Furthermore, the characteristics of electrospun systems such as fiber alignment, size, and porosity play a significant role in protein adsorption [74]. The main advantages are the simplicity of the procedure and the limited damage to fibers and biomolecules. In addition, the interaction could be relatively weak and the release of these biomolecules quick and uncontrolled.

- Chemical immobilization of biomolecules to the surface fibers is realized by the creation of a chemical bonding between functional groups of the components and those of bioactive molecules. Compared to physical adsorption, the covalent surface immobilization of biomolecules results in a more efficient coating; moreover, the bioactive components are retained over a longer period of time, promoting tissue regeneration [75]. In particular, an appropriate choice of polymers—biodegradable or nondegradable—allows the release rate of bioactive components to be controlled.

In this context, great attention is currently devoted to "click chemistry" since "click" reactions offer high reactivity, selectivity, and mild reaction conditions [34,76,77].

The click chemistry approach was first introduced by Sharpless et al. [78], and the overview and classification of "click" reactions are reported in Figure 2.

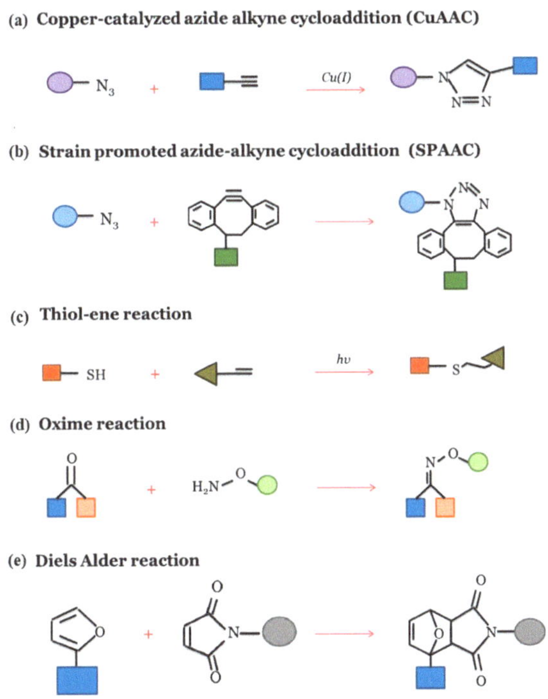

Figure 2. Schematic illustration of representative "click" reactions: (**a**) copper-catalyzed azide-alkyne cycloaddition (CuAAC); (**b**) strain promoted azide-alkyne cycloaddition (SPAAC); (**c**) thiol-ene reaction; (**d**) oxime reaction and (**e**) Diels Alder reaction.

In recent years, the use of 'clickable' building blocks, initiators, monomers, postmodifiers, and cross-linkers has led to the achievement of a wide variety of scaffolds promising in tissue engineering applications [34].

"Clickable" building blocks are common simple molecular subunits possessing one or more "clickable" groups for additional macromolecular construction and required to fabricate polymeric scaffolds with improved functions via "click" reactions. "Clickable" initiators are used to prepare polymers with chain-end 'click' functional groups. Instead, the direct polymerization of "clickable" monomers is needed for the development of polymers with multiple "clickable" sites enabling the further biomodification of scaffold. "Clickable" postmodifiers are recognized as effective tools for the direct functionalization of polymers with 'clickable' groups.

The modification of the electrospun fiber surface with biomolecules by click chemistry can be obtained through the direct electrospinning of functional polymers with reactive and clickable functional groups. For example, Lancuški et al. developed clickable nanofibrous scaffolds from a mixture of PCL-80K and PCL-2K with azide groups by electrospinning [79]. The surface of PCL-2K-azide presented clickable sites for CuAAC reaction (Figure 2a). Then, the obtained nanofibers could easily click onto desired biomolecules containing alkyne groups.

In another work, random and aligned (Dibenzyocyclooctynol) DIBO-terminated PLLA fibers were biofunctionalized with the Tyr-Ile-Gly-Ser-Arg (YIGSR) peptide after electrospinning using SPAAC click reaction (Figure 2b). The results showed that the aligned and biofunctionalized fibers enhanced neurite length and neural expression in mouse embryonic stem cells (mESC) compared to random fibers and to those without YIGSR functionalization [80].

The bioactive peptides and fluorescent molecules were bonded to the surface of electrospun PEU nanofibers by click reaction in aqueous media [81]. In particular, electrospun PEU nanofibers with "clickable" groups such as alkyne, azide, alkene, tyrosine-phenol, and ketone groups were successfully obtained. Depending on the "clickable" group, the functionalization of the PEU nanofibers via CuAAC, thiol-ene (Figure 2c), and oxime reactions (Figure 2d) can be performed [81].

In a recent work, polylactide (PLA)-based copolymers containing furan groups and triethylene glycol (TEG) were synthesized and electrospun to yield nanofibers [82]. After electrospinning, these nanofibers were conjugated with cyclic peptide, cRGDfK- maleimide via Diels-Alder reaction (Figure 2e). The presence of RGD peptide promoted cell adhesion and proliferation of L929 mouse fibroblasts suggesting that the proposed scaffolds were biocompatible and also provided a highly cytocompatible environment [82].

Furthermore, multiple click functional groups on electrospun fibers allow the attachment of different biomolecules. For example, Gly-Arg-Gly-Asp-Ser (GRGDS) and YIGSR peptides were introduced on the electrospun PCL surface via SPAAC and oxime reactions [83]. Modified GRGDS and YIGSR peptides, possessing an azide group and hydroxylamine, respectively, are introduced on the surface DIBO-PLC fibers in a one-pot reaction. The Schwann cell proliferation and attachment measurement suggest that biofunctionalized fibers are promising for peripheral nerve regeneration.

In another research, SPAAC, oxime, and CuAAC reactions were used to sequentially bond GRGDS, BMP-2 peptide, and dopamine on the electrospun DIBO-PCL scaffold surfaces [84]. This approach demonstrated a sequential and well-controlled bioactive component loading which could be useful to mimic the complex native ECM composition and activity in order to promote the regeneration tissue process.

Nowadays, thanks to the advances in click chemistry, clickable polymeric fibers may represent a major opportunity to manufacture suitable systems to bond various biomolecules with different approaches for tissue engineering applications.

4. Customized Functionalization by Click Chemistry of Composite dECM-Based Electrospun Scaffold for Skeletal Tissue Regeneration

Skeletal muscle is an important body-composition component in humans and plays a key role in voluntary movement and locomotion [85]. In addition, skeletal muscle is involved in other physiological processes, including thermogenesis, metabolism, and the secretion of numerous peptides for communication with other tissues [85]. For these reasons, the maintenance of skeletal muscle health is of vital importance. Although skeletal muscle is highly regenerative following injury or disease, endogenous self-regeneration is harshly impaired in the conditions of severe volume traumatic muscle loss (VML). Consequently, a growing demand arose in skeletal muscle TE to fully restore the structure and function of lost muscle [86].

The structure of muscle tissue is composed of oriented muscle fibers (myofibers), which are embedded into an extracellular matrix (ECM) consisting of many components such as collagen, glycoproteins, proteoglycans, and elastin [87].

Collagens form a network of intramuscular connective tissue (IMCT). The IMCT is typically organized in three layers: (i) the endomysium, representing the innermost layer that encloses individual muscle fibers; (ii) the perimysium bundling groups of muscle fibers; (iii) the epimysium enveloping the entire muscle. The IMCT contains various forms of collagens with types I and III being the most abundant [85]. The endomysium interfaces with the myofiber sarcolemma at a specialized basement membrane, which consists primarily of type IV collagen and laminin. Collagen type IV, a helical molecule, produces a network structure that with laminin, constitutes the basis of the basal lamina, directly linked to the sarcolemma of myofibers [85].

In addition, ECM, through its components such as GAGs, bind and store various growth factors which influence cell behavior and regulate cell proliferation, migration, and differentiation [87].

The skeletal muscle engineering approaches focus on the design and development of a fibrous scaffold with appropriate mechanical, morphological, and biofunctional properties to facilitate muscle growth and regeneration [87].

To this end, electrospinning is a forefront method to produce a fibrous scaffold that could mimic the structure and high anisotropic organization of native muscle tissue [88]. In particular, the composite electrospun scaffold could assure biocompatibility, biodegradability, controlled mechanical properties, and high porosity due to the synergistic effects of the different components.

Furthermore, advances in decellularization protocols of muscle tissue allow the preservation of the important components of ECM and the realization of potential dECM-based scaffolds [30,89–91] that could provide the correct support for myofiber development and the appropriate architecture to form muscle regeneration.

Some representative studies on electrospun biomaterials for skeletal muscle regeneration with the related principal outcomes are summarized in Table 2.

Table 2. Some representative studies on electrospun biomaterials for skeletal muscle regeneration, in chronological order.

Electrospun Biomaterials	Experimental Model	Outcomes	Reference
PCL/collagen I	In vitro: Human skeletal muscle cells (hSkMCs)	Aligned PCL/collagen nanofibers significantly induced muscle cell alignment and myotube formation as compared to randomly oriented nanofibers	[47]
PLGA	In vitro: Murine myoblast cells (C2C12)	Aligned PLGA fibers control the myoblast elongation and alignment and encourage myoblast differentiation.	[92]
Chitosan/PCL	In vitro: Murine myoblast cells (C2C12)	Aligned chitosan-PCL nanofibrous scaffolds exhibited superior tensile strength compared to randomly oriented nanofibers and promoted muscle cell proliferation.	[93]
Chitosan/PVA	In vitro: Rabbit's bone marrow (MSCs) In vivo: Adult New Zealand rabbit	Good cell viability, adhesion growth, and significant proliferation with less immune responses when the scaffold was implanted into the leg muscle of rabbit.	[51]
dECM from rabbit skeletal muscle	In vivo: Rabbit	The decellularization protocol of skeletal muscle tissue retains important ECM components. Electrospun scaffold derived completely from skeletal muscle dECM.	[30]
PLGA	In vitro: Murine myoblast cells (C2C12) In vivo: Mdx mice	Aligned PLGA fiber with larger diameter support enhanced alignment, growth, and differentiation of myoblasts. In vivo the optimized scaffolds seeded with primary myoblasts result in the formation of dystrophin-positive myofibers network.	[94]
PCL/dECM from bovine skeletal muscle	In vitro: Rat muscle precursor cells In vivo: C57/BL6 adult mice	Aligned nanofibers support satellite cell growth, myogenic protein expression, and myokine production. In vivo: myofiber regeneration was observed.	[60,95]

Abbreviation: PCL, poly(ε-caprolactone); PLGA, poly(lactic-co-glycolic acid); PVA, poly(vinyl alcohol).

Despite the advantages of using dECM, the attachment of biomolecules onto its surface could be an interesting method of improving its biological potential. For example, the functionalization of dECM platforms with synthetic peptides with different cell adhesive sequences increased in vitro human umbilical vein endothelial cell adhesion compared to nonfunctionalized controls [96]. In addition, dECM was modified with a peptide (QK) mimicking vascular endothelial growth factor in order to enhance angiogenesis of endothelial cells. The functionalization was achieved using CuAAC click reaction, and the resulted triazole linkages seem not to compromise the integrity of dECM [97]. In another study, scaffolds composed of a skeletal muscle-derived decellularized extracellular matrix and myogenic factor insulin growth factor-1 (IGF-1) showed high in vitro cellular activities, including cell viability, proliferation, and differentiation. Moreover, in vivo implantation of dECM-based scaffold in the rabbit muscle defect model showed an acceleration of the novel muscle formation [98].

In addition, dECM scaffolds are known to exhibit poor biomechanical properties and control over their mechanical properties, and degradation is lacking [10,95]. A scaffold composed of dECM and synthetic polymers could be mechanically suitable to support functions of skeletal muscle.

Coupling electrospinning and click reactions to biofunctionalization surface electrospun scaffolds appears to be a valid method to design biofunctionalized fibrous scaffolds promising for skeletal muscle regeneration.

In this context, a promising design strategy concerns the fabrication of composite fiber scaffolds composed of dECM of muscle tissue and synthetic polymer with targeted clickable functional groups. The main steps are reported in Figure 3.

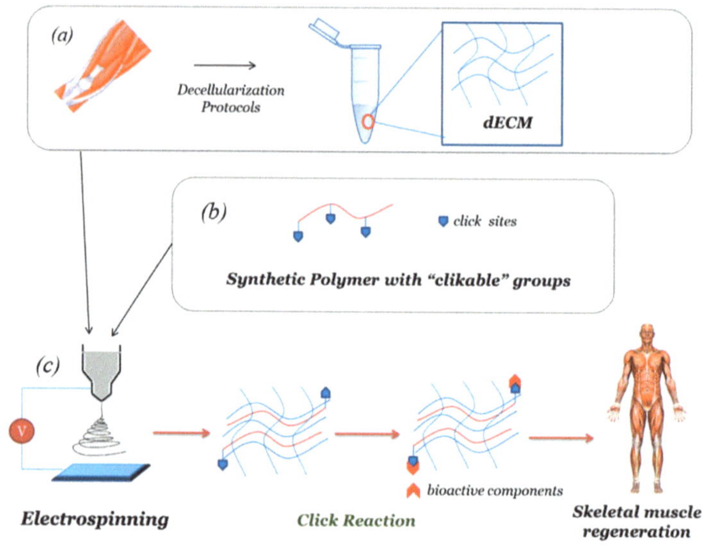

Figure 3. Manufacturing process proposal of composite scaffold for skeletal muscle regeneration: schematic representation. The main steps concern: (**a**) native skeletal muscle is processed to obtain a suitable dECM product; (**b**) synthetic polymers with "clickable" functional groups; (**c**) electrospinning of blend based on dECM (**a**) and the polymeric solution (**b**). The fibrous scaffold is biofunctionalized via click reaction in order to attach bioactive molecules promising for skeletal muscle regeneration.

In particular, scaffolds composed of fibers of decellularized ECM from skeletal muscle could mimic native muscle architecture and provide biological activity. Additionally, combining dECM with a synthetic polymer such as PCL provides mechanical stability to allow the tissue regeneration process [95]. In order to increase the bioactivity of the synthetic polymer component, this could contain clickable sites enabling further binding with appropriate biomolecules in muscle tissue regeneration.

After the electrospinning process, a click reaction such as cycloaddition between an azide and an alkyne is a suitable method to promote bioconjugation between synthetic polymer and biomolecules. It was demonstrated that azide- functionalized PCL [79] or alkyne- functionalized PCL [99] can be obtained.

Azide or alkyne groups are known for their stability to undergo click reactions. The presence of clickable sites on the surface scaffolds allows the binding with various bioactive biomolecules such as alkyne functionalized proteins or azide functionalized growth factors.

The chemical immobilization of azide conjugated epidermal growth factor (EGF) onto dibenzocyclooctyne (DBCO)-modified collagen surface through a strain-promoted azide-alkyne cycloaddition (SPAAC) reaction was reported by Lee et al. [100].

The protocol utilized consists of, first, the introduction of "clickable" functional groups to the collagen and EGF via an N-hydroxysuccinimide (NHS) ester reaction. The chemicals of the DBCO- and azide-containing NHS ester group were utilized for the modification of the collagen surface and EGF, respectively.

Afterwards, SPAAC reaction occurs between the azide group on the EGF and DBCO groups tethered on the collagen surface and generate stable triazole bonding, leading to the covalent immobilization of EGF onto the collagen surface.

This approach demonstrates the potential and useful application of copper-free click chemistry to combine growth factors to the ad hoc modified biomaterial surfaces. However, the selective modification of NHS ester reaction to the N-terminal primary amine group could improve the EGF bioactivity showed after the modification of all primary amine groups.

In an effort to better mimic physiologically relevant environments of skeletal muscle regeneration, growth factors such as FGF, PDGF, HGF and IGF, and short peptides can be used to biofunctionalize the scaffolds [87]. FGF, PDGF, and HGF promote the activation and proliferation of myogenic progenitor cells [101]. Insulin-like growth factor (IGF-I) plays an important role in all phases of satellite cell myogenesis by controlling the migration, proliferation, and differentiation of muscle satellite/progenitor cells [102].

In addition, the Arginine-Glycine-Aspartate (RGD) oligopeptide sequence that specifically recognizes and binds to integrin receptors is used to promote scaffold interaction with the cell surface. Many studies have tested the effectiveness of RGD peptide coated scaffolds for cell adhesion, and their influence on cell behavior with respect to tissue engineering [103].

An attractive alternative concerns the realization of core-shell electrospun scaffolds. In particular, the creating of synthetic polymers-based systems incorporating dECM by the coaxial electrospinning technique could preserve the bioactivity of components dECM. In this approach, the biofunctionalization of the surface scaffold by click chemistry will assure cell adhesion and interaction, and the dECM-based core will assure an appropriate microenvironment for cell activity enabling skeletal muscle regeneration.

5. Conclusions

Electrospinning is a simple and versatile technique to produce polymeric fibrous scaffolds that are capable of mimicking the structure of the natural extracellular matrix (ECM) for a range of tissue engineering applications. In recent years, the decellularized extracellular matrix (dECM) has been demonstrated to be a suitable and promising starting material to fabricate composite electrospun scaffolds. In addition, the biofunctionalization of the electrospun fibers represents an interesting approach for the further exploration of the potential of electrospinning. In particular, the bioactivity of electrospun scaffolds could be assured by the covalent immobilization of biomolecules via click reaction in order to promote cell adhesion and proliferation. The combination of electrospinning and click reactions represents a promising strategy to design and develop leading and smart biomaterials systems for tissue engineering applications. In perspective, this strategy could be a valid starting point for obtaining customizable and intelligent biomaterial systems that are able to successfully guide the tissue regeneration process. In this context, we designed the fabrication of dECM-based composite electrospun scaffolds, and also, we proposed the surface biofunctionalization of these systems via click chemistry in order to modify the surface of the electrospun fibers with selected growth factors or peptides to guide skeletal muscle regeneration.

Author Contributions: Conceptualization, S.P., F.C., and L.T.; writing—original draft preparation, S.P., F.C., A.R., P.D.N., V.M., M.C.A., R.A., S.R., and L.T.; writing—review and editing, S.P., F.C., A.R., P.D.N., V.M., M.C.A., R.A., S.R., and L.T. supervision, F.C., S.R., and L.T.; project administration and funding acquisition, L.T. All authors have read and agreed to the published version of the manuscript.

Funding: This research was funded by Regione Lazio through LazioInnova, Progetto 'SMARTIES' (Smart scaffold ingegnerizzati per la rigenerazione dell'apparato muscolo scheletrico), coordinated by ENEA CR Frascati, Grant Number: 85-2017-15004, AVVISO PUBBLICO PROGETTI DI GRUPPI DI RICERCA Conoscenza e cooperazione per un nuovo modello di sviluppo. Support has been also provided by 'Tor Vergata' University, Grant 'Mission: Sustainability' (DR 2817/2016), Number: E86C18000640005. PROGETTO: BIOMECHANICAL SCALING STIMULA: INDUCING DIFFERENTIATION (BIOMESSID), coordinated by the Dept of Clinical Science and Translational Medicine, University of Rome "Tor Vergata", Italy.

Acknowledgments: We want to express the greatest thanks to Martina Di Cesare (Lazio-innova) for the excellent management of the SMARTIES grant. We would also like to thank Andrea Reale and Flavia Miele (ENEA CR Frascati), for their precious technical assistance to the SMARTIES progetto, and Anna Pagliardini (ENEA CR Frascati), for her administrative valuable support. LT and SR thank the Asian Polymers Association for the 2019 and 2018, respectively, awards for their studies on polymers https://www.asianpolymer.org/distinguished.html.

Conflicts of Interest: The authors declare no conflict of interest.

References

1. Khan, F.; Tanaka, M. Designing smart biomaterials for tissue engineering. *Int. J. Mol. Sci.* **2018**, *19*, 17. [CrossRef] [PubMed]
2. Pérez, R.A.; Won, J.E.; Knowles, J.C.; Kim, H.W. Naturally and synthetic smart composite biomaterials for tissue regeneration. *Adv. Drug Deliv. Rev.* **2013**, *65*, 471–496. [CrossRef] [PubMed]
3. Smoak, M.M.; Mikos, A.G. Advances in biomaterials for skeletal muscle engineering and obstacles still to overcome. *Mater. Today Bio* **2020**, *7*, 100069. [CrossRef] [PubMed]
4. Qazi, T.H.; Duda, G.N.; Ort, M.J.; Perka, C.; Geissler, S.; Winkler, T. Cell therapy to improve regeneration of skeletal muscle injuries. *J. Cachexia. Sarcopenia Muscle* **2019**, *10*, 501–516. [CrossRef]
5. Shin, H.; Jo, S.; Mikos, A.G. Biomimetic materials for tissue engineering. *Biomaterials* **2003**, *24*, 4353–4364. [CrossRef]
6. Carotenuto, F.; Teodori, L.; Maccari, A.M.; Delbono, L.; Orlando, G.; Di Nardo, P. Turning regenerative technologies into treatment to repair myocardial injuries. *J. Cell. Mol. Med.* **2020**, *24*, 2704–2716. [CrossRef]
7. Ma, P.X. Biomimetic materials for tissue engineering. *Adv. Drug Deliv. Rev.* **2008**, *60*, 184–198. [CrossRef]
8. Pina, S.; Ribeiro, V.P.; Marques, C.F.; Maia, F.R.; Silva, T.H.; Reis, R.L.; Oliveira, J.M. Scaffolding Strategies for Tissue Enginnering and Regenerative Medicine Applications. *Materials (Basel)* **2019**, *12*, 1824. [CrossRef]
9. Badylak, S.F. The extracellular matrix as a scaffold for tissue reconstruction. *Semin. Cell Dev. Biol.* **2002**, *13*, 377–383. [CrossRef]
10. Badylak, S.F.; Freytes, D.O.; Gilbert, T.W. Extracellular matrix as a biological scaffold material: Structure and function. *Acta Biomater.* **2009**, *5*, 1–13. [CrossRef]
11. Yao, T.; Baker, M.B.; Moroni, L. Strategies to improve nanofibrous scaffolds for vascular tissue engineering. *Nanomaterials* **2020**, *10*, 887. [CrossRef] [PubMed]
12. Tebyetekerwa, M.; Ramakrishna, S. What Is Next for Electrospinning? *Matter* **2020**, *2*, 279–283. [CrossRef]
13. Li, W.J.; Laurencin, C.T.; Caterson, E.J.; Tuan, R.S.; Ko, F.K. Electrospun nanofibrous structure: A novel scaffold for tissue engineering. *J. Biomed. Mater. Res.* **2002**, *60*, 613–621. [CrossRef] [PubMed]
14. Ramakrishna, S.; Fujihara, K.; Teo, W.E.; Yong, T.; Ma, Z.; Ramaseshan, R. Electrospun nanofibers: Solving global issues. *Mater. Today* **2006**, *9*, 40–50. [CrossRef]
15. Min, L.; Pan, H.; Chen, S.; Wang, C.; Wang, N.; Zhang, J.; Cao, Y.; Chen, X.; Hou, X. Recent progress in bio-inspired electrospun materials. *Compos. Commun.* **2019**, *11*, 12–20. [CrossRef]

16. Ameer, J.M.; Anil Kumar, P.R.; Kasoju, N. Strategies to tune electrospun scaffold porosity for effective cell response in tissue engineering. *J. Funct. Biomater.* **2019**, *10*, 30. [CrossRef]
17. Khorshidi, S.; Solouk, A.; Mirzadeh, H.; Mazinani, S.; Lagaron, J.M.; Sharifi, S.; Ramakrishna, S. A review of key challenges of electrospun scaffolds for tissue-engineering applications. *J. Tissue Eng. Regen. Med.* **2016**, *10*, 715–738. [CrossRef]
18. Seyedmahmoud, R.; Rainer, A.; Mozetic, P.; Maria Giannitelli, S.; Trombetta, M.; Traversa, E.; Licoccia, S.; Rinaldi, A. A primer of statistical methods for correlating parameters and properties of electrospun poly(L-lactide) scaffolds for tissue engineering-PART 1: Design of experiments. *J. Biomed. Mater. Res. Part A* **2015**, *103*, 91–102. [CrossRef]
19. Seyedmahmoud, R.; Mozetic, P.; Rainer, A.; Giannitelli, S.M.; Basoli, F.; Trombetta, M.; Traversa, E.; Licoccia, S.; Rinaldi, A. A primer of statistical methods for correlating parameters and properties of electrospun poly(L-lactide) scaffolds for tissue engineering-PART 2: Regression. *J. Biomed. Mater. Res.-Part A* **2015**, *103*, 103–114. [CrossRef]
20. Ghasemi-Mobarakeh, L.; Prabhakaran, M.P.; Morshed, M.; Nasr-Esfahani, M.H.; Ramakrishna, S. Electrospun poly(ε-caprolactone)/gelatin nanofibrous scaffolds for nerve tissue engineering. *Biomaterials* **2008**, *29*, 4532–4539. [CrossRef]
21. Liang, D.; Hsiao, B.S.; Chu, B. Functional electrospun nanofibrous scaffolds for biomedical applications. *Adv. Drug Deliv. Rev.* **2007**, *59*, 1392–1412. [CrossRef] [PubMed]
22. Rim, N.G.; Shin, C.S.; Shin, H. Current Approaches to Electrospun Nanofibers for Tissue Engineering. *Biomed. Mater.* **2013**, *8*, 014102. [CrossRef] [PubMed]
23. Hanumantharao, S.N.; Rao, S. Multi-functional electrospun nanofibers from polymer blends for scaffold tissue engineering. *Fibers* **2019**, *7*, 66. [CrossRef]
24. Swinehart, I.T.; Badylak, S.F. Extracellular matrix bioscaffolds in tissue remodeling and morphogenesis. *Dev. Dyn.* **2016**, *245*, 351–360. [CrossRef]
25. Porzionato, A.; Stocco, E.; Barbon, S.; Grandi, F.; Macchi, V.; De Caro, R. Tissue-engineered grafts from human decellularized extracellular matrices: A systematic review and future perspectives. *Int. J. Mol. Sci.* **2018**, *19*, 4117. [CrossRef]
26. Krishtul, S.; Baruch, L.; Machluf, M. Processed Tissue–Derived Extracellular Matrices: Tailored Platforms Empowering Diverse Therapeutic Applications. *Adv. Funct. Mater.* **2020**, *30*, 1900386. [CrossRef]
27. Santschi, M.; Vernengo, A.; Eglin, D.; D'Este, M.; Wuertz-Kozak, K. Decellularized matrix as a building block in bioprinting and electrospinning. *Curr. Opin. Biomed. Eng.* **2019**, *10*, 116–122. [CrossRef]
28. Shute, J. Glycosaminoglycan and chemokine/growth factor interactions. *Handb. Exp. Pharmacol.* **2012**, *207*, 307–324. [CrossRef]
29. Heath, D.E. A Review of Decellularized Extracellular Matrix Biomaterials for Regenerative Engineering Applications. *Regen. Eng. Transl. Med.* **2019**, *5*, 155–166. [CrossRef]
30. Smoak, M.M.; Han, A.; Watson, E.; Kishan, A.; Grande-Allen, K.J.; Cosgriff-Hernandez, E.; Mikos, A.G. Fabrication and Characterization of Electrospun Decellularized Muscle-Derived Scaffolds. *Tissue Eng. Part C Methods* **2019**, *25*, 276–287. [CrossRef]
31. Ji, W.; Sun, Y.; Yang, F.; Van Den Beucken, J.J.J.P.; Fan, M.; Chen, Z.; Jansen, J.A. Bioactive electrospun scaffolds delivering growth factors and genes for tissue engineering applications. *Pharm. Res.* **2011**, *28*, 1259–1272. [CrossRef]
32. Yoo, H.S.; Kim, T.G.; Park, T.G. Surface-functionalized electrospun nanofibers for tissue engineering and drug delivery. *Adv. Drug Deliv. Rev.* **2009**, *61*, 1033–1042. [CrossRef] [PubMed]
33. Kalaoglu-Altan, O.I.; Sanyal, R.; Sanyal, A. Reactive and "clickable" electrospun polymeric nanofibers. *Polym. Chem.* **2015**, *6*, 3372–3381. [CrossRef]
34. Zou, Y.; Zhang, L.; Yang, L.; Ding, M.; Lin, F.; Wang, Z.; Li, Y. "Click" chemistry in polymeric scaffolds: Bioactive materials for tissue engineering. *J. Control. Release* **2018**, *273*, 160–179. [CrossRef] [PubMed]
35. Seal, B.L.; Otero, T.C.; Panitch, A. Polymeric biomaterials for tissue and organ regeneration. *Mater. Sci. Eng. R Rep.* **2001**, *34*, 147–230. [CrossRef]
36. Zhang, F.; Zhang, Z.; Zhou, T.; Liu, Y.; Leng, J. Shape Memory Polymer Nanofibers and Their Composites: Electrospinning, Structure, Performance, and Applications. *Front. Mater.* **2015**, *2*, 62. [CrossRef]
37. Liu, X.; Holzwarth, J.M.; Ma, P.X. Functionalized Synthetic Biodegradable Polymer Scaffolds for Tissue Engineering. *Macromol. Biosci.* **2012**, *12*, 911–919. [CrossRef]

38. Soares, R.M.D.; Siqueira, N.M.; Prabhakaram, M.P.; Ramakrishna, S. Electrospinning and electrospray of bio-based and natural polymers for biomaterials development. *Mater. Sci. Eng. C* **2018**, *92*, 969–982. [CrossRef]
39. Sell, S.A.; Wolfe, P.S.; Garg, K.; McCool, J.M.; Rodriguez, I.A.; Bowlin, G.L. The use of natural polymers in tissue engineering: A focus on electrospun extracellular matrix analogues. *Polymers* **2010**, *2*, 522–553. [CrossRef]
40. Hong, M.; Chen, E.Y.X. Chemically recyclable polymers: A circular economy approach to sustainability. *Green Chem.* **2017**, *19*, 3692–3706. [CrossRef]
41. Zha, F.; Chen, W.; Zhang, L.; Yu, D. Electrospun natural polymer and its composite nanofibrous scaffolds for nerve tissue engineering. *J. Biomater. Sci. Polym. Ed.* **2020**, *31*, 519–548. [CrossRef] [PubMed]
42. Zhang, Y.; Ouyang, H.; Chwee, T.L.; Ramakrishna, S.; Huang, Z.M. Electrospinning of gelatin fibers and gelatin/PCL composite fibrous scaffolds. *J. Biomed. Mater. Res.-Part B Appl. Biomater.* **2005**, *72*, 156–165. [CrossRef] [PubMed]
43. Hu, Y.; Feng, B.; Zhang, W.; Yan, C.; Yao, Q.; Shao, C.; Yu, F.; Li, F.; Fu, Y. Electrospun gelatin/PCL and collagen/PCL scaffolds for modulating responses of bone marrow endothelial progenitor cells. *Exp. Ther. Med.* **2019**, *17*, 3717–3726. [CrossRef] [PubMed]
44. Fadaie, M.; Mirzaei, E.; Geramizadeh, B.; Asvar, Z. Incorporation of nanofibrillated chitosan into electrospun PCL nanofibers makes scaffolds with enhanced mechanical and biological properties. *Carbohydr. Polym.* **2018**, *199*, 628–640. [CrossRef] [PubMed]
45. Mochane, M.J.; Motsoeneng, T.S.; Sadiku, E.R.; Mokhena, T.C.; Sefadi, J.S. Morphology and Properties of Electrospun PCL and Its Composites for Medical Applications: A Mini Review. *Appl. Sci.* **2019**, *9*, 2205. [CrossRef]
46. Adeli-Sardou, M.; Yaghoobi, M.M.; Torkzadeh-Mahani, M.; Dodel, M. Controlled release of lawsone from polycaprolactone/gelatin electrospun nano fibers for skin tissue regeneration. *Int. J. Biol. Macromol.* **2019**, *124*, 478–491. [CrossRef]
47. Choi, J.S.; Lee, S.J.; Christ, G.J.; Atala, A.; Yoo, J.J. The influence of electrospun aligned poly(ε-caprolactone)/collagen nanofiber meshes on the formation of self-aligned skeletal muscle myotubes. *Biomaterials* **2008**, *29*, 2899–2906. [CrossRef]
48. Tillman, B.W.; Yazdani, S.K.; Lee, S.J.; Geary, R.L.; Atala, A.; Yoo, J.J. The in vivo stability of electrospun polycaprolactone-collagen scaffolds in vascular reconstruction. *Biomaterials* **2009**, *30*, 583–588. [CrossRef]
49. Shrestha, B.K.; Mousa, H.M.; Tiwari, A.P.; Ko, S.W.; Park, C.H.; Kim, C.S. Development of polyamide-6,6/chitosan electrospun hybrid nanofibrous scaffolds for tissue engineering application. *Carbohydr. Polym.* **2016**, *148*, 107–114. [CrossRef]
50. Aldana, A.A.; Abraham, G.A. Current advances in electrospun gelatin-based scaffolds for tissue engineering applications. *Int. J. Pharm.* **2017**, *523*, 441–453. [CrossRef]
51. Kheradmandi, M.; Vasheghani-Farahani, E.; Ghiaseddin, A.; Ganji, F. Skeletal muscle regeneration via engineered tissue culture over electrospun nanofibrous chitosan/PVA scaffold. *J. Biomed. Mater. Res.-Part A* **2016**, *104*, 1720–1727. [CrossRef] [PubMed]
52. Park, H.K.; Joo, W.; Gu, B.K.; Ha, M.Y.; You, S.J.; Chun, H.J. Collagen/poly(D,L-lactic-co-glycolic acid) composite fibrous scaffold prepared by independent nozzle control multi-electrospinning apparatus for dura repair. *J. Ind. Eng. Chem.* **2018**, *66*, 430–437. [CrossRef]
53. Elmashhady, H.H.; Kraemer, B.A.; Patel, K.H.; Sell, S.A.; Garg, K. Decellularized extracellular matrices for tissue engineering applications. *Electrospinning* **2017**, *1*, 87–99. [CrossRef]
54. Badylak, S.F.; Taylor, D.; Uygun, K. Whole-Organ Tissue Engineering: Decellularization and Recellularization of Three-Dimensional Matrix Scaffolds. *Annu. Rev. Biomed. Eng.* **2011**, *13*, 27–53. [CrossRef]
55. Gilbert, T.W.; Sellaro, T.L.; Badylak, S.F. Decellularization of tissues and organs. *Biomaterials* **2006**, *27*, 3675–3683. [CrossRef]
56. Taylor, D.A.; Sampaio, L.C.; Ferdous, Z.; Gobin, A.S.; Taite, L.J. Decellularized matrices in regenerative medicine. *Acta Biomater.* **2018**, *74*, 74–89. [CrossRef]
57. Baiguera, S.; Del Gaudio, C.; Lucatelli, E.; Kuevda, E.; Boieri, M.; Mazzanti, B.; Bianco, A.; Macchiarini, P. Electrospun gelatin scaffolds incorporating rat decellularized brain extracellular matrix for neural tissue engineering. *Biomaterials* **2014**, *35*, 1205–1214. [CrossRef]

58. Schoen, B.; Avrahami, R.; Baruch, L.; Efraim, Y.; Goldfracht, I.; Elul, O.; Davidov, T.; Gepstein, L.; Zussman, E.; Machluf, M. Electrospun Extracellular Matrix: Paving the Way to Tailor-Made Natural Scaffolds for Cardiac Tissue Regeneration. *Adv. Funct. Mater.* **2017**, *27*, 1700427. [CrossRef]
59. Gao, S.; Guo, W.; Chen, M.; Yuan, Z.; Wang, M.; Zhang, Y.; Liu, S.; Xi, T.; Guo, Q. Fabrication and characterization of electrospun nanofibers composed of decellularized meniscus extracellular matrix and polycaprolactone for meniscus tissue engineering. *J. Mater. Chem. B* **2017**, *5*, 2273–2285. [CrossRef]
60. Patel, K.H.; Dunn, A.J.; Talovic, M.; Haas, G.J.; Marcinczyk, M.; Elmashhady, H.; Kalaf, E.G.; Sell, S.A.; Garg, K. Aligned nanofibers of decellularized muscle ECM support myogenic activity in primary satellite cells in vitro. *Biomed. Mater.* **2019**, *14*, 035010. [CrossRef]
61. Lu, Y.; Huang, J.; Yu, G.; Cardenas, R.; Wei, S.; Wujcik, E.K.; Guo, Z. Coaxial electrospun fibers: Applications in drug delivery and tissue engineering. *Wiley Interdiscip. Rev. Nanomed. Nanobiotechnol.* **2016**, *8*, 654–677. [CrossRef] [PubMed]
62. McClellan, P.; Landis, W.J. Recent Applications of Coaxial and Emulsion Electrospinning Methods in the Field of Tissue Engineering. *Biores. Open Access* **2016**, *5*, 212–227. [CrossRef] [PubMed]
63. Sahoo, S.; Ang, L.T.; Goh, J.C.H.; Toh, S.L. Growth factor delivery through electrospun nanofibers in scaffolds for tissue engineering applications. *J. Biomed. Mater. Res.-Part A* **2010**, *93*, 1539–1550. [CrossRef] [PubMed]
64. Mickova, A.; Buzgo, M.; Benada, O.; Rampichova, M.; Fisar, Z.; Filova, E.; Tesarova, M.; Lukas, D.; Amler, E. Core/shell nanofibers with embedded liposomes as a drug delivery system. *Biomacromolecules* **2012**, *13*, 952–962. [CrossRef] [PubMed]
65. Tian, L.; Prabhakaran, M.P.; Ding, X.; Kai, D.; Ramakrishna, S. Emulsion electrospun nanofibers as substrates for cardiomyogenic differentiation of mesenchymal stem cells. *J. Mater. Sci. Mater. Med.* **2013**, *24*, 2577–2587. [CrossRef]
66. Zhang, H.; Jia, X.; Han, F.; Zhao, J.; Zhao, Y.; Fan, Y.; Yuan, X. Dual-delivery of VEGF and PDGF by double-layered electrospun membranes for blood vessel regeneration. *Biomaterials* **2013**, *34*, 2202–2212. [CrossRef]
67. Park, K.E.; Kim, B.S.; Kim, M.H.; You, H.K.; Lee, J.; Park, W.H. Basic fibroblast growth factor-encapsulated PCL nano/microfibrous composite scaffolds for bone regeneration. *Polymer* **2015**, *76*, 8–16. [CrossRef]
68. Jiang, Y.C.; Wang, X.F.; Xu, Y.Y.; Qiao, Y.H.; Guo, X.; Wang, D.F.; Li, Q.; Turng, L.S. Polycaprolactone Nanofibers Containing Vascular Endothelial Growth Factor-Encapsulated Gelatin Particles Enhance Mesenchymal Stem Cell Differentiation and Angiogenesis of Endothelial Cells. *Biomacromolecules* **2018**, *19*, 3747–3753. [CrossRef]
69. Augustine, R.; Zahid, A.A.; Hasan, A.; Wang, M.; Webster, T.J. Ctgf loaded electrospun dual porous core-shell membrane for diabetic wound healing. *Int. J. Nanomed.* **2019**, *14*, 8573–8588. [CrossRef]
70. Baek, J.; Lee, E.; Lotz, M.K.; D'Lima, D.D. Bioactive proteins delivery through core-shell nanofibers for meniscal tissue regeneration. *Nanomed. Nanotechnol. Biol. Med.* **2020**, *23*, 102090. [CrossRef]
71. Ferrand, A.; Eap, S.; Richert, L.; Lemoine, S.; Kalaskar, D.; Demoustier-Champagne, S.; Atmani, H.; Mély, Y.; Fioretti, F.; Schlatter, G.; et al. Osteogenetic Properties of Electrospun Nanofibrous PCL Scaffolds Equipped With Chitosan-Based Nanoreservoirs of Growth Factors. *Macromol. Biosci.* **2014**, *14*, 45–55. [CrossRef] [PubMed]
72. Decher, G.; Eckle, M.; Schmitt, J.; Struth, B. Layer-by-layer assembled multicomposite films. *Curr. Opin. Colloid Interface Sci.* **1998**, *3*, 32–39. [CrossRef]
73. Jessel, N.; Atalar, F.; Lavalle, P.; Mutterer, J.; Decher, G.; Schaaf, P.; Voegel, J.C.; Ogier, J. Bioactive coatings based on a polyelectrolyte multilayer architecture functionalized by embedded proteins. *Adv. Mater.* **2003**, *15*, 692–695. [CrossRef]
74. Asadian, M.; Rashidi, A.; Majidi, M.; Mehrjoo, M.; Emami, B.A.; Tavassoli, H.; Asl, M.P.; Bonakdar, S. Nanofiber protein adsorption affected by electrospinning physical processing parameters. *J. Iran. Chem. Soc.* **2015**, *12*, 1089–1097. [CrossRef]
75. Koh, H.S.; Yong, T.; Chan, C.K.; Ramakrishna, S. Enhancement of neurite outgrowth using nano-structured scaffolds coupled with laminin. *Biomaterials* **2008**, *29*, 3574–3582. [CrossRef]
76. Xi, W.; Scott, T.F.; Kloxin, C.J.; Bowman, C.N. Click chemistry in materials science. *Adv. Funct. Mater.* **2014**, *24*, 2572–2590. [CrossRef]
77. Costa, A.; Walkowiak, B.; Campanella, L. Tissue Engineering Between Click Chemistry and Green Chemistry. *An Int. J. Hist. Chem.* **2019**, *3*, 29–38. [CrossRef]

78. Kolb, H.C.; Finn, M.G.; Sharpless, K.B. Click Chemistry: Diverse Chemical Function from a Few Good Reactions. *Angew. Chemie Int. Ed.* **2001**, *40*, 2004–2021. [CrossRef]
79. Lancuški, A.; Fort, S.; Bossard, F. Electrospun azido-PCL nanofibers for enhanced surface functionalization by click chemistry. *ACS Appl. Mater. Interfaces* **2012**, *4*, 6499–6504. [CrossRef]
80. Smith Callahan, L.A.; Xie, S.; Barker, I.A.; Zheng, J.; Reneker, D.H.; Dove, A.P.; Becker, M.L. Directed differentiation and neurite extension of mouse embryonic stem cell on aligned poly(lactide) nanofibers functionalized with YIGSR peptide. *Biomaterials* **2013**, *34*, 9089–9095. [CrossRef]
81. Lin, F.; Yu, J.; Tang, W.; Zheng, J.; Xie, S.; Becker, M.L. Postelectrospinning "click" modification of degradable amino acid-based poly(ester urea) nanofibers. *Macromolecules* **2013**, *46*, 9515–9525. [CrossRef]
82. Kalaoglu-Altan, O.I.; Kirac-Aydin, A.; Sumer Bolu, B.; Sanyal, R.; Sanyal, A. Diels-Alder "clickable" Biodegradable Nanofibers: Benign Tailoring of Scaffolds for Biomolecular Immobilization and Cell Growth. *Bioconjug. Chem.* **2017**, *28*, 2420–2428. [CrossRef] [PubMed]
83. Zheng, J.; Kontoveros, D.; Lin, F.; Hua, G.; Reneker, D.H.; Becker, M.L.; Willits, R.K. Enhanced schwann cell attachment and alignment using one-pot "Dual Click" GRGDS and YIGSR derivatized nanofibers. *Biomacromolecules* **2015**, *16*, 357–363. [CrossRef] [PubMed]
84. Zheng, J.; Hua, G.; Yu, J.; Lin, F.; Wade, M.B.; Reneker, D.H.; Becker, M.L. Post-electrospinning "triclick" functionalization of degradable polymer nanofibers. *ACS Macro Lett.* **2015**, *4*, 207–213. [CrossRef]
85. Csapo, R.; Gumpenberger, M.; Wessner, B. Skeletal Muscle Extracellular Matrix–What Do We Know About Its Composition, Regulation, and Physiological Roles? A Narrative Review. *Front. Physiol.* **2020**, *11*, 253. [CrossRef]
86. Gao, L.; Ma, L.; Yin, X.; Luo, Y.; Yang, H.; Zhang, B. Nano- and Microfabrication for Engineering Native-Like Muscle Tissues. *Small Methods* **2020**, *4*, 1900669. [CrossRef]
87. Nakayama, K.H.; Shayan, M.; Huang, N.F. Engineering Biomimetic Materials for Skeletal Muscle Repair and Regeneration. *Adv. Healthc. Mater.* **2019**, *8*, 1801168. [CrossRef]
88. Narayanan, N.; Jiang, C.; Uzunalli, G.; Thankappan, S.K.; Laurencin, C.T.; Deng, M. Polymeric Electrospinning for Musculoskeletal Regenerative Engineering. *Regen. Eng. Transl. Med.* **2016**, *2*, 69–84. [CrossRef]
89. Wolf, M.T.; Daly, K.A.; Reing, J.E.; Badylak, S.F. Biologic scaffold composed of skeletal muscle extracellular matrix. *Biomaterials* **2012**, *33*, 2916–2925. [CrossRef]
90. Teodori, L.; Costa, A.; Marzio, R.; Perniconi, B.; Coletti, D.; Adamo, S.; Gupta, B.; Tarnok, A. Native extracellular matrix: A new scaffolding platform for repair of damaged muscle. *Front. Physiol.* **2014**, *5*, 218. [CrossRef]
91. Porzionato, A.; Sfriso, M.M.; Pontini, A.; Macchi, V.; Petrelli, L.; Pavan, P.G.; Natali, A.N.; Bassetto, F.; Vindigni, V.; De Caro, R. Decellularized human skeletal muscle as biologic scaffold for reconstructive surgery. *Int. J. Mol. Sci.* **2015**, *16*, 14808–14831. [CrossRef] [PubMed]
92. Aviss, K.J.; Gough, J.E.; Downes, S. Aligned electrospun polymer fibres for skeletal muscle regeneration. *Eur. Cells Mater.* **2010**, *19*, 193–204. [CrossRef]
93. Cooper, A.; Jana, S.; Bhattarai, N.; Zhang, M. Aligned chitosan-based nanofibers for enhanced myogenesis. *J. Mater. Chem.* **2010**, *20*, 8904–8911. [CrossRef]
94. Narayanan, N.; Jiang, C.; Wang, C.; Uzunalli, G.; Whittern, N.; Chen, D.; Jones, O.G.; Kuang, S.; Deng, M. Harnessing Fiber Diameter-Dependent Effects of Myoblasts Toward Biomimetic Scaffold-Based Skeletal Muscle Regeneration. *Front. Bioeng. Biotechnol.* **2020**, *8*, 203. [CrossRef] [PubMed]
95. Patel, K.H.; Talovic, M.; Dunn, A.J.; Patel, A.; Vendrell, S.; Schwartz, M.; Garg, K. Aligned nanofibers of decellularized muscle extracellular matrix for volumetric muscle loss. *J. Biomed. Mater. Res.-Part B Appl. Biomater.* **2020**, *108*, 2528–2537. [CrossRef]
96. Aubin, H.; Mas-Moruno, C.; Iijima, M.; Schütterle, N.; Steinbrink, M.; Assmann, A.; Gil, F.J.; Lichtenberg, A.; Pegueroles, M.; Akhyari, P. Customized Interface Biofunctionalization of Decellularized Extracellular Matrix: Toward Enhanced Endothelialization. *Tissue Eng.-Part C Methods* **2016**, *22*, 496–508. [CrossRef]
97. Wang, L.; Zhao, M.; Li, S.; Erasquin, U.J.; Wang, H.; Ren, L.; Chen, C.; Wang, Y.; Cai, C. "Click" immobilization of a VEGF-mimetic peptide on decellularized endothelial extracellular matrix to enhance angiogenesis. *ACS Appl. Mater. Interfaces* **2014**, *6*, 8401–8406. [CrossRef]
98. Lee, H.; Ju, Y.M.; Kim, I.; Elsangeedy, E.; Lee, J.H.; Yoo, J.J.; Atala, A.; Lee, S.J. A novel decellularized skeletal muscle-derived ECM scaffolding system for in situ muscle regeneration. *Methods* **2020**, *171*, 77–85. [CrossRef]

99. Jazkewitsch, O.; Ritter, H. Formation and characterization of inclusion complexes of alkyne functionalized poly(ε-caprolactone) with β-cyclodextrin. Pseudo-polyrotaxane-based supramolecular organogels. *Macromolecules* **2011**, *44*, 375–382. [CrossRef]
100. Lee, H.J.; Fernandes-Cunha, G.M.; Putra, I.; Koh, W.G.; Myung, D. Tethering Growth Factors to Collagen Surfaces Using Copper-Free Click Chemistry: Surface Characterization and in Vitro Biological Response. *ACS Appl. Mater. Interfaces* **2017**, *9*, 23389–23399. [CrossRef]
101. Syverud, B.C.; VanDusen, K.W.; Larkin, L.M. Growth factors for skeletal muscle tissue engineering. *Cells Tissues Organs* **2016**, *202*, 169–179. [CrossRef] [PubMed]
102. Shapiro, L.; Elsangeedy, E.; Lee, H.; Atala, A.; Yoo, J.; Lee, S.; YM, J. In Vitro Evaluation of Functionalized Decellularized Muscle Scaffold for in Situ Skeletal Muscle Regeneration. *Biomed. Mater.* **2019**, *14*. [CrossRef] [PubMed]
103. Bishop, C.J.; Kim, J.; Green, J.J. Biomolecule delivery to engineer the cellular microenvironment for regenerative medicine. *Ann. Biomed. Eng.* **2014**, *42*, 1557–1572. [CrossRef] [PubMed]

© 2020 by the authors. Licensee MDPI, Basel, Switzerland. This article is an open access article distributed under the terms and conditions of the Creative Commons Attribution (CC BY) license (http://creativecommons.org/licenses/by/4.0/).

nanomaterials

Correction

Correction: Niu, Y.; Galluzzi, M. Hyaluronic Acid/Collagen Nanofiber Tubular Scaffolds Support Endothelial Cell Proliferation, Phenotypic Shape and Endothelialization. *Nanomaterials* 2021, *11*, 2334

Yuqing Niu [1,*] and Massimiliano Galluzzi [2]

[1] Department of Pediatric Surgery, Guangdong Provincial Key Laboratory of Research in Structural Birth Defect Disease, Guangzhou Women and Children's Medical Center, Guangzhou Medical University, Guangzhou 510623, China

[2] Materials Interfaces Center, Shenzhen Institutes of Advanced Technology, Chinese Academy of Sciences, Shenzhen 518055, China; galluzzi@siat.ac.cn

* Correspondence: niuyuqing@gwcmc.org

Citation: Niu, Y.; Galluzzi, M. Correction: Niu, Y.; Galluzzi, M. Hyaluronic Acid/Collagen Nanofiber Tubular Scaffolds Support Endothelial Cell Proliferation, Phenotypic Shape and Endothelialization. *Nanomaterials* 2021, *11*, 2334. *Nanomaterials* **2024**, *14*, 1203. https://doi.org/10.3390/nano14141203

Received: 12 June 2024
Accepted: 19 June 2024
Published: 16 July 2024

Copyright: © 2024 by the authors. Licensee MDPI, Basel, Switzerland. This article is an open access article distributed under the terms and conditions of the Creative Commons Attribution (CC BY) license (https://creativecommons.org/licenses/by/4.0/).

Error in Figures

In the original publication [1], there were mistakes in Figures 1B and 2A. The SEM micrograph of the cross-section view of the HA/collagen sample before cross-linking in the upper panel of Figure 1B was erroneously displayed for a wall thickness of ~0.35 mm, and the correct image was originally omitted. In Figure 2A, the SEM micrographs of the surface of cross-linked, electrospun HA/collagen nanofibers and collagen nanofibers samples were inadvertently displayed before cross-linking images during the figure arrangement. The corrected upper panel for Figure 1B and the corrections for Figure 2A are shown below.

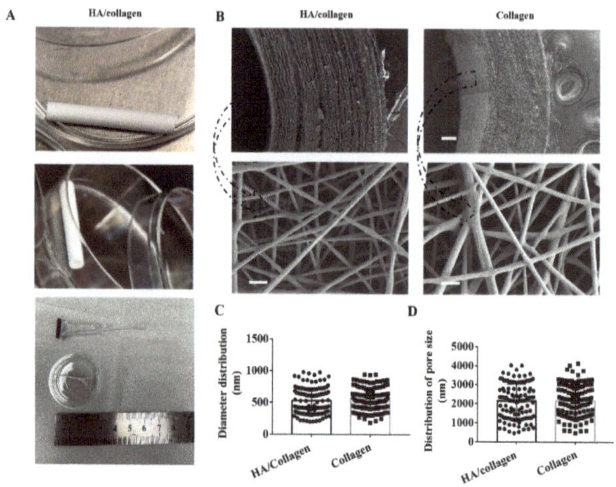

Figure 1. Physical characteristics of electrospun nanofibers. (**A**) The lateral (upper panel), cross-section (middle panel), and strip nanofiber concentric axis film (after cutting along the axis) in the image of an HA/collagen nanofiber tube. (**B**) SEM micrographs of the cross-section (upper panel) views and the inner wall surface (lower panel) of HA/collagen and collagen nanofibers before cross-linking. Scale bars: 100 μm (upper panel), 2 μm (lower panel). Statistical data of the diameter (**C**), pore size, and (**D**) distribution of various nanofibers.

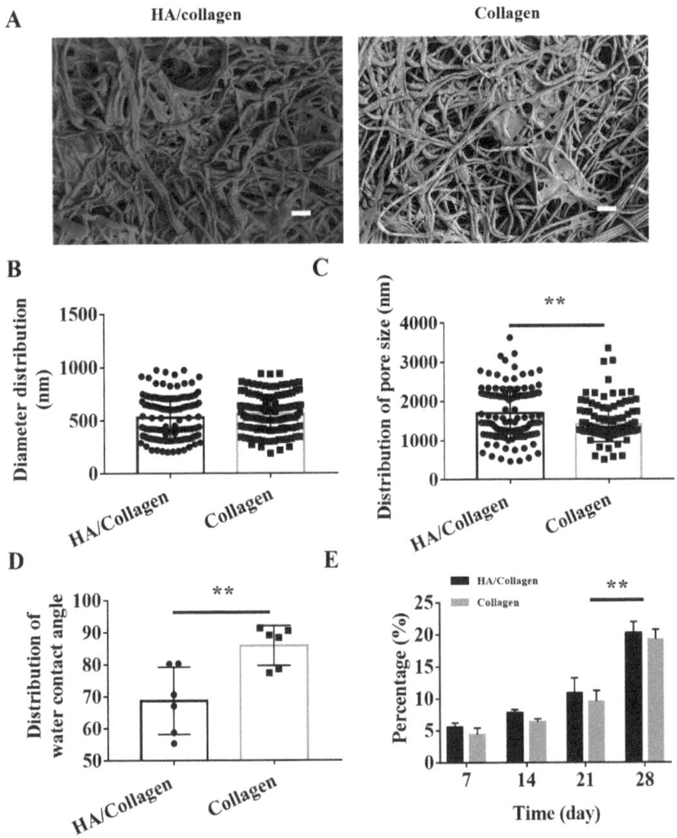

Figure 2. Morphology, hydrophilicity and degradability of cross-linked electrospun nanofibers in vitro. (**A**) SEM micrographs of the surface of cross-linked electrospun HA/collagen and collagen nanofibers. Scale bars: 4 μm. Statistical data of the diameter (**B**) and pore size (**C**) distribution of cross-linked nanofibers ($n = 120$). Average water contact angle (**D**) and degradability (**E**) of different nanofibers ($n = 6$). ** $p < 0.01$.

In the published version of Figure 5A, the fluorescence micrograph of the collagen sample in the right lower panel in Figure 5A is incorrectly represented. The corrected figure is displayed below.

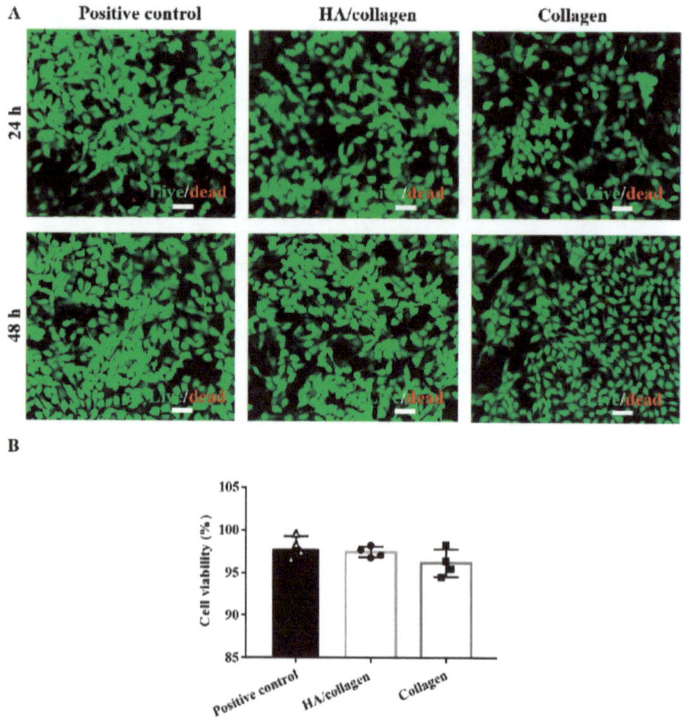

Figure 5. In vitro cytocompatibility analysis. (**A**) Fluorescence microscopy graphs of stained mouse PAECs with and without crosslinked nanofiber films after 24 and 48 h of culture. Scale bars, 50 μm. (**B**) The cell viability of mouse PAECs with and without crosslinked nanofiber films after 48 h of culture.

The authors apologize for any inconvenience caused and state that the scientific conclusions are unaffected. This correction was approved by the Academic Editor. The original publication has also been updated

Reference

1. Niu, Y.; Galluzzi, M. Hyaluronic Acid/Collagen Nanofiber Tubular Scaffolds Support Endothelial Cell Proliferation, Phenotypic Shape and Endothelialization. *Nanomaterials* **2021**, *11*, 2334. [CrossRef]

Disclaimer/Publisher's Note: The statements, opinions and data contained in all publications are solely those of the individual author(s) and contributor(s) and not of MDPI and/or the editor(s). MDPI and/or the editor(s) disclaim responsibility for any injury to people or property resulting from any ideas, methods, instructions or products referred to in the content.

nanomaterials

MDPI

Perspective

Information-Driven Design as a Potential Approach for 3D Printing of Skeletal Muscle Biomimetic Scaffolds

Silvia Baiguera [1,2,3,*,†], Costantino Del Gaudio [4,†], Felicia Carotenuto [1,2,3], Paolo Di Nardo [2,3,5] and Laura Teodori [1,3,*]

1 Departmet of Fusion and Technologies for Nuclear Safety and Security, Diagnostic and Metrology (FSN-TECFIS-DIM), ENEA, 00196 Roma, Italy; carotenuto@med.uniroma2.it

2 Department of Clinical Science and Translational Medicine, University of Rome "Tor Vergata", 00133 Roma, Italy; dinardo@uniroma2.it

3 Interdepartmental Center of Regenerative Medicine (CIMER), University of Rome "Tor Vergata", 00133 Roma, Italy

4 E. Amaldi Foundation, Via del Politecnico snc, 00133 Rome, Italy; costantino.delgaudio@fondazioneamaldi.it

5 L.L. Levshin Institute of Cluster Oncology, I. M. Sechenov First Medical University, 119991 Moscow, Russia

* Correspondence: Silvia.Baiguera@uniroma2.it (S.B.); laura.teodori@enea.it (L.T.); Tel.: +39-06-7259-4246 (S.B.); +39-06-9400-5642 (L.T.)

† These authors contributed equally to this work.

Received: 12 September 2020; Accepted: 2 October 2020; Published: 8 October 2020

Abstract: Severe muscle injuries are a real clinical issue that still needs to be successfully addressed. Tissue engineering can represent a potential approach for this aim, but effective healing solutions have not been developed yet. In this regard, novel experimental protocols tailored to a biomimetic approach can thus be defined by properly systematizing the findings acquired so far in the biomaterials and scaffold manufacturing fields. In order to plan a more comprehensive strategy, the extracellular matrix (ECM), with its properties stimulating neomyogenesis and vascularization, should be considered as a valuable biomaterial to be used to fabricate the tissue-specific three-dimensional structure of interest. The skeletal muscle decellularized ECM can be processed and printed, e.g., by means of stereolithography, to prepare bioactive and biomimetic 3D scaffolds, including both biochemical and topographical features specifically oriented to skeletal muscle regenerative applications. This paper aims to focus on the skeletal muscle tissue engineering sector, suggesting a possible approach to develop instructive scaffolds for a guided healing process.

Keywords: decellularized extracellular matrix; engineered muscular construct; stereolithography; volumetric muscle loss

---

1. Introduction: the Current Scenario

Severe muscle injuries (muscle loss greater than 20%) are beyond the endogenous regenerative capacity of skeletal muscle (SM) tissue, and the current gold standard approach (autologous tissue transfer) is often associated with donor site morbidity, functional deficiency and graft failure [1,2]. Therefore, alternative and effective treatments, such as engineered muscle tissues, become a necessity.

Tissue engineering has usually been considered a promising applicative option to address urgent and critical clinical needs. Moreover, systems biology techniques can further support the collection of the desired output by developing physiologically relevant bio-engineered tissues [3]. The potential of these multidisciplinary approaches can pave the way to the definition of novel and original medical and surgical protocols capable of overcoming most of the drawbacks elicited by conventional therapies, such as functional impairment, reduced movement ability, severe immune response risks and high costs [1,2]. Such an expectation has driven almost all the investigations published so far aimed to design suitable scaffolds to actually promote muscular regeneration, closely mimicking the morphological, biochemical and functional properties of the specific natural extracellular matrix (ECM) [4–6]. This is a keypoint of the classic tissue engineering paradigm in which the scaffold acts as a temporary ECM, providing all the instructive cues guiding the formation of the autologous tissue, and is capable of remodeling according to the physiological processes of the host. Clearly, this outcome should avoid the typical limitations related to tissue/organ transplantation and conventional implantable prostheses, such as the lack of growth and integration [7,8].

In this regard, the challenge of tissue engineering is highly demanding, and the design of ad hoc scaffolds has to be accurately performed in order to offer a valid alternative. In particular, SM comprises bundles of long, highly aligned, striated, multinucleated fibers organized in a hierarchical manner within the ECM. Thus, a parallel-aligned structure, which implies dealing with an anisotropic tissue that contributes to influencing cellular orientation and maturation, should provide an overall replication of the physiologic microenvironment, as this is pivotal when seeking to develop an engineered biomimetic muscle. Thus, the selection of both peculiar materials and fabrication techniques that support the realization of a three-dimensional microarchitecture similar to that of the SM ECM plays a major role is providing functional engineered muscle tissues. However, these starting design considerations are just a partial solution to a large-scale problem. It is well-known that microstructure, stiffness and porosity, to mention only the most common and controllable features, represent the first input for dictating cell response. However, it is also known that this condition may be insufficient to promote the desired biological outcome. Dealing with a comprehensive tissue engineering protocol underlines, therefore, the necessity of developing a biomimetic approach to preparing an integrated materials/cells/signaling construct that orchestrates a univocal result. For this aim, several proposals targeted at muscle regeneration have already been presented, focusing on (i) the materials, synthetic and natural ones, and their combinations [5,6,9], (ii) the fabrication techniques to evaluate their role on scaffold morphology and properties [10–12], (iii) the surface functionalization [13], (iv) the cells to be used, stem cells or already committed [6,10], (v) the addition of different nanomaterials to deal with constructs responsive to electrical, magnetic and photothermal stimulation [14] and (vi) the methods to specifically induce vascularization, innervation and contractility [10,13]. More than interesting and promising findings have been reported; however, the proper cellular microenvironment to bioengineer the SM construct has not, to date, been found, and there is still a clinical need to introduce new strategies that can facilitate safe and large muscle tissue repair and regeneration [13]. This work, therefore, proposes a possible technical/functional approach for designing and preparing suitable biomimetic scaffolds for SM regeneration, starting from the experimental findings collected so far and focusing on decellularized ECM to be used as a starting biomaterial for bioprinting and stereolithography.

2. Moving Towards Biomimetic Engineered Muscular Constructs

Biomimetic scaffolds, usually recalled as potential three-dimensional structures for an effective healing protocol, should be accurately designed to replicate one or more attributes of a biological environment produced by a living organism. This expected outcome involves several key parameters that need to be critically assessed and selected since it is unlikely that all the ECM characteristics can be reproduced. Biomaterials are pivotal, as their interaction with the seeded cells will determine the final result of the tissue-engineered construct. For this aim, dealing with biologically derived materials seems to be a reasonable option, if the most relevant biochemical cues are preserved [9,15,16].

Noteworthy, additive manufacturing represents a recent generation of appealing means to develop bioactive scaffolds, largely recapitulating biochemical and functional features of the tissue to be healed. However, for most of the 3D printing technologies, the resolution that can be achieved is not comparable to the characteristic dimensions of the structural elements of the ECM to be resembled, and this may impair the desired result related to the natural microarchitecture. In this context, stereolithography could instead usefully support the fabrication of biomimetic scaffolds for muscle repair, as it is characterized by a higher resolution level (the highest among 3D printing techniques for tissue engineering), and thus capable of fabricating aligned structural elements with the muscle fiber characteristic size. This approach represents a promising scenario if the processing solution components are properly selected. Decellularized ECM (dECM) can be a suitable choice to address this issue, as it can be modified to act as a photopolymerizable hydrogel, intrinsically providing a number of biological cues to guide cell response. Herein, after an overview of dECMs used as bioinks and in stereolithographic setups, a design-oriented process, starting from histological images and SM dECM applications, is suggested in order to propose biomimetic SM dECM-based scaffolds for muscular regeneration.

2.1. SM dECM-Based Hydrogel: A Material for Biomimetic Engineered Muscular Constructs?

The bioink is a printable hydrogel containing nutrients, ECM components and cells in various permutations that can be crosslinked in order to acquire stability and tailored structural properties. This allows dealing with a three-dimensionally seeded scaffold, which is one of the goals to be reached for a real tissue-engineered scaffold. An ideal bioink should (i) provide tissue-specific biological, physical and mechanical cues, (ii) be biocompatible and (iii) have good printability and biodegradability [17]. Possible limitations for the fabrication process can be related to the rheological characteristics of the hydrogel, the shear stress values experienced by cells during the process and the photopolymerization needed to crosslink the scaffold, which in the end may negatively affect cell viability [12].

Different and specific bioinks, both natural and synthetic, have already been considered for bioprinting SM constructs [12,18]. Synthetic bioinks are characterized by tunable mechanical properties and crosslinking capacity, but also by insufficient cellular adhesion. In contrast, natural bioinks provide biological cues but have poor mechanical properties [19]. Moreover, a large number of them could not actually represent the complexity of natural tissues and thus are inadequate to recreate a suitable microenvironment for cell engraftment, survival and function. dECM, retaining specific tissue components, including collagen, glycosaminoglycans, growth and differentiation factors, can recapitulate most of the features of natural ECM, if not all. Thus, dealing with bioinks made of only dECM, providing cells with their natural microenvironment, could represent a further step towards the design of a real tissue-mimicking scaffold. In this context, dECM bioinks have been derived from different organs [17,20,21]. The obtainment of dECM-based bioinks for 3D printing involves different steps [17,22,23]. The first is the decellularization of the tissue of interest by removing the cells while preserving the ECM. Several physical, chemical, enzymatic and biological processes can be adopted; the selection depends on several factors, such as tissue and organ thickness, density and lipid content [20,21]. To prepare a printable material, dECM needs to be processed as a hydrogel: ECM is usually crushed into a powder form and solubilized in a physiological buffer solution (generally, enzymatic acidic digestion with a pH adjusted to accommodate cells, if to be included in the final formulation) in order to deal with a solution at 10 °C and a gel at 37 °C. dECM gels may have the intrinsic mechanical strength to be directly printed or need to be mixed with a crosslinking agent, reacting during the printing process [17,20]. Moreover, a carrier polymer can also be used to increase the solubility, tune the viscosity, or induce/enhance the post-cross-linking of the bioink [17,22].

Focusing on engineered SM constructs, recent studies based on additive manufacturing approaches using SM dECM as a bioink have shown promising results. Choi et al. demonstrated that an SM dECM bioink had sufficient printability for producing different shapes of 3D freestanding volumetric muscle constructs, indicating that this bioink can be usefully considered to design and fabricate the original structures of injured muscles prior to implantation [24]. Moreover, cells were encapsulated in a microenvironment similar to that of the native ECM, offering an additional advantage for myogenesis [24]. The same group demonstrated that prevascularized 3D printed muscle constructs co-axially fabricated with muscle and vascular dECM bioinks successfully mimicked the hierarchical architecture of vascularized muscles with improved de novo fiber formation, vascularization, innervations and 85% functional recovery in volumetric muscle loss injuries [25]. SM dECM has also been chemically modified to impart photo-crosslinkable properties and thus to enhance mechanical stability. By employing a multi-crosslinking approach, Skardal et al. achieved a soft, extrudable material for bioprinting, which could be further stabilized after deposition in order to reach the elastic modulus of a targeted material and thus match that of the tissue of interest [26]. Recently, Kim et al. chemically modified SM dECM and fabricated a uniaxially aligned/micro-topographical SM structure, suggesting that this construct may lead to efficient functional recovery of damaged or diseased SM tissues [27]. Indeed, the biochemical cues given by SM dECM accelerated myogenic differentiation, while the topographical ones allowed achieving a suitable cellular orientation [27].

These 3D printed tissue models represent a critical proof of concept for the use of dECMs as a proper bioink to prepare engineered muscular constructs. Moreover, it has been demonstrated that ECM degradation releases growth factors (such as basic fibroblast growth factor and vascular endothelial growth factor) and low molecular weight matricryptic oligopeptides to recruit and influence stem or endogenous progenitor cells that play a role in the constructive and functional remodeling process, including vasculature and innervation formation [28,29].

Although further studies are needed, the bioactive features of natural materials can support the generation of an innovative class of implantable devices capable of recapitulating the main functional characteristics of the tissue to be healed. However, this is a part of the requested solution, as ad hoc manufacturing methods should be selected or developed in order to correctly deal with the desired scaffold.

2.2. *Stereolithography: A Suitable Fabrication Technique for Engineered Muscular Constructs?*

Stereolithography is a 3D printing technique based on the layer photopolymerization of specific materials by means of light irradiation, usually UV, leading to a final stacked scaffold as the build stage is vertically translated. This fabrication method is characterized by an intrinsic high resolution, the highest among all the additive manufacturing techniques (about 6 μm), which allows finely controlling external geometry and internal architecture of the designed scaffold to realize complex geometries [30]. In addition, a reliable modification can be further considered to include additional functional cues [31].

Several working principles can be implemented, acting, for instance, on the computer-controlled movement of the light source to polymerize each layer of the structure, or to directly polymerize the whole layer by means of an array of thousands of micro-mirrors reflecting light in a predefined spatial pattern, i.e., the digital micro-mirror device [32].

The intrinsic high resolution can be further improved by two-photon stereolithography, reaching about 200 nm, by focusing two consecutive photons within the focal volume of a laser beam, typically less than 1 μm^3 [33]. This experimental approach can allow the fabrication of scaffolds with nanoscale features similar to those of natural ECM, thus leading to assessing cell behavior in this kind of environment [34].

To overcome stereolithography limitations, such as the necessity to use photopolymerizable solutions containing potentially cytotoxic UV-activated photo-initiators, different chemical and biological functional agents have been conjugated to dECM [26,27,35,36]. The collected results suggested that these photo-crosslinking molecules have minimal negative effects on cell viability and function [26,27,35,36]. In particular, vitamin B2, activated by long-wavelength UV radiation, showed to be non-toxic, significantly increased tissue rigidity and provided therapeutic outputs, proving that it is a completely biocompatible photo-crosslinking agent [37,38]. Moreover, it has been recently demonstrated that the vitamin B2-UVA crosslinking approach is an effective strategy to enhance ECM-hydrogel performance for tissue engineering and regenerative medicine applications, without inducing significant effects on cell viability and apoptosis, and therefore, being a potentially attractive and tunable means for scaffolding strategies [39–41]. To date, muscle-derived dECM has not yet been assessed for the preparation of biomimetic scaffolds by means of stereolithography. According to the findings already reported and including tissue-specific ECM, a possible strategy could then be developed in order to support the definition of ad hoc therapeutic protocols for large muscle defects.

3. Proposal of a Biomimetic Scaffold for the Treatment of Severe Muscle Injuries

The treatment of severe muscle injuries can be an investigational field in which the synergic potential of dECM and stereolithography might support the development of an effective biomimetic scaffold. Indeed, stereolithography could close the loop, allowing the fabrication of a scaffold with the characteristic dimensions, alignment and orientation of the muscle fibers. A similar approach can pave the way to the development of scaffolds substantially made of the tissue-specific "biomaterial". Moreover, it can be assumed that such a solution could be characterized by extreme biocompatibility and valuable matching properties with the biological environment. Furthermore, with the use of physiological datasets (histological images, SEM micrographs or medical images) and computer-aided design (CAD) to model scaffold fabrication, stereolithography can allow creating scaffolds with precise tissue architectures [42]. The output of medical imaging can effectively support the accurate definition of a scaffold resembling the target tissue in order to provide, e.g., stability and flow transport as peculiar properties of the extracellular microenvironment. The integration of an information-driven design and CAD-based micromanufacturing approach has been previously highlighted, reviewing a number of solutions from different fabrication techniques, always referring to the ECM characteristics to be reproduced [43]. This approach, for instance, was considered for preparing 3D printed devices for neural regeneration starting from magnetic resonance imaging, computed tomography, and/or 3D virtual visualization to acquire topological data for the tailored production of structures that match the injured microenvironment, a promising methodology for patient-specific implants [44]. Similar considerations can be presented regarding the orthopedic field in which imaging data can pave the way for personalized treatment [45]. However, it should be underlined that most of the solutions proposed so far are generally focused on large defects treated with scaffolds that may not finely reproduce the ultrastructure of the tissue to be healed. In this regard, the use of morphological images with micro/nano details of the tissue of interest (such as histological images and/or SEM micrographs) [46] could allow the development of biomimetic structures, resembling the natural ECM and can actually provide a step forward in the definition of novel therapeutic protocols. In this regard, the design-oriented process implementing stereolithography can start, for instance, from histological images or scanning electron micrographs of muscle fibers to be considered as the architectural structure to be 3D printed. Following this suggestion, Figure 1 shows the planning approach to prepare a scaffold aimed at resembling the fiber morphology and packing density of an SM structure. In this regard, the proposal elaborates histological observations to design aligned fibers reproducing the muscle cross-section. Moreover, considering that SM fibers appear striated due to the sarcomeres alignment, two other scaffolds could be considered: a scaffold composed of organized parallel, aligned and striated fibers (as control) (Figure 1B) and a scaffold composed of striated fibers organized as in the SM structure (Figure 1C). These latter options can represent a simplified version of the SM scaffold, which can further support the optimization of the design process towards the final applicative proposal.

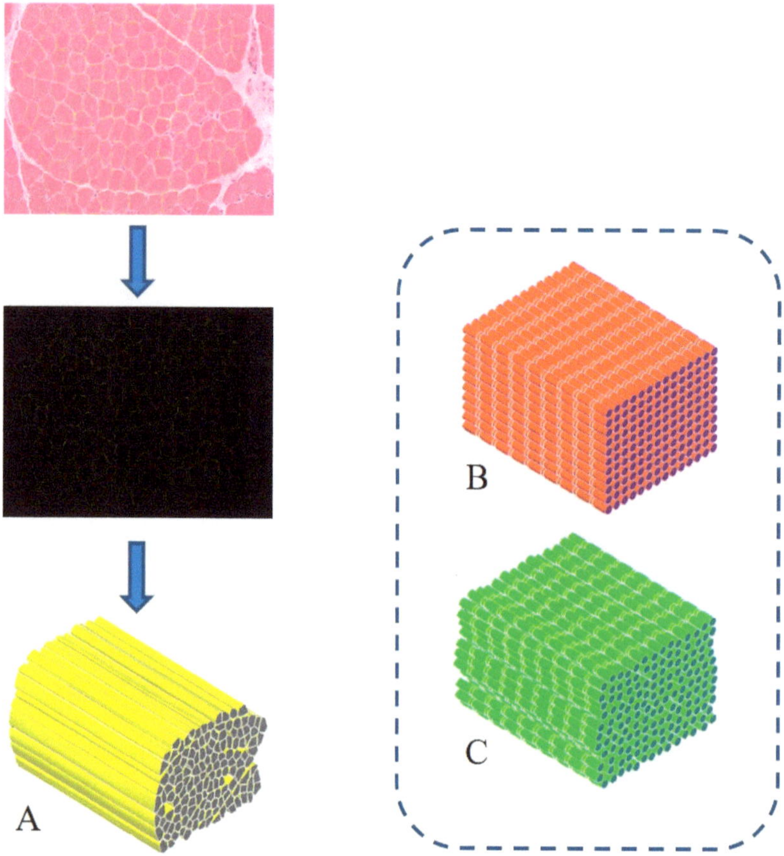

Figure 1. Biomimetic design of scaffold for skeletal muscle (SM) tissue engineering. A three-dimensional bioactive construct can be fabricated thanks to stereolithography, resembling the muscle structure and including SM decellularized extracellular matrix (dECM) (**A**). Similarly, a simple (**B**) and a more realistic (**C**) control case can be developed, mimicking the striated structure.

In addition, to complete the design stage, computational models can be proposed as a means to investigate and predict either the fluid dynamics and the mechanical response in a simulated relevant environment, e.g., a bioreactor or physiological conditions. However, several requirements should be accomplished in order to deal with a realistic solution in terms of, for instance, experienced shear stress, mass transport phenomena and local stiffness properties. This implies an accurate validation step, which depends on the implementation of biological data as input and numerical physiological matching output, supported by high-performance computing resources.

Such an approach can lead to an effective tissue-engineered construct combining most of the requirements needed for a substantial biomimetic investigation. The possibility of resembling the tissue microarchitecture according to physiological datasets, of modeling the printing input to be fabricated by means of tissue-specific biomaterials and ad hoc 3D printing techniques capable of assuring the scaffold resolution can then actually allow to mimic SM tissue and address the crucial issue of severe muscle injuries. More refined solutions can obviously be considered, but the different manufacturing alternatives within the stereolithography sector can usefully and promptly realize this kind of substrate. Hydrogel formulation, light source type and power, exposure time and patterning strategy concur to modulate the final mechanical characteristics to match those of the physiological site of implantation.

Therefore, a rational and comprehensive plan should be developed that includes and further exploits all the keypoints here addressed: (i) the selection of the material(s) to formulate the hydrogel to be processed, combining dECM, cells and possibly other polymers; (ii) the CAD input for the stereolithography setup, taking advantage of the resolution and accuracy of the technique that can actually allow collecting scaffolds morphologically and functionally similar to the natural ECM; and (iii) the implementation of the most suitable fabrication methodology, e.g., single- or two-photon stereolithography.

Author Contributions: Conceptualization, S.B., C.D.G. and L.T.; original draft preparation and writing, S.B. and C.D.G.; review and editing, S.B., C.D.G., F.C., P.D.N. and L.T.; project administration and funding acquisition, L.T. All authors have read and agreed to the published version of the manuscript.

Funding: This research was funded by Regione Lazio through LazioInnova, Progetto 'SMARTIES' (Smart scaffold ingegnerizzati per la rigenerazione dell'apparato muscolo scheletrico), coordinated by ENEA CR Frascati, Grant Number: 85-2017-15004, AVVISO PUBBLICO PROGETTI DI GRUPPI DI RICERCA Conoscenza e cooperazione per un nuovo modello di sviluppo. S.B. fellowship is supported by the research project 'SMARTIES'.

Conflicts of Interest: The authors declare that the research was conducted in the absence of any commercial or financial relationships that could be construed as potential conflicts of interest.

References

1. Mase, V.J.; Hsu, J.R.; Wolf, S.E.; Wenke, J.C.; Baer, D.G.; Owens, J.; Badylak, S.F.; Walters, J.T. Clinical Application of an Acellular Biologic Scaffold for Surgical Repair of a Large, Traumatic Quadriceps Femoris Muscle Defect. *Orthopedics* **2010**, *33*, 511. [CrossRef] [PubMed]
2. Turner, N.; Badylak, S.F. Regeneration of skeletal muscle. *Cell Tissue Res.* **2011**, *347*, 759–774. [CrossRef] [PubMed]
3. Rajagopalan, P.; Kasif, S.; Murali, T.M. Systems Biology Characterization of Engineered Tissues. *Ann. Rev. Biomed. Eng.* **2013**, *15*, 55–70. [CrossRef] [PubMed]
4. Vigodarzere, G.C.; Mantero, S. Skeletal muscle tissue engineering: Strategies for volumetric constructs. *Front. Physiol.* **2014**, *5*, 362. [CrossRef]
5. Grasman, J.M.; Zayas, M.J.; Page, R.L.; Pins, G.D. Biomimetic scaffolds for regeneration of volumetric muscle loss in skeletal muscle injuries. *Acta Biomater.* **2015**, *25*, 2–15. [CrossRef]
6. Mulbauer, G.D.; Matthew, H. Biomimetic Scaffolds in Skeletal Muscle Regeneration. *Discoveries* **2019**, *7*, e90. [CrossRef]
7. Orlando, G.; Soker, S.; Stratta, R.J. Organ Bioengineering and Regeneration as the New Holy Grail for Organ Transplantation. *Ann. Surg.* **2013**, *258*, 221–232. [CrossRef]
8. Trevisan, C.; Maghin, E.; Dedja, A.; Caccin, P.; De Cesare, N.; Franzin, C.; Boso, D.; Pesce, P.; Caicci, F.; Boldrin, F.; et al. Allogenic tissue-specific decellularized scaffolds promote long-term muscle innervation and functional recovery in a surgical diaphragmatic hernia model. *Acta Biomater.* **2019**, *89*, 115–125. [CrossRef]
9. Urciuolo, A.; De Coppi, P. DecellularizedTissue for Muscle Regeneration. *Int. J. Mol. Sci.* **2018**, *19*, 2392. [CrossRef]
10. Qazi, T.H.; Mooney, D.J.; Pumberger, M.; Geißler, S.; Duda, G.N. Biomaterials based strategies for skeletal muscle tissue engineering: Existing technologies and future trends. *Biomaterials* **2015**, *53*, 502–521. [CrossRef]
11. Lee, H.; Kim, W.; Lee, J.; Yoo, J.J.; Kim, G.H.; Lee, S.J. Effect of Hierarchical Scaffold Consisting of Aligned dECM Nanofibers and Poly(lactide-co-glycolide) Struts on the Orientation and Maturation of Human Muscle Progenitor Cells. *ACS Appl. Mater. Interfaces* **2019**, *11*, 39449–39458. [CrossRef] [PubMed]
12. Ostrovidov, S.; Salehi, S.; Costantini, M.; Suthiwanich, K.; Ebrahimi, M.; Sadeghian, R.B.; Fujie, T.; Shi, X.; Cannata, S.; Gargioli, C.; et al. 3D Bioprinting in Skeletal Muscle Tissue Engineering. *Small* **2019**, *15*, e1805530. [CrossRef] [PubMed]
13. Liu, J.; Saul, D.; Böker, K.O.; Ernst, J.; Lehman, W.; Schilling, A.F. Current Methods for Skeletal Muscle Tissue Repair and Regeneration. *BioMed Res. Int.* **2018**, *2018*, 1–11. [CrossRef] [PubMed]
14. Farr, A.C.; Hogan, K.J.; Mikos, A.G. Nanomaterial Additives for Fabrication of Stimuli-Responsive Skeletal Muscle Tissue Engineering Constructs. *Adv. Health Mater.* **2020**, 2000730. [CrossRef]

15. Gilpin, A.; Yang, Y. Decellularization Strategies for Regenerative Medicine: From Processing Techniques to Applications. *BioMed Res. Int.* **2017**, *2017*, 1–13. [CrossRef]
16. Sarrafian, T.L.; Bodine, S.C.; Murphy, B.; Grayson, J.K.; Stover, S.M. Extracellular matrix scaffolds for treatment of large volume muscle injuries: A review. *Vet. Surg.* **2018**, *47*, 524–535. [CrossRef]
17. Choudhury, D.; Tun, H.W.; Wang, T.; Naing, M.W. Organ-Derived Decellularized Extracellular Matrix: A Game Changer for Bioink Manufacturing? *Trends Biotechnol.* **2018**, *36*, 787–805. [CrossRef]
18. Popov, A.; Malferrari, S.; Kalaskar, D.M. 3D bioprinting for musculoskeletal applications. *J. 3D Print. Med.* **2017**, *1*, 191–211. [CrossRef]
19. Gopinathan, J.; Noh, I. Recent trends in bioinks for 3D printing. *Biomater. Res.* **2018**, *22*, 11. [CrossRef]
20. Dzobo, K.; Motaung, K.S.; Adesida, A.B. Recent Trends in Decellularized Extracellular Matrix Bioinks for 3D Printing: An Updated Review. *Int. J. Mol. Sci.* **2019**, *20*, 4628. [CrossRef]
21. Kabirian, F.; Mozafari, M. Decellularized ECM-derived bioinks: Prospects for the future. *Methods* **2020**, *171*, 108–118. [CrossRef]
22. Ji, S.; Guvendiren, M. Recent Advances in Bioink Design for 3D Bioprinting of Tissues and Organs. *Front. Bioeng. Biotechnol.* **2017**, *5*, 23. [CrossRef]
23. Pati, F.; Cho, D.-W. Bioprinting of 3D Tissue Models Using Decellularized Extracellular Matrix Bioink. *Breast Cancer* **2017**, *1612*, 381–390. [CrossRef]
24. Choi, Y.-J.; Kim, T.G.; Jeong, J.; Yi, H.-G.; Park, J.W.; Hwang, W.; Cho, D.-W. 3D Cell Printing of Functional Skeletal Muscle Constructs Using Skeletal Muscle-Derived Bioink. *Adv. Health Mater.* **2016**, *5*, 2636–2645. [CrossRef] [PubMed]
25. Choi, Y.-J.; Jun, Y.-J.; Kim, D.Y.; Yi, H.-G.; Chae, S.-H.; Kang, J.; Lee, J.; Gao, G.; Kong, J.-S.; Jang, J.; et al. A 3D cell printed muscle construct with tissue-derived bioink for the treatment of volumetric muscle loss. *Biomaterials* **2019**, *206*, 160–169. [CrossRef] [PubMed]
26. Skardal, A.; Devarasetty, M.; Kang, H.; Mead, I.; Bishop, C.; Shupe, T.; Lee, S.J.; Jackson, J.; Yoo, J.; Soker, S.; et al. A hydrogel bioink toolkit for mimicking native tissue biochemical and mechanical properties in bioprinted tissue constructs. *Acta Biomater.* **2015**, *25*, 24–34. [CrossRef]
27. Kim, W.; Lee, H.; Lee, J.; Atala, A.; Yoo, J.J.; Lee, S.J.; Kim, G. Efficient myotube formation in 3D bioprinted tissue construct by biochemical and topographical cues. *Biomaterials* **2019**, *230*, 119632. [CrossRef]
28. Teodori, L.; Costa, A.; Marzio, R.; Perniconi, B.; Coletti, D.; Adamo, S.; Gupta, B.; Tárnok, A. Native extracellular matrix: A new scaffolding platform for repair of damaged muscle. *Front. Physiol.* **2014**, *5*, 218. [CrossRef]
29. Dziki, J.; Badylak, S.; Yabroudi, M.; Sicari, B.; Ambrosio, F.; Stearns, K.; Turner, N.; Wyse, A.; Boninger, M.L.; Brown, E.H.P.; et al. An acellular biologic scaffold treatment for volumetric muscle loss: Results of a 13-patient cohort study. *NPJ Regen. Med.* **2016**, *1*, 16008. [CrossRef]
30. Tamay, D.G.; Usal, T.D.; Alagoz, A.S.; Yucel, D.; Hasirci, N.; Hasirci, V. 3D and 4D Printing of Polymers for Tissue Engineering Applications. *Front. Bioeng. Biotechnol.* **2019**, *7*, 164. [CrossRef]
31. Ciocci, M.; Mochi, F.; Carotenuto, F.; Di Giovanni, E.; Prosposito, P.; Francini, R.; De Matteis, F.; Reshetov, I.V.; Casalboni, M.; Melino, S.; et al. Scaffold-in-Scaffold Potential to Induce Growth and Differentiation of Cardiac Progenitor Cells. *Stem Cells Dev.* **2017**, *26*, 1438–1447. [CrossRef] [PubMed]
32. VijayaVenkataRaman, S.; Yan, W.-C.; Lu, W.F.; Wang, C.-H.; Fuh, J.Y.H. 3D bioprinting of tissues and organs for regenerative medicine. *Adv. Drug Deliv. Rev.* **2018**, *132*, 296–332. [CrossRef] [PubMed]
33. Spangenberg, A.; Hobeika, N.; Stehlin, F.; Malval, J.; Wieder, F.; Prabhakaran, P.; Baldeck, P. Recent advances in two-photon stereolithography. In *Updates in Advanced Lithography*; Hosaka, S., Ed.; IntechOpen: London, UK, 2013; pp. 35–64. [CrossRef]
34. Skoog, S.A.; Goering, P.L.; Narayan, R.J. Stereolithography in tissue engineering. *J. Mater. Sci. Mater. Electron.* **2013**, *25*, 845–856. [CrossRef] [PubMed]
35. Jang, J.; Kim, T.G.; Kim, B.S.; Kim, S.-W.; Kwon, S.M.; Cho, D.-W. Tailoring mechanical properties of decellularized extracellular matrix bioink by vitamin B2-induced photo-crosslinking. *Acta Biomater.* **2016**, *33*, 88–95. [CrossRef]
36. Jang, J.; Park, H.-J.; Kim, S.-W.; Kim, H.; Park, J.Y.; Na, S.J.; Kim, H.J.; Park, M.N.; Choi, S.H.; Park, S.H.; et al. 3D printed complex tissue construct using stem cell-laden decellularized extracellular matrix bioinks for cardiac repair. *Biomaterials* **2017**, *112*, 264–274. [CrossRef]

37. Wollensak, G.; Spoerl, E.; Seiler, T. Stress-strain measurements of human and porcine corneas after riboflavin–ultraviolet-A-induced cross-linking. *J. Cataract. Refract. Surg.* **2003**, *29*, 1780–1785. [CrossRef]
38. Wollensak, G.; Spoerl, E.; Seiler, T. Riboflavin/ultraviolet-a-induced collagen crosslinking for the treatment of keratoconus. *Am. J. Ophthalmol.* **2003**, *135*, 620–627. [CrossRef]
39. Stachon, T.; Wang, J.; Song, X.; Langenbucher, A.; Seitz, B.; Szentmáry, N. Impact of crosslinking/riboflavin-UVA-photodynamic inactivation onviability, apoptosis and activation of human keratocytes in vitro. *J. Biomed. Res.* **2015**, *29*, 321–325. [CrossRef]
40. Ahearne, M.; Coyle, A. Application of UVA-riboflavin crosslinking to enhance the mechanical properties of extracellular matrix derived hydrogels. *J. Mech. Behav. Biomed. Mater.* **2016**, *54*, 259–267. [CrossRef]
41. Arrizabalaga, J.H.; Nollert, M.U. Riboflavin-UVA crosslinking of amniotic membranes and its influence on the culture of adipose-derived stem cells. *J. Mech. Behav. Biomed. Mater.* **2020**, *106*, 103729. [CrossRef]
42. Bajaj, P.; Chan, V.; Jeong, J.H.; Zorlutuna, P.; Kong, H.; Bashir, R. 3-D biofabrication using stereolithography for biology and medicine. *IEEE Eng. Med. Biol. Soc.* **2012**, *2012*, 6805–6808. [CrossRef]
43. Steier, A.; Muñiz, A.; Neale, D.; Lahann, J. Emerging Trends in Information-Driven Engineering of Complex Biological Systems. *Adv. Mater.* **2019**, *31*, e1806898. [CrossRef] [PubMed]
44. Joung, D.; Lavoie, N.S.; Guo, S.; Park, S.H.; Parr, A.M.; McAlpine, M.C. 3D Printed Neural Regeneration Devices. *Adv. Funct. Mater.* **2019**, *30*, 1906237. [CrossRef] [PubMed]
45. Wong, K.C. 3D-printed patient-specific applications in orthopedics. *Orthop. Res. Rev.* **2016**, *8*, 57–66. [CrossRef]
46. Teodori, L.; Crupi, A.; Costa, A.; Diaspro, A.; Melzer, S.; Tárnok, A. Three-dimensional imaging technologies: A priority for the advancement of tissue engineering and a challenge for the imaging community. *J. Biophotonics* **2016**, *10*, 24–45. [CrossRef]

© 2020 by the authors. Licensee MDPI, Basel, Switzerland. This article is an open access article distributed under the terms and conditions of the Creative Commons Attribution (CC BY) license (http://creativecommons.org/licenses/by/4.0/).

MDPI AG
Grosspeteranlage 5
4052 Basel
Switzerland
Tel.: +41 61 683 77 34

Nanomaterials Editorial Office
E-mail: nanomaterials@mdpi.com
www.mdpi.com/journal/nanomaterials

Disclaimer/Publisher's Note: The title and front matter of this reprint are at the discretion of the Guest Editors. The publisher is not responsible for their content or any associated concerns. The statements, opinions and data contained in all individual articles are solely those of the individual Editors and contributors and not of MDPI. MDPI disclaims responsibility for any injury to people or property resulting from any ideas, methods, instructions or products referred to in the content.

www.ingramcontent.com/pod-product-compliance
Lightning Source LLC
Chambersburg PA
CBHW041359210926
51850CB00013B/3128